First published by Earthscan in the UK and USA in 2006 for and on behalf of the United Nations Human Settlements Programme (UN-HABITAT).

United Nations Human Settlements Programme (UN-HABITAT)
P.O. Box 30030, Nairobi, Kenya
Tel: +254 20 7621 234
Fax: +254 20 7624 266/7
Website: www.unhabitat.org

DISCLAIMER

The designations employed and the presentation of the material in this report do not imply the expression of any opinion whatsoever on the part of the Secretariat of the United Nations concerning the legal status of any country, territory, city or area, or of its authorities, or concerning delimitation of its frontiers or boundaries, or regarding its economic system or degree of development. The analysis, conclusions and recommendations of this report do not necessarily reflect the views of the United Nations Human Settlements Programme or its Governing Council.

HS/814/06E (paperback)
HS/815/06E (hardback)

ISBN-10: 1-84407-378-5 (Earthscan paperback)
ISBN-13: 978-1-84407-378-8 (Earthscan paperback)
ISBN-10: 1-84407-379-3 (Earthscan hardback)
ISBN-13: 978-1-84407-379-5 (Earthscan hardback)

92-1-131811-4 (UN-HABITAT paperback)
978-92-1-131811-1 (UN-HABITAT paperback)
92-1-131812-2 (UN-HABITAT hardback)
978-92-1-131812-8 (UN-HABITAT hardback)

Design and layout by Michael Jones Software, Nairobi, Kenya.
Printed and bound in Malta by Gutenberg Press Ltd.

For a full list of Earthscan publications contact:

Earthscan
8–12 Camden High Street
London, NW1 0JH, UK
Tel: +44 (0)20 7387 8558
Fax: +44 (0)20 7387 8998
E-mail: earthinfo@earthscan.co.uk
Website: www.earthscan.co.uk

22883 Quicksilver Drive, Sterling, VA 20166-2012, USA

Earthscan is an imprint of James & James (Science Publishers) Ltd and publishes in association with the International Institute for Environment and Development.

A catalogue record for this book is available from the British Library.

Library of Congress Cataloging-in-Publication Data has been applied for.

Foreword

Since the adoption of the Millennium Declaration in 2000, many Governments have made significant progress toward reducing poverty by expanding primary education, promoting gender equality and improving people's access to basic services, among other important efforts. But those gains have been unevenly distributed, not only among and within countries, but also among and within cities.

As this third edition of the *State of the World's Cities* makes clear, much more needs to be done to secure the well-being of people living in poverty in the world's cities. Governments and development agencies have traditionally emphasized the improvement of rural areas, because that is where the vast majority of the world's poor live. But as rapid urbanization continues, similar energies are needed in urban areas. The problem is not urbanization *per se,* but the fact that urbanization in many developing regions has not resulted in greater prosperity or a more equitable distribution of resources. Indeed, efforts to improve the lives of the urban poor have not kept up with the rate of urbanization.

Rural poverty has long been the world's most common face of destitution. But urban poverty can be just as intense, dehumanizing and life-threatening. New UN-HABITAT data on slum populations reported here for the first time reveal startling facts about life in impoverished urban settlements, where parents must choose between paying rent and buying food for their children, where decrepit school facilities and costly materials deprive young people of a basic education, and where lack of sanitation, clean water and ventilation increase the risk of disease, particularly among women and children.

The world's cities are simultaneously places of great human progress and deep human destitution. This Report illustrates how very different the lives of the urban rich and the urban poor can be, and explains why the international community must concentrate more of its efforts on improving the lives of the urban poor if the Millennium Development Goals are to be achieved. Slum settlements are already home to almost one billion people, or one-third of the world's urban population, and over the next two decades the cities of the developing world are expected to absorb 95 per cent of the world's urban population growth. Without a renewed commitment to the needs of our urban era, matched by resources, the world's urban transition will see a further expansion and entrenchment of slums, and the spread of urban ills. This Report offers information and analysis that can help us do better, and I commend it to a wide global audience.

Kofi A. Annan
Secretary-General
United Nations

Introduction

The 2006/2007 edition of the *State of the World's Cities* marks two important milestones: the dawn of the urban millennium in 2007 and the 30th anniversary of the first Habitat Conference held in Vancouver in June 1976, which placed "urbanization" on the global development agenda. This publication also marks a less triumphal moment in history: thirty years after the world's governments first pledged to do more for cities, almost one-third of the world's urban population lives in slums, without access to decent housing or basic services and in neighbourhoods where disease, illiteracy and crime are rampant.

Since its establishment in 1978, the United Nations Human Settlements Programme (UN-HABITAT) has continued to highlight the important role and contribution of cities in fostering economic and human development. Understanding the complex social, cultural and economic dynamics of cities and urbanization is more important now than ever before as we strive to attain internationally agreed development goals. In a rapidly urbanizing world, attaining these goals will require policies and strategies based on clear and accurate data on the human settlements conditions and trends in each country.

This edition of the *State of the World's Cities Report* advances this objective by breaking new ground in the area of urban data collection, analysis and dissemination. For the first time in the history of the United Nations, urban data is reported here at *slum* and *non-slum* levels, going far beyond the traditional urban-rural dichotomy. UN-HABITAT's intra-urban data analysis – involving disaggregated data for more than 200 cities around the world – takes this work further and provides detailed evidence of urban inequalities in the areas of health, education, employment and other key indicators. The implications are significant for the attainment of the Millennium Development Goals as we can no longer assume that the urban poor are better off than their rural counterparts, or that all urban dwellers are able to benefit from basic services by virtue of proximity.

UN-HABITAT has led the drive for urban indicators since 1991 by working with other United Nations agencies and external partners to consistently refine methods for data collection and analysis and to better inform our common quest for "adequate shelter for all" and "sustainable human settlements development in an urbanizing world" - the twin goals of the Habitat Agenda adopted by the world's governments in Istanbul in 1996. With the adoption of the Millennium Declaration by the world's leaders in 2000, much of this work is now focused on monitoring progress in attaining Millennium Development Goal 7, target 11 on improving the lives of at least 100 million slum dwellers by 2020. This task requires a deeper analysis of how well cities are doing and of the actual living conditions of the urban poor. Data for this Report comes primarily from Phase III of UN-HABITAT's Urban Indicators Programme that compiles global, regional, country and household-level data of specific relevance to the Habitat Agenda and the Millennium Development Goals.

This Report clearly shows how shelter conditions have a direct impact on human development, including child mortality, education and employment. The correlation between a poor living environment, characterized by one or more shelter deprivations, and poor performance on key indicators of the Millennium Development Goals underscores the assertion that "where we live matters". The findings of this Report are unfolding a new urban reality that needs to be urgently addressed by pro-poor and gender-sensitive urban policies and legislation.

Finally, as the international community celebrates Vancouver + 30, it should also reflect on the important lessons learned in urban development and the need to reduce inequalities within cities. Cities present an unparalleled opportunity for the simultaneous attainment of most, if not all, of the internationally agreed development goals. Interventions in, for example, pro-poor water and sanitation, have immediate positive knock-on effects in terms of improved health, nutrition, disease prevention and the environment. However, unless such concerted action is taken to redress urban inequalities, cities may well become the predominant sites of deprivation, social exclusion and instability worldwide.

Anna K. Tibaijuka
Under-Secretary-General and Executive Director
UN-HABITAT

Acknowledgements

Core Team

Overall coordination: Don Okpala
Director: Nefise Bazoglu
Task Manager: Eduardo López Moreno
Statistical Adviser: Gora Mboup
Editor: Rasna Warah

Principal authors

Nefise Bazoglu, Tanzib Chowdhury, Eduardo López Moreno, Gora Mboup and Rasna Warah

Support Team

Graphs: Natsuo Ito
Maps: Iris Knabe and Martin Raithelhuber
Editorial Assistance: Darcy Varney
Research Assistance: Martha Mathenge and Raymond Otieno
Statistical Annex: Julius Majale, Philip Mukungu and Ezekiel Ngure
Cover design and page layout: Mike Jones
Cover photographs © Still Pictures

Contributors

Cecilia Andersson, Christine Auclair, Francis Dodoo, Alex Ezeh, Anna Alvazzi del Frate, Meg Holden, Asa Jonsson, Sunita Kapila, Wendy Mendes, Jaana Mioch, Luc Mougeot, Shipra Narang, Karen A. Stanecki, Darcy Varney, Ananda Weliwita and Eliya Zulu

UN-HABITAT Advisory and Technical Support

Graham Alabaster, Alioune Badiane, Daniel Biau, William Cobbett, Szilard Fricska, Iole Issaias, Anne Klen, Anantha Krishnan, Dinesh Mehta, Mutinta Munyati, Jane Nyakairu, Roman Rollnick, Wandia Seaforth, Sharad Shankardass, Farouk Tebbal, Raf Tuts, Francisco Vasquez, Satyanarayana Vejella, Chris Williams and Habitat Programme Managers in select countries.

International Advisory Board

Jo Beall, Joep Bijlmer, Andrew Boraine, Edesio Fernandes, Ilona Kickbusch, Miloon Kothari, Susan Loughhead, Patricia L. McCarney, Luc Mougeot, Kalpana Sharma and Molly O'Meara Sheehan

This Report is dedicated to the memory of Tanzib Chowdhury, UN-HABITAT Human Settlements Officer, who conceptualized the global scorecard on slums and analysed the slum upgrading and prevention policies described in Part Four.

Contents

Part 1: Cities, Slums and the Millennium Development Goals

Boxes

Tables

Figures

Maps

+

+

=
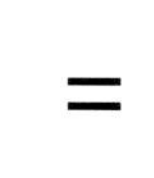

Part 2: The State of the World's Slums

Boxes

Tables

Figures

Maps

Part 3: Where We Live Matters

Boxes

Figures

Part 4: Policies and Practices that have Worked

Boxes

Figures

City Stories

Statements

Overview

Sometimes it takes just one human being to tip the scales and change the course of history. In the year 2007, that human being will either move to a city or be born in one. Demographers watching urban trends will mark it as the moment when the world entered a new urban millennium, a period in which, for the first time in history, the majority of the world's people will live in cities.

The year 2007 will also see the number of slum dwellers in the world cross the one billion mark – when one in every three city residents will live in inadequate housing with no or few basic services. This statistic may be reported in newspaper headlines, but it is still not yet clear how it will influence government policies and actions, particularly in relation to Millennium Development Goal 7, target 11: by 2020, to have improved the lives of at least 100 million slum dwellers.

Cities, Slums and the Millennium Development Goals

Three important trends characterize the urbanization process in this new urban era. Firstly, the biggest cities in the world will be found mainly in the developing world. "Metacities" – massive conurbations of more than 20 million people, above and beyond the scale of megacities – are now gaining ground in Asia, Latin America and Africa. These cities are home to only 4 per cent of the world's population and most have grown at the relatively slow rate of about 1.5 per cent annually. However, the sheer size of these urban agglomerations points to the growth of city-regions and "metropolitanization" that call for more polycentric forms of urban governance and management and stronger inter-municipal relations. The scale of environmental impact of metacities and megacities on their hinterlands is also significant and is likely to be a cause for concern in coming decades.

Secondly, despite the emergence of metacities, the majority of urban migrants will be moving to small towns and cities of less than one million inhabitants. Already, more than half of the world's urban population lives in cities of fewer than 500,000 inhabitants, and almost one-fifth lives in cities of between 1 and 5 million inhabitants. These intermediate cities are predicted to grow at a faster rate than any other type of city. Natural population increase, rather than rural-to-urban migration, is becoming a more significant contributor to urban growth in many regions, as is reclassification of rural areas into urban areas. However, the relative absence of infrastructure, such as roads, water supply and communication facilities, in many small and intermediate-sized cities makes these cities less competitive locally, nationally and regionally, and leads to a lower quality of life for their citizens.

Thirdly, cities of the developing world will absorb 95 per cent of urban growth in the next two decades, and by 2030, will be home to almost 4 billion people, or 80 per cent of the world's urban population. After 2015, the world's rural population will begin to shrink as urban growth becomes more intense in cities of Asia and Africa, two regions that are set to host the world's largest urban populations in 2030, 2.66 billion and 748 million, respectively. Urban poverty and inequality will characterize many cities in the developing world, and urban growth will become virtually synonymous with slum formation in some regions. Asia is already home to more than half of the world's slum population (581 million) – followed by sub-Saharan Africa (199 million) and Latin America and the Caribbean (134 million). Sub-Saharan Africa has both the highest annual urban growth rate and the highest slum growth rate in the world, 4.58 per cent and 4.53 per cent, respectively, more than twice the world average. The continued threat of conflict in several African countries is a significant contributing factor in the proliferation of slums in the region's urban areas. The prolonged crisis in southern Sudan, for instance, has led to the mass exodus of rural communities to the capital Khartoum, which accommodated almost half of the more than 6 million internally displaced persons in the country in the late 1990s. These trends will most likely concern policymakers in the developing world as they confront the reality of growing inequality and poverty in their cities.

The good news is that urbanization can also be a positive force for human development; countries that are highly urbanized tend to have higher incomes, more stable economies, stronger institutions and are better able to withstand the volatility of the global economy. In both developed and developing countries, cities generate a disproportionate share of gross domestic product and provide extensive opportunities for employment and investment. However, evidence suggests that despite the enormous potential of cities to bring about prosperity, the wealth generated by cities does not automatically lead to poverty reduction; on the contrary, in many cities, inequalities between the rich and the poor have grown, as have the sizes and proportions of slum populations.

Although poverty remains a primarily rural phenomenon, urban poverty is becoming a severe, pervasive – and largely unacknowledged – feature of urban life. Large sections of the population in urban areas are suffering from extreme levels of deprivation that are often even more debilitating than those experienced by the rural poor. UN-HABITAT analyses reflected in this Report show that the incidence of disease and mortality is much higher in slums than in non-slum urban areas, and in some cases, such as HIV prevalence and other health indicators, is equal to or even higher than in rural areas. These disparities are often not reflected in national statistics, which mask the deprivation experienced in poor urban neighbourhoods. Inequality in access to services, housing, land, education, health care and employment opportunities within cities have socio-economic, environmental and political repercussions, including rising violence, urban unrest, environmental degradation and underemployment, which threaten to diminish any gains in income and poverty reduction.

This edition of the *State of the World's Cities Report* provides an overview of a range of issues that link cities, slums and the Millennium Development Goals. It makes clear that the global fight against poverty – encapsulated in the Millennium Development Goals – is heavily dependent on how well cities perform. The Report highlights three inter-related issues:

- The Millennium Development Goals provide an apt framework for linking the opportunities provided by cities with improved quality of life;
- The achievement of the Goals depends on governments' capacity to speed up progress in reducing urban poverty and inequality and in reversing current trends in slum formation;
- Improving the living conditions of slum dwellers (housing, tenure, infrastructure and access to basic services) will automatically have a positive impact on the attainment of most of the Goals and their related targets.

Where We Live Matters: The Social and Health Costs of Living in a Slum

For as long as governments have been monitoring the human development performance of their countries, achievements in various sectors have tended to focus on only two geographical areas: rural and urban. In general, statistics show that urban populations are better off than those living in villages: they tend to enjoy more access to services and generally perform well on a range of human development indicators, including life expectancy and literacy. However, evidence suggests that in many developing countries, urban poverty is becoming as severe and as dehumanizing as rural poverty. This Report presents for the first time data disaggregated at urban, rural, slum and non-slum levels. The findings show remarkable similarities between slums and rural areas:

- In low-income countries, such as Bangladesh, Ethiopia, Haiti, India, Nepal and Niger, 4 out of every 10 slum children are malnourished, a rate that is comparable to rural areas of those countries.
- Likewise, in some cities, such as Khartoum and Nairobi, the prevalence of diarrhoea is much higher among slum children than among rural children. In slums, child deaths are attributed not so much to lack of immunization against measles, but inadequate living conditions, such as lack of access to water and sanitation or indoor air pollution, which lead to water-borne and respiratory illnesses among children.
- Malnutrition and hunger in slums is almost the same as in villages in some countries. In India, for instance, slum dwellers suffer slightly more from malnutrition than the rural population of the country.
- Recent data on HIV/AIDS shows that in various sub-Saharan African countries, HIV prevalence is significantly higher in urban areas than in rural areas, and is also higher in slums than in non-slum urban areas. Moreover, slum women are particularly at risk, with HIV prevalence rates that are higher than that of both men and rural women.
- Age pyramids for slum and rural populations in several countries show similar patterns: both groups tend to be younger and generally die sooner than non-slum urban populations, which tend to have the lowest child mortality rates and the highest life expectancy rates.

The above examples show that slum populations are not benefiting from the advantages and opportunities offered by cities. Studies have also shown that children living in a slum within a city are more likely to die from pneumonia, diarrhoea, malaria, measles or HIV/AIDS than those living in a non-slum area within the same city; many of these diseases are the result of poor living conditions prevalent in slums rather than the absence of immunization coverage or lack of health facilities. In many cases, poverty, poor sanitation and indoor air pollution make children and women living in slums more vulnerable to respiratory illnesses and other infectious diseases than their rural counterparts. For many slum dwellers, overcrowding, housing located in hazardous areas and the threat of eviction affects other livelihood issues, such as employment. Some studies have also found a strong correlation between where people live and their chances of finding a job. One such study in France showed that job applicants residing in poor neighbourhoods were less likely to be called for interviews than those who lived in middle- or high-income neighbourhoods. Another study in Rio de Janeiro found that living in a *favela* (slum) was a bigger barrier to gaining employment than being dark skinned or female, a finding that confirms that "where we live matters" when it comes to health, education and employment.

These findings reveal "a tale of two cities within one city". Thus, policymakers, governments, development practitioners and funding agencies should no longer see the city as one homogenous entity. Slums are not only a manifestation of poor housing standards, lack of basic services and denial of human rights, they are also a symptom of dysfunctional urban societies where inequalities are not only tolerated, but allowed to fester.

This Report unfolds a new urban reality where slum dwellers die earlier, experience more hunger, have less education, have fewer chances of employment in the formal sector and suffer more from ill-health than the rest of the urban population.

The international community cannot afford to ignore slum dwellers because, after rural populations, they represent the second largest target group for development interventions – and their size is set to grow as the developing world becomes more urbanized. The Millennium Development Goals thus have to target this disadvantaged and vulnerable group of people; if they are ignored, it is very likely that the Millennium Development Goals will not be achieved.

The State of the World's Slums

The growth of slums in the last 15 years has been unprecedented. In 1990, there were nearly 715 million slum dwellers in the world. By 2000 – when world leaders set the target of improving the lives of at least 100 million slum dwellers by 2020 – the slum population had increased to 912 million. Today, there are approximately 998 million slum dwellers in the world. UN-HABITAT estimates that, if current trends continue, the slum population will reach 1.4 billion by 2020.

One out of every three city dwellers lives in slum conditions. Some slums become less visible or more integrated into the urban fabric as cities develop and as the incomes of slum dwellers improve. Others become permanent features of urban landscapes. Both types of slums have carved their way into modern-day cities, making their mark as a distinct category of human settlement that needs to be looked at over and above the traditional rural-urban dichotomy.

Slum dwellers often live in difficult social and economic conditions that manifest different forms of deprivation – material, physical, social and political. Throughout this Report, UN-HABITAT uses an operational definition of slums – one with measurable indicators at household level. Four of the five indicators measure physical expressions of slum conditions: lack of water; lack of sanitation; overcrowding; and non-durable housing structures. These indicators – known also as shelter deprivations – focus attention on the circumstances that surround slum life, depicting deficiencies and casting poverty as an attribute of the environments in which slum dwellers live. The fifth indicator – security of tenure – has to do with legality, which is not as easy to measure or monitor, as the status of slum dwellers often depends on *de facto* or *de jure* rights – or lack of them. By knowing how many slum dwellers there are in cities and what shelter deprivations they suffer most from, it becomes possible to design interventions that target the most vulnerable urban populations.

Not all slums are homogeneous and not all slum dwellers suffer from the same degree of deprivation. In this Report, UN-HABITAT presents an analysis of the degrees of shelter deprivation in some selected countries and regions. This type of information helps to connect monitoring information to policy, making more rigorous and systematic the development of programmes and interventions that are better attuned to specific locations and situations.

The *State of the World's Cities Report 2006/7* provides an overview of the state of the world's slums with regards to the five indicators. The following provides a summary of the main findings.

Lack of durable housing

It is estimated that 133 million people living in cities of the developing world lack durable housing. Non-durable or non-permanent housing is more prevalent in some regions than in others; over half the urban population living in non-permanent houses resides in Asia, while Northern Africa has the least numbers of people living in this kind of housing. However, UN-HABITAT analysis shows that global figures on housing durability are highly underestimated due to the fact that durability is based primarily on permanence of individual structures, not on location or compliance with building codes. Moreover, estimates are made taking into account only the nature of the floor material, since information on roof and wall materials is collected in very few countries. For instance, figures indicate that over 90 per cent of the world's urban dwellings have permanent floors, but when estimates are made combining floor, roof and wall materials, this figure drops dramatically in several countries. In Bolivia, for instance, when only floor material is considered, 83.8 per cent of the urban population is counted as having durable housing, but when wall and roof materials are taken into account, this figure drops to 27.7 per cent. Statistical analysis presented in this Report shows that when more physical structure variables are combined, the results provide a more realistic image of housing durability.

Lack of sufficient living area

Overcrowding is a manifestation of housing inequality and is also a hidden form of homelessness. In 2003, approximately 20 per cent of the developing world's urban population – 401 million people – lived in houses that lacked sufficient living area (with three or more people sharing a bedroom). Two-thirds of the developing world's urban population living in overcrowded conditions resides in Asia; half of this group, or 156 million people, reside in Southern Asia. This Report shows how living conditions, including overcrowding and poor ventilation, are related to rates of illness, child mortality and increase in negative social behaviors. It stresses that the risk of disease transmission and multiple infections becomes substantially higher as the number of people crowded into small, poorly ventilated spaces increases. After presenting overcrowding data by region, the Report highlights some of the local variances of the definition.

Lack of access to improved water

Although official statistics reflect better water coverage in urban areas than in rural areas, various surveys show that in many cities, the quantity, quality and affordability of water in low-income urban settlements falls short of acceptable standards. Improved water provision in the world's urban areas was reported to be as high as 95 per cent in 2002. This statistic, however, presents an overly optimistic picture since "improved" provision of water does not always mean that the provision is safe, sufficient, affordable or easily accessible. For example, further analysis reveals that getting water from a tap is a luxury enjoyed by only two-third of the world's urban population; less than half of this group (46 per cent) have piped water within their dwelling; 10 per cent rely on public taps, while 8 per cent have access only to manually pumped water or protected wells. Inter-regional differences indicate that Africa has the lowest proportion (38.3 per cent) of urban households with access to piped water, while the Latin American and Caribbean region has the highest (89.3 per cent). Sometimes, even when water is available, it may not be affordable or safe to drink. In Addis Ababa, Ethiopia, a UN-HABITAT survey showed that the proportion of low-income urban residents with access to water supply dropped to 21 per cent from 89 per cent when the operational definition of "access" included variables such as cost and quality. Poor access to water in urban areas has a direct bearing on rates of water-borne or water-related diseases in urban areas, a phenomenon that is explored in some depth in the latter part of the Report.

Lack of access to improved sanitation

Over 25 per cent of the developing world's urban population – or 560 million city residents – lack adequate sanitation. Asia alone accounts for over 70 per cent of this group, mainly because of the large populations of China and India; in 2000, lack of sanitation coverage in Chinese cities was reported to be approximately 33 per cent. UN-HABITAT analysis shows that while cities in South-Eastern Asia and Southern Asia have made significant progress in recent years to improve sanitation coverage in urban areas, access lags far behind in sub-Saharan Africa and Eastern Asia, where 45 per cent and 31 per cent of the urban population still lacks access to improved sanitation, respectively. However, some countries in Southern Asia have extremely low coverage, notably Afghanistan, where only 16 per cent of the urban population has access to a proper toilet. Lack of access to an adequate toilet not only violates the dignity of the urban poor, but also affects their health. Every year, hundreds of thousands of people die as a result of living conditions made unhealthy by lack of clean water and sanitation. The number of deaths attributable to poor sanitation and hygiene alone may be as high as 1.6 million per year – five times as many people who died in the 2004 Indian Ocean tsunami. A disproportionate share of the labour and health burden of inadequate sanitation falls on women, who have to wait for long periods to gain access to public toilets or have to bear the indignity of defecating in the open. Inadequate sanitation is therefore something of a "silent tsunami" causing waves of illness and death, especially among women and children. As this Report shows, mortality rates are quite often linked to whether or not children or their mothers have access to adequate sanitation facilities; in the city of Fortaleza in Brazil, for instance, child mortality rates dropped dramatically when sanitation coverage increased.

Lack of secure tenure

Mass evictions of slum and squatter settlements in various cities in recent years suggest that security of tenure is becoming increasingly precarious, particularly in cities of sub-Saharan Africa and Asia, where evictions are often carried out to make room for large-scale infrastructure or city "beautification" programmes. A global survey in 60 countries found that 6.7 million people had been evicted from their homes between 2000 and 2002, compared with 4.2 million in the previous two years. Many of these evictions were carried out without legal notice or without following due process. Improving the tenure of urban households could go a long way in preventing evictions, but operationalizing security of tenure for the purpose of global monitoring remains difficult. At present, it is neither possible to obtain household-level data on secure tenure in most countries, nor to produce global comparative data on various institutional aspects of secure tenure, as data on secure tenure is not regularly collected by censuses or household surveys. However, non-empirical information suggests that between 30 per cent and 50 per cent of urban residents in the developing world lack security of tenure. Although home ownership is regarded as the most secure form of tenure, evidence from around the world also suggests that ownership is not the norm in both the developed and the developing world, and is not the only means to achieve tenure security. In fact, informal – or illegal – growth has become the most common form of housing production in the developing world, where gaining access to housing through legal channels is the exception rather than the rule for the majority of urban poor households. UN-HABITAT and its partners are currently working on the preparation of a global monitoring system that could in the future provide a framework to assist governments at local and national levels to produce estimates at household level on how many people have secure tenure, using an agreed-upon methodology in terms of definitions, indicators and variables.

30 Years of Shaping the "Habitat" Agenda: Policies and Practices that have Worked

Since the first UN Conference on Human Settlements (Habitat I) took place in Vancouver in 1976, governments and the international community have adopted and implemented a range of human settlements policies and programmes with mixed results. Many programmes were unsuccessful; others, while successful at the pilot stages, could not be scaled up and remained small "islands of success" that did not have a significant impact on urban poverty levels or slum growth rates. Few interventions had an economic or social impact on urban poor populations.

Getting urban poverty on the development agenda has been a struggle in the last thirty years. Silence or neglect have characterized most policy responses. However, with the adoption of the Millennium Declaration in 2000, urban poverty is now being brought to the centre stage of the global development agenda. As part of its mandate to assess the performance of countries on Millennium Development Goal 7, target 11 – to improve the lives of at least 100 million slum dwellers by 2020 – UN-HABITAT built a broad architecture for global monitoring and reporting. As part of this process, the organization has evaluated the performance of more than 100 countries to see if they were "on track", "stabilizing", "at risk" or "off track" vis-à-vis the slum target. Three criteria were used to rate countries: annual slum growth rate; slum percentage; and slum population.

Analysis of the results revealed some interesting findings: countries that had successfully reduced slum growth rates, slum proportions and slum populations in the last 15 years shared many attributes: their governments had shown long-term political commitment to slum upgrading and prevention; many had undertaken progressive pro-poor land and housing reforms to improve the tenure status of slum dwellers or to improve their access to basic services; most used domestic resources to scale up slum improvements and prevent future slum growth; and a significant number had put in place policies that emphasized equity in an environment of economic growth. In many countries, improvements in just one sector, such as sanitation, had a significant impact on slum reduction, particularly in cities where inhabitants suffered from only one or two shelter deprivations.

Another major finding of this analysis of country performance on the slum target showed that those countries doing well in managing slum growth had highly centralized systems and structures of governance; even in cases where decentralized systems existed, policy actions for slum prevention and upgrading were implemented through centralized interventions. This was possible because central governments – having command and control – could put in place measures and resources to ensure cohesiveness in the design and implementation of slum upgrading projects. Central governments had the capacity to put forward legislation and pro-poor policy reforms to tackle basic shelter deprivations – reforms that require political support at the national level before being filtered downward to local levels of government. These central governments have been able to set up the institutional arrangements, allocate important budgets, and execute projects to effectively meet their targets and commitments. In countries such as Brazil, Egypt, Mexico, South Africa, Thailand and Tunisia, implementation of inclusive policies, land reforms, regularization programmes and commitment to improve the lives of the urban poor by the top leadership were key to the success of slum upgrading or prevention programmes. These countries developed either specific slum upgrading and prevention policies or have integrated slum upgrading and prevention as part of broader poverty reduction policies and programmes. They have done this not only to respond to social imperatives, but also to promote national economic development. Central governments in these countries, among others, have played a critical role, not just in the physical improvement of slums, but also in ensuring that investments are made in other sectors as well, such as education, health, sanitation and transport, which have benefited slum communities.

This perhaps is a prelude to a change in governance paradigms, in which central governments and local authorities would develop a more coordinated approach in the development and implementation of policies, with central governments taking the lead in urban poverty reduction programmes as they would have the power and authority to institute pro-poor reforms and the mandate and ability to allocate resources to various priority sectors. On the other hand, local authorities would be able to locally coordinate operational actions bringing together different actors.

This Report also clearly shows that not all countries struggling to cope with high slum growth rates have shied away from committing to change. Some sub-Saharan African countries, namely Burkina Faso, Senegal and Tanzania, have in recent years shown promising signs of growing political support for slum upgrading and prevention that includes reforms in policies governing land and housing.

Some low- or middle-income countries that are starting to stabilize or reverse slum growth rates, including Colombia, El Salvador, Philippines, Indonesia, Myanmar and Sri Lanka, did not wait to achieve important milestones in economic growth in order to address slums. These countries have managed to prevent slum formation by anticipating and planning for growing urban populations – by expanding economic and employment opportunities for the urban poor, by investing in low-cost, affordable housing for the most vulnerable groups and by instituting pro-poor reforms and policies that have had a positive impact on low-income people's access to services. These countries give hope and direction to other low-income countries by showing that it is possible to prevent slum formation with the right policies and practices.

What comes out clearly in this Report is that slum formation is neither inevitable nor acceptable. "Running the poor out of town" – through evictions or discriminatory practices – is not the answer: rather, helping the poor to become more integrated into the fabric of urban society is the only long-lasting and sustainable solution to the growing urbanization of poverty. Ultimately, as the developing world becomes more urban and as the locus of poverty shifts to cities, the battle to achieve the Millennium Development Goals will have to be waged in the world's slums.

STATE OF THE WORLD'S cities 2006/7

The Millennium
Development Goals and
Urban Sustainability:
30 Years of Shaping
the Habitat Agenda

Part One

Cities, Slums and the Millennium Development Goals

This Part highlights the major urbanization trends in the world and provides a global and regional overview of slums. It also presents the first findings of country performance on the Millennium Development Goals' slum target using a new global scorecard developed by UN-HABITAT. The Part concludes by emphasizing the importance of implementing the Millennium Development Goals at the city level, and more importantly, in slums. A table linking cities and slums to the Millennium Development Goals is also presented.

Bangkok, Thailand JEARANAIKUL/UNEP/STILL PICTURES

ยอร์ค... คุ้มค่าไปถึงวันหน้า
YORK

1.1 'City-zens' of the World: Urban Trends in the 21st Century

Havana, Cuba RASNA WARAH

2007: The dawn of the urban millennium

The year 2007 will mark a turning point in human history: the world's urban population will for the first time equal the world's rural population. Although it is difficult to predict on which day or month this radical transformation will occur, what is certain is that this milestone will herald the advent of a new urban millennium: a time when one out of every two people on the planet will be a "city-zen".

Cities, whether small municipalities of 2,000 inhabitants or massive agglomerations of 10 million people or more, are becoming a widespread phenomenon. The global urban population has quadrupled since 1950, and cities of the developing world now account for over 90 per cent of the world's urban growth.

In 2005, the world's urban population was 3.17 billion out of world total of 6.45 billion. Current trends predict the number of urban dwellers will keep rising, reaching almost 5 billion by 2030. Between 2005 and 2030, the world's urban population is expected to grow at an average annual rate of 1.78 per cent, almost twice the growth rate of the world's total population. As more and more people occupy cities, the population of rural settlements around the globe will begin to contract after 2015, decreasing at an average annual rate of -0.32 through 2030 – a decrease of more than 155 million people over 15 years.[1]

Asia and Africa will host the largest urban populations

Whereas Europe, North America and Latin America experienced intense urbanization – the increased concentration of people in cities rather than in rural areas – and rapid urban growth through the mid-20th century, the trend has now shifted to the developing regions of Asia and Africa. In-migration, reclassification and natural population increase are contributing to a rapid urban transformation of these regions. Annual urban growth rates are highest in sub-Saharan Africa (4.58 per cent), followed by South-Eastern Asia (3.82 per cent), Eastern Asia (3.39 per cent), Western Asia (2.96 per cent), Southern Asia (2.89 per cent) and Northern Africa (2.48 per cent). The developed world's cities are growing at a slower pace, averaging 0.75 per cent a year.

Latin America is the most urbanized region in the developing world, with 77 per cent of its population – 433 million people – living in cities.[2] The urbanization of Latin America has yet to reach its peak; by 2015, it is predicted that 81 per cent of its population will reside in urban areas. In terms of sheer numbers, however, Asia has the largest urban population (with more than 1.5 billion people inhabiting its cities) even though slightly less than 40 per cent of its population is urbanized. The total population of cities in the developing regions of the world already exceeds that of cities in all of the developed regions (by 1.3 billion people). If predictions prove accurate, by 2030, nearly 4 billion people – 80 per cent of the world's urban dwellers – will live in cities of the developing world.

Asia and Africa will continue to dominate global urban growth through 2030. Currently the least urbanized regions in the world, with 39.9 per cent and 39.7 per cent of their populations living in cities in 2005, respectively, by 2030, both regions will become predominantly urban, Asia with 54.5 per cent of its population living in cities, and Africa with 53.5 per cent of its population urban. Asia alone will account for more than half the world's urban population (2.66 billion out of a global urban population of 4.94 billion); and the urban population of Africa (748 million) will by 2030 be larger than the total population of Europe at that time (685 million).

Defining "Urban"

The United Nations defines an ***urban agglomeration*** as the built-up or densely populated area containing the city proper, suburbs and continuously settled commuter areas. It may be smaller or larger than a metropolitan area; it may also comprise the city proper and its suburban fringe or thickly settled adjoining territory.

A ***metropolitan area*** is the set of formal local government areas that normally comprise the urban area as a whole and its primary commuter areas.

A ***city proper*** is the single political jurisdiction that contains the historical city centre.

However, an analysis of countries shows that different criteria and methods are currently being used by governments to define "urban":

- 105 countries base their urban data on **administrative** criteria, limiting it to the boundaries of state or provincial capitals, municipalities or other local jurisdictions; 83 use this as their sole method of distinguishing urban from rural.
- 100 countries define cities by **population size or population density**, with minimum concentrations ranging broadly, from 200 to 50,000 inhabitants; 57 use this as their sole urban criterion.
- 25 countries specify **economic** characteristics as significant, though not exclusive, in defining cities – typically, the proportion of the labour force employed in non-agricultural activities.
- 18 countries count the availability of **urban infrastructure** in their definitions, including the presence of paved streets, water supply systems, sewerage systems, or electric lighting.
- 25 countries provide **no definition** of "urban" at all.
- 6 countries regard their **entire populations** as urban.

Sources: United Nations: Principles and Recommendations for Population and Housing Censuses (1998) and World Urbanization Prospects: The 2003 Revision.

Small and intermediate cities will absorb most urban growth

Small cities with less than 500,000 inhabitants and intermediate cities with between 1 and 5 million inhabitants, not megacities (defined as cities with 10 million or more people), will continue to absorb most of the urban population around the world well into the future. More than 53 per cent of the world's urban population lives in cities of fewer than 500,000 inhabitants, and another 22 per cent of the global urban population lives in cities of 1 to 5 million inhabitants. These cities are significant sites of social and economic activity, often serving as centres of trade and destinations for rural migrants.[3] They are often the first places where the social urban transformation of families and individuals occurs; by offering economic linkages between rural and urban environments, they can provide a "first step" out of poverty for impoverished rural populations and a gateway to opportunities in larger cities. In Eastern Africa, South-Eastern Asia, the Caribbean and Europe, cities of fewer than 500,000 are particularly prevalent, hosting approximately two-thirds of those regions' urban residents.

FIGURE 1.1.1 PROPORTION OF URBAN POPULATION BY REGION, 1950-2030

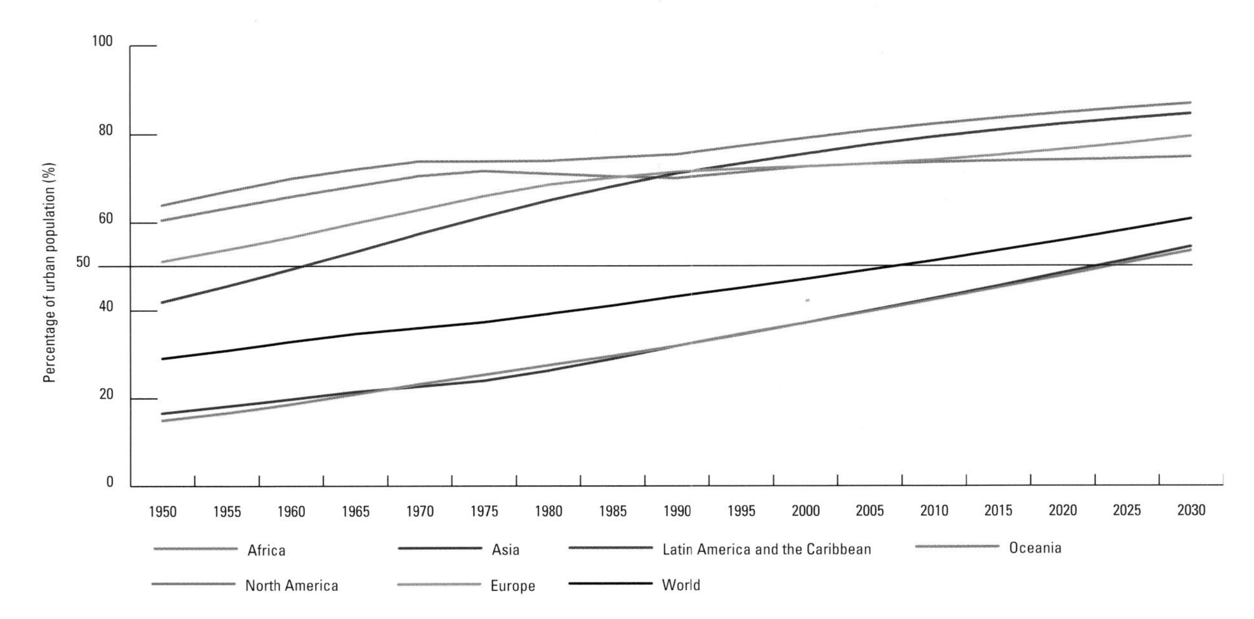

Source: United Nations, World Urbanization Prospects: The 2003 Revision.

Most of the world's urban population will continue to live in small cities over the next decade, but intermediate cities are predicted to grow at a faster rate: between 2000 and 2015, cities of fewer than 500,000 will likely increase their populations by 23 per cent, while cities of 1 to 5 million are predicted to increase their populations by 27 per cent. By 2015, Asia will have gained 37 cities of 1 to 5 million people, rising to a total of 253; Africa will have gained 20, totaling 59; and Latin America and the Caribbean will have gained 16, rising to a total of 65.

The emergence of the "metacity"

Although "megacities" of more than 10 million inhabitants have been around since the 1950s, when New York and Tokyo were the largest cities in the world, "metacities"[4] – massive conurbations of more than 20 million people – are now gaining ground in Asia, Latin America and Africa.

Called "hypercities" by some,[5] cities of more than 20 million inhabitants represent a new type of settlement above and beyond the scale of megacities. Spurred by economic development and increased population, they gradually swallow up rural areas, cities and towns, becoming multi-nuclear entities counted as one. The world has never before known so many cities as large as these metacity agglomerations. Many of these cities have populations larger than entire countries; the population of Greater Mumbai (which will soon achieve metacity status), for instance, is already larger than the total population of Norway and Sweden combined.

Contrary to common perception, however, the world's largest cities are home to only 4 per cent of the world's total population, and most have grown at the relatively slow rate of about 1.5 per cent annually. Although new research techniques that combine population statistics with satellite imagery reveal that these huge urban agglomerations may already be home to 7 per cent of the world's population, they still represent just a small minority of cities worldwide.[6]

The first metacity came into being in the mid-1960s when Tokyo's population crossed the 20 million inhabitant threshold. Tokyo continues to be the only metacity in the world today, with a population in excess of 35 million people – more than the total population of Canada. In less than a decade, however, Mumbai, Delhi and Mexico City will have joined the league of metacities, closely followed by São Paulo, New York, Dhaka, Jakarta and Lagos, each with more than 17 million inhabitants. By 2020, all of these cities are expected to attain metacity status.[7] Lagos is experiencing an exceptional growth rate – more than 5 per cent per year through 2005 – and is expected to continue growing faster than the other largest cities of the world through 2020.

On average, the largest cities grow more slowly than cities of 5 to 10 million people, and they account for only about 9 per cent of the world's urban population. The impact of the largest cities on their regions, however, is great. Between 1950 and 2020, the New York-Newark metropolitan area is expected to have increased its population by only 40 per cent, whereas Mumbai will have grown by 88 per cent, and Dhaka will have grown to more than 50 times its size in 1950. Throughout the past half-century, the largest cities of the developing world have had to absorb astounding increases in their urban populations while many cities in the developed regions have grown considerably less, or, as in the case of London, decreased in population.

FIGURE 1.1.2 RURAL AND URBAN POPULATION BY REGION IN 2005 AND 2030

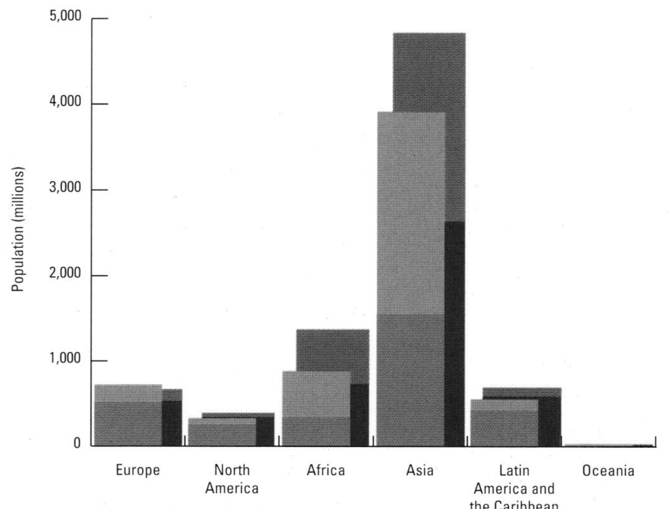

Africa is the least urbanized continent but by 2030 its urban population will exceed the total population of Europe.

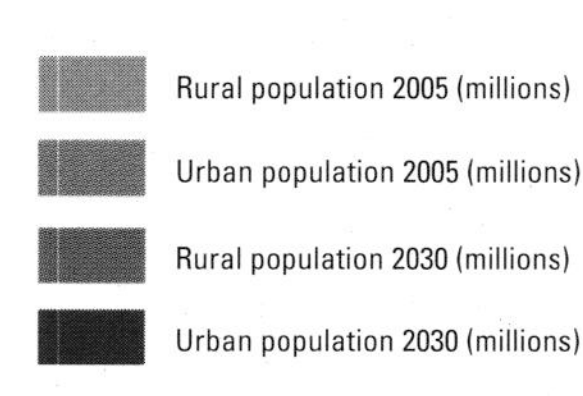

Source: United Nations, World Urbanization Prospects: The 2003 Revision.

Street Scene, Morocco NATSUO ITO

More than 53 per cent of the world's urban population lives in cities of fewer than 500,000 inhabitants.

FIGURE 1.1.3 POPULATION BY CITY SIZE, 2000-2015

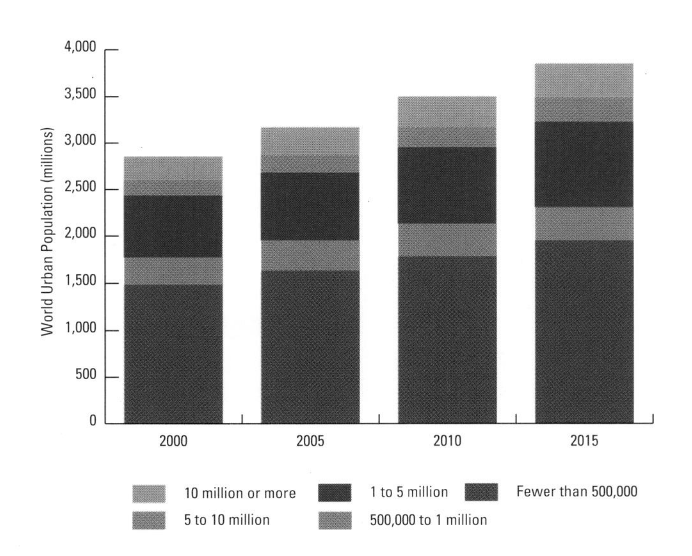

Source: United Nations, World Urbanization Prospects: The 2003 Revision.

City-regions will require new forms of coordinated management for sustainability

By 2020, all but 4 of the world's largest cities will be in developing regions, 12 of them in Asia alone. While still few in number, these metacities point to new forms of urban planning and management, leading to the growth of city-regions and "metropolitanization".

As cities increase in size, metropolitanization is becoming a progressively more dominant mode of urbanization, particularly in megacities and newly emerging metacities. Metropolitanization can take various forms: it may involve densely settled regions in which villagers or people living in suburbs commute to work in the nearby cities but where many of the production and service activities are located in rural areas and suburbs; it may mean that a stagnating and declining population and economic base in the core of a city shifts to nearby secondary cities; or it may refer to the development of inter-connected systems of cities that create city-regions linked by manufacturing and other activities, such as the Hong Kong/Pearl River Delta region of China.[8]

Metropolitanization calls for new, innovative and more decentralized forms of governance. Already, many large cities are decentralizing governance to the appropriate levels with more municipalities and boroughs managing different parts of the city. This calls for better inter-municipal coordination, more intermediate metropolitan levels of governance, more civil society participation and more autonomy for various parts of this new organism called the metacity.

The scale of environmental impact of metacities and megacities on their hinterlands is also significant and is likely to be a cause for concern in coming decades. For instance, in China, rapid economic growth and urbanization, combined with inconsistent implementation of industrial emissions standards and increased use of motor vehicles, have had a negative impact on the urban environment; the country is the largest producer of greenhouse gases, after the United States, and hosts 16 of the 20 most polluted cities in the world. Managing environmental sustainability, economic sustainability and socio-political sustainability – the three pillars of sustainable urbanization – will require more polycentric forms of governance, more environmentally-friendly legislation and a regional approach to planning and management of human settlements.

Competition between "world cities" will intensify

Cities are more than simply concentrations of people and resources. As hubs of trade, culture, information and industry, cities also articulate and mediate major functions of the global economy. In developed countries, cities generate over 80 per cent of national economic output, while in developing countries, urban economic activity contributes significantly to national revenue, generating up to 40 per cent of gross domestic product.[9] Wealthy world cities are also increasingly operating like city-states and city-regions, independent of regional or national mediation.[10]

Global urban economies are increasingly reliant on advanced producer services for their income: advertising, finance, banking, insurance, law, management consultancy, and other service-based businesses.

Today, several major cities play pivotal roles in global networks, not only producing goods and services and hosting institutions, but also generating related economic and civil society activity in other cities. These "world cities" provide economies of scale and access to resources of local and global significance. Connectivity, economic production and cultural innovation have long kept London, New York, Paris, and Tokyo at the top of the world-city scale, closely followed by cities such as Frankfurt, Hong Kong, Amsterdam, Singapore, São Paulo and Shanghai, which are emerging as trend-setters on the global financial scene. Other cities, such as Dubai and Rotterdam, are becoming global transport hubs, while Bangalore, Seattle and Silicon Valley have emerged as world leaders in the area of information technology.

In the new urban millennium, world city status is beginning to extend to several key cities in developing regions as well, based on their new roles in the global economy and their capacity for linking resources with populations in need. Cities such as Istanbul and Mumbai are already establishing the cultural trends in their countries and regions, and this influence could cross international borders through films, literature and satellite television networks and entertainment. Large cities in the developing world, including Nairobi, Addis Ababa and Bangkok, among others, are increasingly bringing together major national and international partners by hosting international agencies and development partners, and offering avenues for constructive peer exchange, mediation and diplomacy.

At the same time, cities are becoming more competitive with each other. No longer are only the world's most highly recognized cities jockeying for the honour of hosting major international events and corporate headquarters; growing cities of the developing regions are also competing with each other to become important regional, corporate and development centres. The primary economic rival of India's financial capital of Mumbai, for example, is Shanghai in neighbouring China, which has similar global aspirations, but has a much smaller proportion of its population living in poverty.

FIGURE 1.1.4 URBAN GROWTH IN THE WORLD'S LARGEST CITIES, 1950-2020

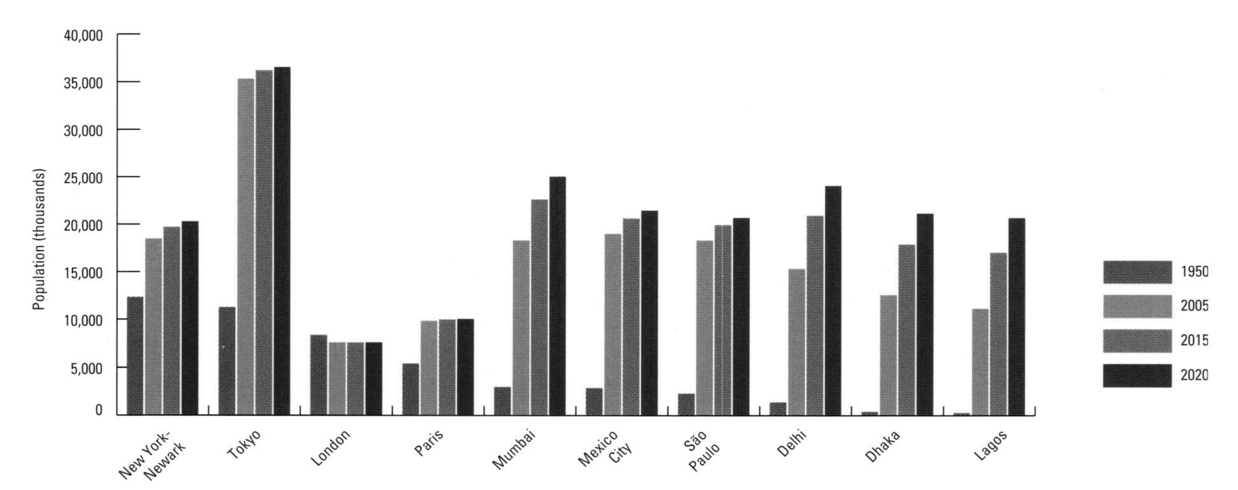

Source: United Nations, World Urbanization Prospects: The 2003 Revision.
Note: Population in 2020 was estimated from population in 2010 and 2015 assuming that trends for these years remain the same.

In developing countries, informal employment comprises one-half to three-quarters of non-agricultural employment. In many of these countries, the informal sector provides more employment opportunities than the formal sector.

Simit Seller, Turkey DARCY VARNEY

The urban economy in the developing world will be largely informal

In the developing world, there has been a trend toward "informalization" of the urban economy, with increasing shares of incomes earned in unregulated employment.[11] In many developing countries, particularly in Africa and Asia, the formal sector has not been able to provide adequate jobs for rapidly growing urban populations, leading to the proliferation of the urban informal sector. In Latin America and the Caribbean, 7 out of 10 new jobs in urban areas are created in the informal sector.[12] Two main processes have significantly contributed to the rise in urban informal activities. One is the failure of the formal sector to provide adequate jobs and income-generating opportunities for a rapidly growing urban population; the other is the growing tendency of the formal sector to contract services out to secondary labour markets, which are mainly in the informal sector.

In developing countries, informal employment comprises one-half to three-quarters of non-agricultural employment. In many of these countries, the informal sector provides more employment opportunities than the formal sector. In sub-Saharan Africa, for instance, the informal sector accounts for about 78 per cent of all non-agricultural employment.[13] In Kenya alone, there are an estimated 5.5 million informal sector workers compared with only about 1.7 million wage-earners in formal establishments.[14] In Asia, 65 per cent of all the non-agricultural employment is in the informal sector. In Latin America and Northern Africa, it is 51 and 48 per cent, respectively. If informal employment in the agricultural sector is also included, the proportion of informal employment in the total labour force is even larger. It is common to attribute the high proportion of informal sector workers in these regions to migration of unskilled workers from rural areas to urban areas. However, in many regions, informal sector employment is the only option available to skilled and educated people who are not absorbed by the formal labour market. In addition to its contribution to employment, the sector also contributes significantly to national economies in terms of income. The sector's share of gross domestic product (GDP) is approximately 41 per cent in sub-Saharan Africa, 31 per cent in Asia, 29 per cent in Latin America and 27 per cent in North Africa.[15]

Women account for a disproportionately larger share of the informal labour force than men, particularly in sub-Saharan Africa and Asia. In developing countries as a whole, more than 60 per cent of women are engaged in informal employment in the non-agricultural sector. In sub-Saharan Africa, about 84 per cent of women are employed in non-agricultural informal activities compared with 63 per cent of men. In Latin America, 58 per cent of women are engaged in non-agricultural informal sector activities compared with 48 per cent of men. Women's participation in both the informal and formal sectors of the urban economy has had a positive impact on their social mobility and political involvement in urban affairs. A study in Dhaka, Bangladesh, for instance, found a positive correlation between women's employment in factories and their level of political participation.[16]

Insecurity will be a growing concern in cities of the developed world

Since the attacks on New York and Washington on 11 September 2001, cities of the developed world have become increasingly concerned about their vulnerability to acts of terrorism. These concerns have been magnified by recent attacks on cities such as London, Madrid, Bali, New Delhi, Nairobi and Dar es Salaam. Because of their dense populations and intricate infrastructure, cities are deeply affected by attacks, which exact heavy physical, psychological and financial costs. It is estimated that the city of New York lost $110 billion in infrastructure, buildings and jobs as a result of the 11 September attacks.[17]

Although terrorism affects both developed and developing countries, the former have in recent years put in place measures to respond to the crisis by increasing their budgets for security and surveillance apparatus and tightening immigration policies. In some cities, such as New York and London, this has meant visible changes in urban form: loss of use of public space, restriction of movement within the city, weakening of popular participation and increased use of security and surveillance equipment in strategic locations. Terrorism and insecurity create fear and change perceptions in cities, leading to situations in which "a bag is no longer a bag, but a bomb". A 2003 report by the City of London Police found that almost one in ten Londoners worried about the threat of terrorism on a daily basis. London authorities are said to have increased police patrols and installed more surveillance cameras in strategic locations, making the city one of the most closely surveilled spaces in the United Kingdom, and perhaps the world.[18] Some are concerned that the implementation of these measures could mean that less money will be available for social services, such as health and education, which could become a source of friction and conflict in the future.

While terrorism dominates the concerns of cities of the developed world, most developing country cities are contending with other forms of insecurity that threaten their lives and livelihoods. The security of the urban poor, in particular, is affected by their health status, which influences both their ability to work and their ability to escape poverty. The HIV/AIDS pandemic has particular implications for urban security as it leads to loss of household income, growth in the phenomenon of orphaned street children and disintegration of the family unit. Many urban poor families also face the constant threat of eviction. Insecurity is exacerbated by insecure tenure with respect to both housing and land.

Divided cities: Cities are, and will continue to be, sites of extreme inequality

In recent years, an increasing number of countries have opened up markets and expanded political freedoms. Others have made impressive gains in economic growth. But democracy and economic growth have not helped reduce inequalities in much of the world; the wealthiest 20 per cent of the world's people account for 86 per cent of private consumption, while the poorest account for just 1 per cent. In the past six years, 23 million more Latin Americans slipped into poverty, and most African countries – with the exception of Botswana and Egypt – are poorer today than they were in the 1970s. Despite impressive economic growth rates in both China and India in the last decade, these two countries have not been able to join the ranks of "high-income" countries, and have been unable to bridge the income gap between rural and urban populations and between the urban rich and the urban poor.[19]

These inequalities manifest themselves most starkly in cities. In many cities of both the developed and the developing world, economic growth has not resulted in prosperity for all. On the contrary, intra-city inequalities have risen as the gap between the rich and the poor has widened. Although poverty remains a primarily rural phenomenon, large sections of the population in urban areas are suffering from extreme levels of deprivation.

Cairo ©ALESSANDRO BOLIS. IMAGE FROM BIGSTOCKPHOTO.COM

One out every three city dwellers – nearly one billion people – lives in a slum.

The highest levels of income inequality exist in Africa and Latin America, the least and most urbanized developing regions, respectively. This inequality is most stark in urban areas, and particularly in large cities. Although the proportion of poor people in rural areas is larger than the proportion in urban areas, there are more poor people living in Latin America's cities than its rural areas. In 1999, for example, only 77 million of the region's 211 million poor lived in rural areas, while the remaining 134 million lived in urban areas, although the proportion of rural poor was much greater than that of the urban poor – at 64 per cent and 34 per cent, respectively. In Africa, the proportion of people living in poverty in rural areas is 59 per cent, compared with 43 per cent in cities, a gap that is likely to shrink in an environment of economic decline.[20] Sub-Saharan African countries have some of the world's highest levels of urban poverty, extending to more than 50 per cent of the urban populations in Chad, Niger and Sierra Leone. Countries of Northern Africa and Western Asia have urban poverty levels near or below 20 per cent. In Asia, India has the highest urban poverty levels, at 30 per cent.

In Latin America and the Caribbean, levels of urban poverty vary widely, from 8 per cent of the urban population in Colombia to 57 per cent in Honduras. In some countries around the world – notably Nigeria, Egypt and Trinidad and Tobago – urban and rural poverty percentages are almost equal.[21] However, income-based statistics should be viewed with caution as the true extent of urban poverty is likely higher than they suggest. The high cost of non-food items, such as transport, health, education, and water in cities – and poor living conditions, including inadequate housing and poor access to water and sanitation – impact the ability of the urban poor to rise out of poverty. When these items are included to measure poverty, poverty estimates for urban areas are likely to rise significantly.

■ Slums are emerging as a dominant and distinct type of settlement in cities of the developing world

One out every three city dwellers – nearly one billion people – lives in a slum. The vast majority of slums – more than 90 per cent – are located in cities of the developing world, which are also absorbing most of the world's urban growth.[22]

Urbanization has become virtually synonymous with slum growth, especially in sub-Saharan Africa, Western Asia and Southern Asia, where annual slum and urban growth rates are almost identical. Annual slum and urban growth rates are highest in sub-Saharan Africa, 4.53 per cent and 4.58 per cent respectively, nearly twice those of Southern Asia, where slum and urban growth rates are 2.2 per cent and 2.89 per cent, respectively. In Western Asia, slums and cities are growing at a similar pace, 2.71 per cent and 2.96 per cent respectively. Northern Africa is the only sub-region where slum growth rates are declining, largely due to positive measures taken by individual countries to address the plight of slum dwellers. Eastern Asia, South-Eastern Asia and Latin America and the Caribbean are also regions where annual slum growth rates have not kept pace with annual urban growth rates. Nonetheless, these regions continue to have large numbers of their urban populations residing in slums.

Sub-Saharan Africa has the highest prevalence of slums in the world – 71.8 per cent of its urban population lives in slums – and in the last 15 years, the number of slum dwellers in the region has almost doubled, from 101 million in 1990 to 199 million, in 2005. Given the high slum growth rate in the sub-region, the number of slum dwellers is projected to double by 2020, reaching nearly 400 million, and overtaking the slum populations of both Southern Asia and Eastern Asia, where slum populations are projected to rise to 385 million and 299 million, respectively. In terms of absolute numbers, Asia still has the largest share of the world's slum

Charcoal vendor, Kibera, Nairobi HIROSHI SATO

Sub-Saharan African countries have some of the world's highest levels of urban poverty.

population; in 2005, the region housed more than half the world's slum dwellers, or 581 million people.

Slums in many cities are no longer just marginalized neighbourhoods housing a relatively small proportion of the urban population; in many cities, they are the dominant type of human settlement, carving their way into the fabric of modern-day cities, and making their mark as a distinct category of human settlement that now characterizes so many cities in the developing world. Although slums do not directly denote levels of urban poverty, their prevalence in a city can be an indicator of urban inequality. UN-HABITAT projections indicate that the number of slum dwellers in the world will rise to 1.4 billion by 2020 if no remedial action is taken.

Endnotes

1. All urban population statistics are drawn from the United Nations Department of Economic and Social Affairs, Population Division, *World Urbanization Prospects: The 2003 Revision,* unless otherwise noted.
2. Of the more-developed regions, Australia and New Zealand have the highest proportion of their populations living in cities: 91.6 per cent in 2005.
3. Satterthwaite & Tacoli 2003.
4. Term coined by UN-HABITAT for cities with populations of more than 20 million.
5. Davis, 2004.
6. The Center for International Earth Science Information Network 2005.
7. This prediction is based on a linear trend model, using the population growth between 2010 and 2015 as the baseline to extend the population to 2020.
8. UN-HABITAT/DFID 2002.
9. UN-HABITAT 2002c.
10. Taylor 2005.
11. Cohen 2004.
12. Inter-American Development Bank 2004.
13. International Labour Organization 2002.
14. Bindra 2005.
15. International Labour Organization 2002.
16. Sachs 2005.
17. Cohen 2004.
18. Coaffee 2004.
19. United Nations 2005a.
20. Ibid.
21. World Bank 2002 estimates.
22. All slum data drawn from UN-HABITAT Global Urban Observatory. For a definition of "slum", see Chapter 1.2.

PETER HALL

WHY SOME CITIES FLOURISH WHILE OTHERS LANGUISH

There is a vital question for cities, but the answer eludes urbanists: at a time when the knowledge economy is becoming all-pervasive, what makes cities innovative? Why do cities flourish creatively, but then languish? Why are Athens or Florence no longer leading creative cities? Why have Guangzhou and Shanghai taken the places of Manchester and Detroit? How do a few cities, including London and New York, manage to retain their edge?

London MJS

Economists approach this question through regression equations to try to identify the critically important ingredient in successful cities. But this may not provide a robust explanation for all or most cities over time. There is another way, through economic and social history: carefully dissecting the sequence and combination of causes that facilitated the emergence of creative cities. The danger with this approach, too, is that it may fail to generate good general explanations.

Yet it need not. In my own work on creative cities, common themes emerge. All were leading cities of their age economically; invariably they were the centres of trading empires. They were at the economic forefront, or near it. Thus, they became magnets for people with ability, who migrated from far corners of their far-flung empires. It is no accident that key roles were played by outsiders: non-citizen Metics in Periclean Athens, Jews in the Vienna of the early 1900s, artists like Pablo Picasso in Paris shortly after. The immigrants considered themselves half inside, half outside the established societies in which they lived. They became creative lightning rods of a sort, illuminating the underlying tensions inside these societies. Occasionally, as in Berlin in the 1920s, such tensions tore those societies apart. Generally, the results were happier.

Technological innovation shows similar features – but also interesting differences. Manchester in 1780, Detroit in 1910 and Silicon Valley in 1960 were upstart cities or city-regions, egalitarian places that welcomed new talent, stressing individual self-improvement and mutual education. Their people engaged in extraordinary chains of innovation, through networks that – paradoxically – were simultaneously competitive and cooperative. There are amazing parallels between Lancashire during the period between 1760 and 1830 and Silicon Valley since 1960: one innovation stimulated another, in long and complex chains. Regions like these became creative because of an extraordinary process of mutual learning and mutual stimulation.

Very well, policymakers may say; what are the lessons for the places now competing to become the 21st century equivalents? One point is clear: the candidates are no longer individual cities but the megacity-regions of Southeast England, the Northeastern seaboard of the United States and the Yangtze River Delta in China, clustered around global cities such as London, New York and Hong Kong. Here we find the 21st century equivalents of 18th century Manchester or 19th century Berlin. Centres of concentrated innovative power, now diffusing into neighbouring cities and towns through networks of information exchange: London to Reading and Milton Keynes, Shanghai to Suzhou.

The key is information. Knowledge is the new production factor, and the truly vital information is exchanged face-to-face, brain-to-brain, through local networks in quite concentrated downtown areas, such as the City of London or Downtown Manhattan. So the critical question now concerns the pattern of diffusion and reconcentration, and the limits to that process, within these multi-centred city-regions, set by some critical time limit from the central city.

Information, the raw material of the knowledge economy, is manipulated by workers in the advanced producer services, who generate incomes that trigger a dazzling array of consumer services. Megacity-regions compete in a winner-takes-all contest. The most successful city-regions grow faster, reinforced by advantages in communications, such as a major international airport or train hub. It is also no accident that London and New York, the great English-speaking cities, represent the twin peaks of this global information economy.

Yet in the coming century, this may change. The 21st century will clearly be the Asian century. China and India are already racing ahead, aiming to regain the leading positions they occupied in past centuries. Their resurgence will come through their great cities – Shanghai, Beijing, Mumbai – and their surrounding megacity-regions. This will be an ongoing East-West economic Olympic Games – but the stadiums, where the prizes will be won, will be urban.

Sir Peter Hall is Professor of Planning at the Bartlett School of Architecture and Planning, University College London. He is the author and editor of over 30 books, including *Cities of Tomorrow* (1988) and *Cities in Civilization* (1998).

China's rising cities

Recent economic reforms coupled with modernization policies have improved the living conditions of millions of people in China. In the last two decades, the world's most populous country has witnessed annual economic growth rates of more than 9 per cent while the proportion of people living on less than $1 a day dropped dramatically from 634 million in 1981 to 212 million in 2001.

The impact of economic growth is most evident in urban areas. China's cities are not only doing better than its rural areas but are largely responsible for the country's economic boom, the effects of which are concentrated in the larger cities. In 2001, per capita disposable income for urban residents was $829 compared to $278 for rural residents. In 1987, the income of the average urban household was almost twice that of the average rural household; today it is almost three times higher.

Some coastal cities, such as Shanghai, have skewed these figures even further. In 2001, China's largest city with a population of 12.7 million had a gross domestic product (GDP) of $4,510 per capita, almost five times the national average. The opening up of the Chinese economy has made Shanghai China's most modern city and a favourite for foreign investment: in the 1990s, foreign investment in the city totalled $45.6 billion. The city, which had only one skyscraper in 1988, today has more than 300. A mass transit system, first-rate sea and river ports, well-developed railway and road networks and two international airports have increased the investment potential of the city and made it a leading centre of international commerce and finance.

The prosperity of China's cities is largely a result of economic reform policies that have a pro-urban focus. Although China already hosts 4 of the 30 largest urban agglomera-

Sources: Linch & Zhi 2003; UN-HABITAT 2005a; TIME 2005; United Nations 2005a; Economy 2005; Worldwatch Institute 2006b; WHO/UNICEF 2000.

tions in the world – Shanghai, Beijing, Tianjin and Hong Kong – since the mid-1980s, the country has been pursuing an aggressive urbanization policy as a means of stimulating both rural and urban economic development. The policy aims to absorb the hundreds of millions of farmers who are flocking to cities as a result of economic reforms and easing of previously strictly enforced "urban residency permits". The residency control system is likely to be completely eradicated in coming years as capital investments are made to improve the infrastructure and economy of urban areas. The aim of China's pro-urban policies is to focus on the development of towns and secondary cities to ease congestion in the larger cities.

Equity grants

Economic growth has also led to growing urban disparities. Prior to the economic reforms, the system made it difficult for villagers to migrate to cities, with the result that slum formation was controlled, whenever possible. But economic reforms saw a significant increase in migration of unemployed workers and farmers to cities, with the result that some inner-city and peri-urban areas have been suffering from a gradual deterioration of living conditions. In 2000, for instance, an estimated one-third of the urban population in the country lacked adequate sanitation. While the economic boom experienced by Chinese cities induced investment in high and middle segments of the housing market, it posed problems of affordability and accessibility for families with limited income and savings.

Until the early 1980s, China's urban housing market was almost entirely the purview of state-owned enterprises that were responsible for investing in and allocating housing within a strict command-and-control economy. High rates of urbanization and economic growth in the last two decades led to major macroeconomic reforms geared towards a "socialist economy based on market principles" and to the liberalization of the urban housing market in the late 1990s.

To facilitate low-income people's access to the housing market, Chinese cities have been practising a policy of stimulating supply and demand through the use of equity grants for people living in sub-standard housing. While land remains the property of the state, leases are auctioned to developers to supply housing on a home ownership basis. Low-income families living in slums or substandard housing are thus provided with once-in-a-lifetime equity grants based on the market value of their existing housing, which enables them to access mortgage instruments. Developers, on the other hand, are provided incentives in the form of tax reductions or exemptions.

The use of equity grants, combined with incentives for housing developers to provide affordable housing, led to the production of more than 20 million housing units in the last five years. Chinese cities are hoping to avert the proliferation of urban ghettos and slums by providing more affordable housing. In large housing estate developments, many of which attract foreign direct investment, a new level of self-governance has also emerged, with residents electing committees to oversee and manage urban safety and security, environmental conservation and the needs of youth and the elderly.

Prosperity and pollution

China's recent gains in economic growth and industrialization have in many cases exacerbated environmental problems in its cities. Economic growth has increased consumer purchasing power, with the result that Chinese cities, such as Beijing – once the bicycle capital of the world – are now teeming with motor vehicles, a leading cause of air pollution. There are 1.3 million private cars in Beijing alone, an increase of 140 per cent since 1997. Experts believe that China's skyrocketing private car ownership and lax implementation of industrial emission regulations could threaten the recent gains it has achieved on the economic front. China's manufacturing-based economy has made it one of the world's largest consumers: in 2005, the country used 26 per cent of the world's crude steel, 32 per cent of its rice and 47 per cent of its cement.

According to the World Bank, China is home to 16 of the 20 most polluted cities on the planet. China is also the second largest producer of greenhouse gases, after the United States. Environmental degradation robs the nation of up to 12 per cent of its GDP, and every year some 400,000 Chinese die prematurely of respiratory illnesses and some 30,000 children die from diarrhoea caused by drinking unclean water. Towns and villages along China's most polluted rivers are also reporting more cases of cancer and miscarriages. According to the Yellow River Conservancy Commission, river pollution costs the country $.1.9 million annually. If China is to sustain its remarkable economic growth, it must also ensure that its cities are sustainable.

1.2 Putting Slums on the Map: A Global and Regional Overview

Global trends

In the last decade, an increasing number of governments around the world have enlarged democratic space within their countries and opened up their markets in response to the demands of a globalizing world. But democracy and market economies have not had the desired effect of reducing inequalities within and among the world's regions. On the contrary, from 1960 to 1999, the incomes of the richest countries grew to exceed those of the poorest by 35 times.[1] Economic growth, it turns out, does not automatically result in prosperity for all. In many countries, national gross domestic product (GDP) rates have risen much more quickly than national poverty rates have fallen; the effect has been a growing gap between the rich and the poor, a gap that is most evident in cities of the developing world.

Inequality has a direct bearing on patterns of urbanization. The rich in most countries live a world apart from the poor, with homes in protected urban enclaves and access to the latest technology, the best services and the most comfort. The rest, especially slum dwellers, live in the most deprived neighbourhoods, struggling to gain access to adequate shelter and basic services, such as water and sanitation. Many slum dwellers also live under the constant threat of eviction. Such stark differences and divisions can be found among regions and countries, but also within countries and cities. Especially in the developing world, urban zones of poverty and despair commonly skirt modern cosmopolitan zones of plenty. If current trends are not reversed, cities will become more and more spatially divided, with high- and middle-income residents living in the better-serviced parts of the city,

TABLE 1.2.1 POPULATION OF SLUM AREAS AT MID-YEAR, BY REGION; 1990, 2001, 2005 AND ANNUAL SLUM GROWTH RATE

Region	% slum 1990	Slum Population (thousand) 1990	% slum 2001	Slum Population (thousand) 2001	% slum 2005	Slum Population (thousand) 2005	Slum annual growth rate (%)
WORLD	**31.3**	**714,972**	**31.2**	**912,918**	**31.2**	**997,767**	**2.22**
Developed regions	**6.0**	**41,750**	**6.0**	**45,191**	**6.0**	**46,511**	**0.72**
EURASIA (Countries in CIS)	10.3	18,929	10.3	18,714	10.3	18,637	-0.10
European countries in CIS	6.0	9,208	6.0	8,878	6.0	8,761	-0.33
Asian countries in CIS	30.3	9,721	29.4	9,836	29.0	9,879	0.11
Developing regions	**46.5**	**654,294**	**42.7**	**849,013**	**41.4**	**933,376**	**2.37**
Northern Africa	37.7	21,719	28.2	21,355	25.4	21,224	-0.15
Sub-Saharan Africa	72.3	100,973	71.9	166,208	71.8	199,231	4.53
Latin America and the Caribbean	35.4	110,837	31.9	127,566	30.8	134,257	1.28
Eastern Asia	41.1	150,761	36.4	193,824	34.8	212,368	2.28
Eastern Asia excluding China	25.3	12,831	25.4	15,568	25.4	16,702	1.76
Southern Asia	63.7	198,663	59.0	253,122	57.4	276,432	2.20
South-Eastern Asia	36.8	48,986	28.0	56,781	25.3	59,913	1.34
Western Asia	26.4	22,006	25.7	29,658	25.5	33,057	2.71
Oceania	24.5	350	24.1	499	24.0	568	3.24

Source: UN-HABITAT 2005, Global Urban Observatory, Urban Indicators Programme, Phase III.
Note: % slum indicates the proportion of the urban population living in slums; 2005 figures are projections.

Slum overlooking Mumbai, India R. A. ACHARYA/UNEP/STILL PICTURES

Low-income housing in Mexico City, Mexico RON GILING/STILL PICTURES

TABLE 1.2.2 URBAN AND SLUM GROWTH RATES BY REGION

	Regions	Urban growth rate	Slum growth rate
Urban growth significantly higher than slum growth	Latin America and the Caribbean	2.21	1.28
	Northern Africa	2.48	-0.15
	Eastern Asia	3.39	2.28
	South-Eastern Asia	3.82	1.34
Urban and slum growth similar	Western Asia	2.96	2.71
	Southern Asia	2.89	2.20
	Sub-Saharan Africa	4.58	4.53
Developed world		**0.75**	**0.72**
World		**2.24**	**2.22**

Source: UN-HABITAT 2005, Global Urban Observatory.

In 2005, there were 998 million slum dwellers in the world; if current trends continue, the slum population will reach 1.4 billion by 2020.

and the poor living in spatially or socially segregated slums with few services or none at all.

Not all of the world's urban poor live in slums; poverty in cities has various social and economic dimensions that have little to do with the physical structure of the houses or the environments in which people live. Conversely, not all those who live in slums are poor – many people who have risen out of income poverty choose to continue living in slums for various reasons ranging from lack of affordable housing in better parts of the city to proximity to family and social networks. However, if the quality of housing and the existence of basic services are used as criteria to determine poverty levels, then slums represent a physical dimension of poverty. This aspect of urban poverty is the focus of this Report.

What is a slum?

At an Expert Group Meeting in November 2002, UN-HABITAT and its partners came up with a provisional definition of "slum": a settlement in an urban area in which more than half of the inhabitants live in inadequate housing and lack basic services. Developing an operational definition – one with measurable indicators – required further refinement, recognizing that slums can be geographically contiguous or isolated units. UN-HABITAT therefore focuses on the household as the basic unit of analysis.

A single operational definition of slums is used throughout this Report.*

A slum household is a group of individuals living under the same roof in an urban area who lack *one or more* of the following five conditions:

Durable housing: A house is considered "durable" if it is built on a non-hazardous location and has a structure permanent and adequate enough to protect its inhabitants from the extremes of climatic conditions, such as rain, heat, cold and humidity.

Sufficient living area: A house is considered to provide a sufficient living area for the household members if *not more than three people* share the same room.

Access to improved water: A household is considered to have access to improved water supply if it has a *sufficient amount of water* for family use, at an *affordable price*, available to household members without being subject to *extreme effort*, especially on the part of women and children.

Access to sanitation: A household is considered to have adequate access to sanitation if an excreta disposal system, either in the form of a *private toilet* or a *public toilet shared with a reasonable number of people*, is available to household members.

Secure tenure: Secure tenure is the right of all individuals and groups to effective protection against forced evictions. People have secure tenure when there is *evidence of documentation* that can be used as proof of secure tenure status or when there is either *de facto* or *perceived protection against forced evictions.*

** This definition may be amended according to the situation in a specific city. For example, in Rio de Janeiro, living area is insufficient for both the middle classes and the slum population and is not a good indicator. It could either be omitted, or it could be combined with another indicator to denote two or more shelter deprivations.*

The findings presented in the following chapters represent a new approach to measuring and understanding slums, developed by UN-HABITAT in response to the international community's recognition in 2000 that slums cannot be considered an unfortunate by-product of urbanization, but instead need to be addressed comprehensively as a major development issue. Approaching slums as a specific type of human settlement with discernable characteristics and impacts on the people who live in them provides a framework for moving toward Millennium Development Goal 7, target 11: *by 2020, to have achieved a significant improvement in the lives of at least 100 million slum dwellers*

FIGURE 1.2.1 SLUM POPULATIONS, 1990-2020

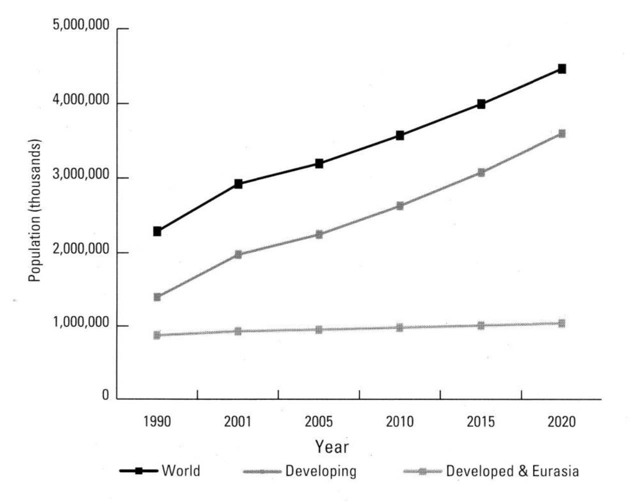

Source: UN-HABITAT, Global Urban Observatory 2005.

Slums: The emerging human settlements of the 21st century?

The word "slum" first appeared in 19th century London, when the burgeoning urban working classes moved into overcrowded and poorly serviced tenements, living close to the factories and industrial plants that employed them. The term referred to what was initially designated "a room of low repute", but over time took on the generic definition, "a squalid and overcrowded urban area inhabited by very poor people".[2]

Although slums continued to grow over the course of the last two centuries, their evolution was particularly swift in the latter half of the 20th century as the developing world became more urbanized. Today's slums are much larger and have many more residents than the slums prevalent in 19th century Europe and North America. The slum population of Rio de Janeiro, for example, is almost the same size as the total population of Helsinki. Mumbai's more than 5 million slum dwellers exceed the total population of Nairobi. Slum dwellers now live primarily in the cities of Africa, Asia and Latin America, although a smaller number also live in cities of the developed world.

Slum housing ranges from crowded tenement buildings in Hong Kong to mud-and-tin shacks in Cape Town. Slums can be inner-city tenements in cities of the developed world, shanty towns on the periphery of large cities or densely packed neighbourhoods bordering high-income areas. Individual households located in high- or middle-income neighbourhoods may also fit the definition of slums. In some parts of the developing world, gradations of slums are common, with each variation having a different name. For instance, in India, a *chawl* (a densely packed block of one-room "apartments" with shared toilets and bathrooms) is quite different from a *zopadpatti* (a shack made of non-durable materials, often located in a crowded slum settlement within or on the outskirts of a city). Yet, both types of housing could fall under UN-HABITAT's definition of slum households if they lack one or more of the five conditions that are necessary to deem a house "adequate" (see box).

One out of every three city dwellers – nearly one billion people – lives in a slum today. For many years, governments and local authorities viewed slums as transient settlements that would disappear as cities developed and as the incomes of slum dwellers improved. However, evidence shows that slums are growing and becoming permanent features of urban landscapes. Slums have carved their way into the fabric of modern-day cities, making their mark as a distinct category of human settlement that constitutes a space between "rural" and "urban". Given the proliferation of slums around the world and the growth of city-regions, in which larger cities act as centres for smaller cities and towns and a large rural hinterland, the simple duality that exists in the traditional framing of human settlement patterns no longer suffices to describe the reality of people's lives. Slums in many cities are no longer just marginalized neighbourhoods housing a relatively small proportion of the urban population: in some cities, particularly in Southern Asia and sub-Saharan Africa, slums host significantly large proportions of the urban population and slum growth is virtually synonymous with urbanization; this calls for new ways of looking at cities and the slums within them.

Will slums become a predominant type of settlement in the 21st century? If no preventive or remedial action is taken, they may indeed come to characterize cities in many parts of the developing world.

The vast majority of slums, more than 90 per cent, are located in cities of the developing world, where urbanization has become virtually synonymous with slum formation. This is especially so in sub-Saharan Africa, Southern Asia and Western Asia, where urban growth over the last 15 years has been accompanied by a commensurate growth in slums.

Slum and urban growth rates are highest in sub-Saharan Africa, 4.53 per cent and 4.58 per cent per year, respectively – nearly twice those of Southern Asia, where slum and urban growth rates are 2.2 per cent and 2.89 per cent per year, respectively. In Western Asia, annual slum and urban growth rates are quite similar, at 2.71 per cent and 2.96 per cent respectively, while in Eastern Asia and Latin America, slum growth rates are significantly lower than urban growth rates, although slum growth rates are relatively high in both regions: 2.28 per cent and 1.28 per cent per year, respectively. South-Eastern Asia and Northern Africa are two regions where slum growth has not kept pace with urbanization; in both regions the proportion of slum dwellers has actually declined in recent years from over 36 per cent of the urban population in 1990 to approximately 25 per cent in 2005. Eastern Asia and Latin America and the Caribbean also have urban growth rates that are higher than slum growth rates. This suggests that countries within these regions have in recent years taken active steps to reduce the number of slum dwellers or prevent slum formation.

Kibera, Nairobi HIROSHI SATO

At the global level, 31.2 per cent of all urban dwellers lived in slums in 2005, a proportion that has not changed significantly since 1990. However, in the last 15 years, the magnitude of the problem has increased substantially: 283 million more slum dwellers have joined the global urban population. In 1990, there were nearly 715 million slum dwellers in the world.[3] By 2000 – when world leaders set the target of improving the lives of at least 100 million slum dwellers by 2020 – the slum population had increased to 912 million. In 2005, there were almost 1 billion (998 million) slum dwellers in the world; if current trends continue, UN-HABITAT estimates that the slum population will reach 1.4 billion by 2020.

Trends in developing regions

Slum trends in Africa

An interesting disparity exists between Northern Africa and sub-Saharan Africa in terms of slum growth and slum prevalence: while the former is experiencing negative slum growth, the latter is experiencing the opposite trend, with extremely high slum growth rates of 4.53 per cent per year – the highest in the world.[4]

Northern Africa achieved a reduction in both the number and proportion of slum dwellers between 1990 and 2005. The share of slum dwellers in the region fell from 37.7 per cent to 25.4 per cent, with the absolute number of people living in slums decreasing by half a million, to just over 21 million. The reduction may be attributed to the relatively low levels of slum prevalence in the region in general, as well as to the implementation of policies aimed at reducing the number of slum dwellers within countries. Countries such as Morocco and Tunisia have been very successful in improving the lives of slum dwellers, while Egypt, where slums exist on a much larger scale, was able to address the problem with pro-poor policies and substantial investments in improving the shelter conditions of people living in cities.

Sub-Saharan Africa, on the other hand, has been unable to manage or reduce slum growth. This is partly attributed to the declining economies of some countries in the region, coupled with its disproportionate share of HIV prevalence and conflicts, both of which have exacerbated slum formation and worsened living conditions in cities. Slums in cities such as Khartoum, for instance, have grown remarkably in the last decade, largely due to an influx of internally displaced persons (IDPs) from Southern Sudan, which has been suffering from a protracted civil war.

Sub-Saharan Africa has the highest proportion of slum dwellers in the world: 71.8 per cent. In terms of absolute numbers, it is home to the third most populous slum population among the regions of the developing world, after Southern Asia and Eastern Asia. In the last 15 years, the number of slum dwellers in sub-Saharan Africa has almost doubled, from 101 million in 1990 to 199 million in 2005. Urban growth in the region is almost identical to slum growth – a trend that is also prevalent to a lesser extent in other regions. Given the high slum growth rate in the region, the number of slum dwellers will likely double by 2020, reaching nearly 400 million, and overtake the slum populations of both Southern Asia and Eastern Asia, which are estimated to rise to 385 million and 299 million, respectively.

In many of the region's countries, notably Angola, Ethiopia, Mali, Mauritania, Sudan, and Tanzania, slum populations are expected to double within the next 15 years. Slum households are the norm rather than the exception in many cities. For instance, UN-HABITAT's Urban Inequities Survey,[5] conducted in the capital of Ethiopia, Addis Ababa, has shown that less than 10 per cent of that city's inhabitants live in non-slum areas.

When slums constitute the largest proportion of a city, differentials between, even within, slums also become apparent.[6] As in Ethiopia, several other primarily rural sub-Saharan African countries in the first stages of their urban transition – including Chad, and the Central African Republic – have very high proportions of slums in their cities. Not surprisingly, the living conditions in slums within these countries are also

MAP 1 URBAN POPULATION AND SLUM PROPORTION IN AFRICAN COUNTRIES, 2001

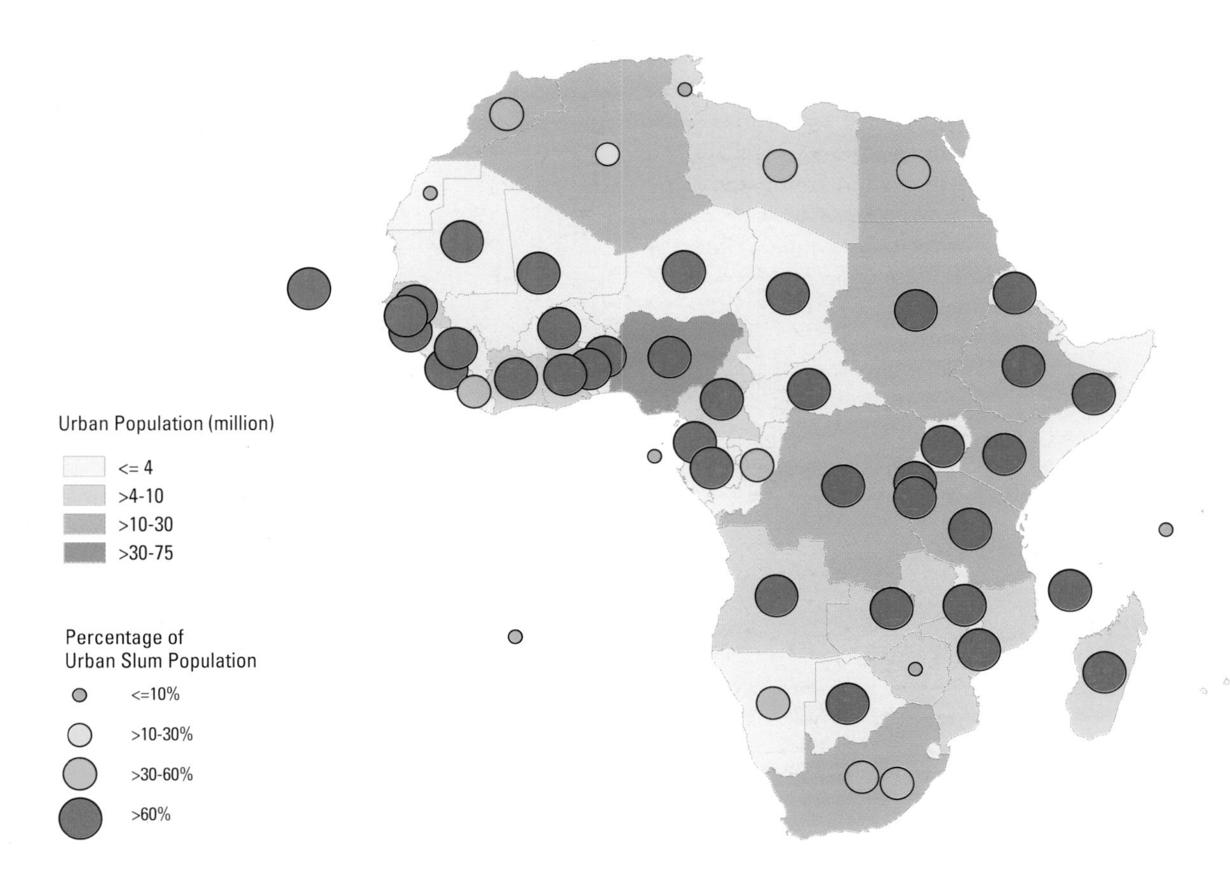

Source: UN-HABITAT, Global Urban Observatory 2005.

extremely severe, as 70 per cent to 90 per cent of households are deprived of more than two basic shelter needs, such as water or sufficient living space. This trend applies to much of sub-Saharan Africa, where slum households are likely to lack water, sanitation, durable housing, sufficient living space, and secure tenure altogether, or a combination of at least three of these indicators of adequate housing.

Slum trends in Asia

In absolute numbers, Asia has the largest share of the world's slum population – in 2005, the region was home to more than half the world's total slum population, or about 581 million people. Some sub-regions within Asia are faring worse than others. Eastern and Southern Asia harbour 80 per cent of the slum dwellers in the region, with Southern Asia hosting nearly half the region's slum population. These figures are largely attributable to China and India, which are the most populous countries in the world and have significant proportions of their urban populations living in slum conditions. Although China hosts the world's largest slum population – almost 196 million people – its slum prevalence in 2001 was lower than that of India; UN-HABITAT estimates that 38 per cent of China's urban residents lived in slum conditions that year, compared with India's 56 per cent.

Most of the slum dwellers in **Southern Asia** – 63 per cent, or almost 170 million people – reside in India. The share of Southern Asia's slum dwellers constitutes 27 per cent of the global total; India alone accounts for 17 per cent of the world's slum dwellers. India has pioneered many best practices and good policies in recent years that are having some impact on the lives of slum dwellers, but they have not reached a sufficient scale to ameliorate the proliferation of slums. Although the country has seen remarkable economic growth rates in recent years and has managed to reduce extreme poverty by 10 per cent in the last decade, the impact of poverty reduction is still not being felt in cities. Unless more radical policies are pursued in India, the global target for improving the lives of slum dwellers will not be reached.

Other countries that need to address this challenge urgently are Bangladesh and Pakistan which, along with India, have among the highest urban poverty rates and the largest urban populations in the sub-region. UN-HABITAT data shows that Bangladesh was home to 30 million slum dwellers in 2001, and 85 per cent of its urban population lived in poverty that year; 74 per cent of Pakistan's urban population lived in poverty in 2001 – more than 35 million people. Through successful initiatives such as the Orangi project,[7] Pakistan has demonstrated how the lives of slum dwellers can be improved at the local level. Three decades on, however, the project has not been able to scale up its interventions to have a national impact. Bangladesh's development campaigns through the Bangladesh Rural Advancement Corporation and its Grameen Bank initiative have focused on alleviating rural poverty, so have had negligible impact in urban areas.

Ninety per cent – or 195.7 million people – of **Eastern Asia's** slum dwellers live in China. Chinese slum dwellers account for 20 per cent of the world's total. It is important to note that since 1990, China has been held up as a success in increasing the scale of low-cost housing schemes, thus preventing slums before they even form.[8] Despite such measures, the country suffers from high levels of slum prevalence. There could be two reasons for this, other than the fact that the total population of China constitutes one fifth of the world's population. One is the need for a lapse in time for slum prevention policies to have an impact on the ground. The other could be the mismatch between UN-HABITAT and national definitions of what constitutes a slum. While UN-HABITAT considers the *de facto* status of dwellings in the cities of China, irrespective of their legal status, national authorities do not consider people who live outside the *de jure* residential area or those who do not possess residency permits as bona fide residents of a city.[9]

It is interesting to note that real success stories in the region, in terms of decreasing slum growth significantly, have occurred in **South-Eastern Asia,** in countries such as Thailand, where policies implemented even before the 1990s have had a strong

Mumbai RASNA WARAH

impact on both the magnitude and the proportion of slums. The main reasons behind Thailand's ability to reduce slum growth are a strong political commitment by its leadership, accompanied by a tradition of strategic planning and monitoring development efforts, which have been an integral part of the development tradition for the last 30 years. Many of the sub-region's countries also have an active civil society. Although the high slum growth rates in Cambodia and Lao People's Democratic Republic suggest a less optimistic future, there is evidence that they might be able to curb slum growth, as they have recently initiated slum prevention policies.

Western Asia, on the other hand, lags far behind the other sub-regions in terms of slum prevention. Slum and urban growth rates in the sub-region are almost the same, reaching nearly 3 per cent per year. The countries of Western Asia have made little progress on any of the Millennium Development Goal indicators and have not been able to sustain the momentum of development they gained between 1980 and 1990, as the region has in recent years been engulfed in political turmoil that has exacerbated the refugee crisis and worsened conditions in cities. In countries such as Jordan, slums have grown at the rate of 4.3 per cent per year, and Lebanon has also experienced an increase in its slum population. Both countries have relatively small populations, so the reduction or increase in slums there might not make a dent in the overall slum figures by 2020, but progress is still important, as it would indicate greater stability in the sub-region, accompanied by better social indicators. Slum growth in the largest country of the sub-region, Turkey, declined radically between 1990 and 2001, from 23.3 per cent to 17.9 per cent, primarily because of an effective policy of decentralization, which empowered the municipal governments to borrow directly from international financial institutions to build or upgrade water and sanitation networks.

Most of the slum dwellers in Southern Asia – 63 per cent, or almost 170 million people – reside in India.

MAP 2 URBAN POPULATION AND SLUM PROPORTION IN ASIAN COUNTRIES, 2001

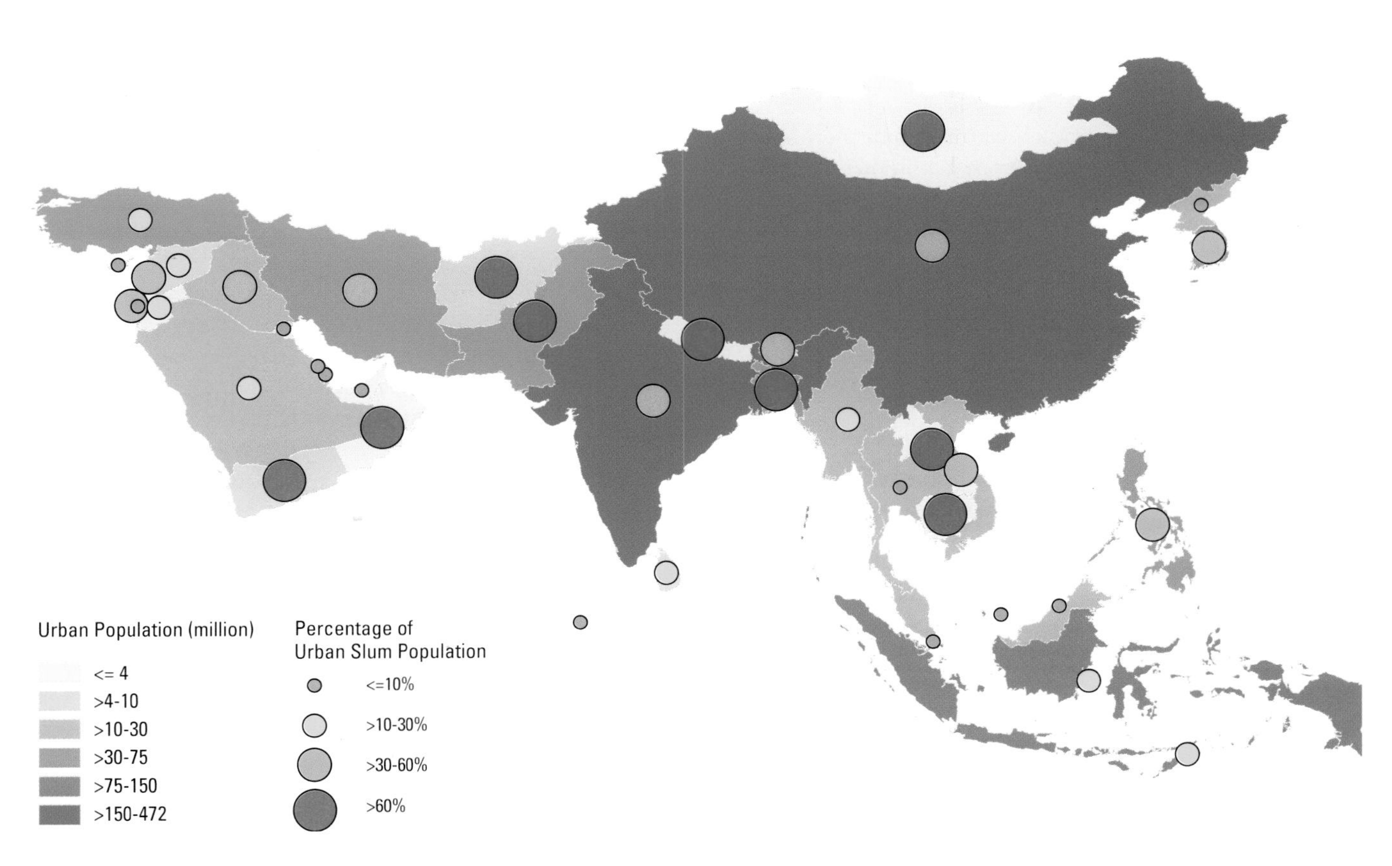

Source: UN-HABITAT, Global Urban Observatory 2005.

Mumbai's quest for 'world city' status

Mumbai, the capital city of the state of Maharashtra and India's most important financial capital, has a population of 18.3 million people, making it the fourth largest urban agglomeration in the world, after Tokyo, Mexico City and New York-Newark. The city hosts one of the world's largest slum populations: more than 5 million of the city's residents are slum dwellers. More people live in Mumbai's slums than in the entire country of Norway.

Despite its large slum population – or, as some would argue, because of it – Mumbai has emerged as one of India's leading commercial and cultural centres, home to the country's hugely successful film industry and a booming stock exchange. The city alone pays almost 40 per cent of the nation's taxes.

There is a perception among many of the country's policy-makers, however, that Mumbai's progress is being hampered by its image as a city of slum dwellers, which severely erodes its ambitions to become the "Shanghai of India" with clean streets, gleaming skyscrapers and a modern and efficient transport and communications network. This perception was reinforced in late 2004 when the Indian government embarked on a campaign to make Mumbai a "world-class city" in response to a call by the Indian Prime Minister in October 2004 to transform Mumbai into an international hub for trade and commerce. The call was supported by Maharashtra's chief minister, who submitted an ambitious four-year, $8 billion proposal for modernizing Mumbai, which included the building of new roads, a subway system and a large-scale public housing project. The modernization proposal followed an earlier Slum Rehabilitation Scheme in the mid-1990s that aimed to improve the lives of 4 million slum dwellers through public-private partnerships that involved builders in the private sector and the authorities.

Demolition drive

In late 2004, despite progressive slum improvement and tenure regularization policies and programmes, the government of Maharashtra began a slum demolition drive aimed at removing slums and shanty towns in the city. Between December 2004 and March 2005, more than 90,000 shanties were torn down, in violation of poll promises, international covenants to which India is a signatory and a 2001 Slum Areas Act, which protected all slums built prior to 1995. (The Act stipulates that all slum dwellers who could establish that their names were on the electoral roll on 1 January 1995 were protected, to the extent that their homes could not be demolished without rehabilitation.)

Amid public outcry and pressure from the ruling Congress party, the demolitions were halted in February 2005, but many believe that the plan to make Mumbai a world-class city is still very much on the cards. "The reality of course is that a new Mumbai cannot be built on the corpses of its poor, the very people who hold up this city," argued journalist Kalpana Sharma of *The Hindu* newspaper.

Jockin Arputham, founder of India's National Slum Dwellers Federation, has consistently argued that it is Mumbai's poor, who allow the city to flourish by providing cheap labour and services. "The poor work as refuse collectors, construction labourers, handcart pullers, vegetable vendors, factory workers, domestic workers and so on. They provide goods and services at rates that most of the city's people can afford. But when it comes to their housing, the city turns its back on them."

Dire slum conditions

Despite the active role played by non-governmental organizations (NGOs) and slum federations in the city, the situation of slum dwellers in Mumbai remains dire. A recent survey for the Mumbai Sewerage Disposal Project found that 42 per cent of slum dwellings in the city had an area of less than 10 square metres and only 9 per cent had an area of more than 20 square metres. Almost half of the households in slums got their water from shared standpipes and only 5 per cent had direct access to water through individual taps. The city's sanitation situation was even more alarming: 73 per cent of the city's slum households – housing 3.86 million residents – depended exclusively on public toilets. Moreover, overuse and poor maintenance had made public toilets a health hazard, especially in areas where the user group was undefined. Less than one per cent of the slum population had access to individual toilets or to pay-per-use toilets constructed by private agencies or NGOs.

Despite the daunting conditions in its slums, Mumbai is a magnet for Indians, not only from neighbouring cities and villages, but also from the rest of the country. According to "Vision Mumbai", a 2003 report by the private consultancy firm McKinsey & Co, the city urgently needs to build at least 1.1 million affordable housing units in the next decade for current

Sources: Moreau and Mazumdar 2005; Risbud, 2003; YUVA and Montgomery Watson Consultants 2001; Sharma 2005a and 2005b; Burra 2005; Arputham 2001; Mehta 2004; Kothari 2005; United Nations Department of Economic and Social Affairs, Population Division 2004.

Slum children, Mumbai MARK EDWARDS/STILL PICTURES

and future generations of slum dwellers and migrants. (Currently, only 58,000 new low-cost housing units are available for pre-1995 slum dwellers.) The report provides the framework for the city's urban renewal scheme, which, if implemented, will cost upwards of $40 billion over the next decade. City authorities are already looking into how the funds can be raised from federal and local governments and from international lending institutions.

Author Suketu Mehta feels that Mumbai cannot escape the demands of globalization, including the pressures to become "world class". In many ways, he writes, the city has already parted with the rest of India as the gap between the haves and the have-nots widens: "In the Bayview Bar of the Oberoi Hotel you can order a bottle of Dom Pérignon for one and a half times the average annual income … . People are still starving to death in other parts of India. In Bombay, there are several hundred slimming clinics." (Bombay was officially renamed Mumbai in 1996, but like many diehard Bombayites, Mehta prefers to call it by its old name.)

Others are cynical about Mumbai's attempts to reach world-class status. "One hardly needs to emphasize that the world over, cities with good and affordable public transport are also the most liveable … . The best cities in the world are also the ones that have affordable housing for all classes," writes Sharma, who is the author of a book on Dharavi, Mumbai's largest slum. "We need to put aside our obsession with becoming 'world class'. Let us make our cities liveable for all the people. That itself is a big agenda for the future."

Slums: The shelter dimension of urban poverty

Current debates over the multidimensional aspects of poverty recognize that income-based poverty measurements do not capture the scale or range of poor living conditions experienced by people around the world. The link between income and levels of deprivation is weak and misleading, as many who live above the poverty line may suffer from serious deprivations in other areas, while those below the poverty line may suffer from income poverty, but may not be "poor" in other aspects.

These debates are highly applicable to measurements of urban poverty. UN-HABITAT and others have consistently argued that understanding the various dimensions and degrees of urban poverty is important in order to construct pro-poor policies that have a tangible impact on the living conditions of the urban poor. UN-HABITAT is convinced that neither the food basket nor the one-dollar-a-day indicator can accurately reflect the diverse experiences of people living in poverty in both rural and urban areas.

The problem with current measurements of urban poverty, and consequent policy discussions, is the division of urban populations into the "poor" and the "non-poor" with little recognition of the diversity within the "poor" and the "almost poor" with regard to their deprivations, vulnerabilities and needs. This measurement also fails to recognize that people have a variety of assets, which may or may not translate into income or cash, but which nonetheless play an important role in determining levels of poverty. These include: human assets (such as skills and good health), natural assets (such as land), physical assets (such as access to roads and other infrastructure), financial assets (such as access to savings and credit) and social assets (such as networks of family and other contacts that can be called upon in times of need). Capturing the depth and magnitude of urban poverty is particularly important when monitoring the achievement of Millennium Development Goal 7, target 11. It is also important to have better data to (i) guide the day-to-day work for those working on the ground and (ii) in order to lobby for adequate sources/funding to match the size of the problem, and to get urban poverty as a higher national priority (eg in PRSPs).

Urban poverty and slums

In this Report, UN-HABITAT attempts to analyse a particular dimension of urban poverty that has not been adequately captured in either national statistics or in United Nations data – that of shelter deprivation. This shelter dimension of urban poverty is measured using five key indicators: *access to water; access to sanitation; durability of housing; sufficient living area; and secure tenure.*

In cities, poverty is quite often physically and spatially visible in slums, which suffer from poor quality, insecure, hazardous and overcrowded housing and lack infrastructure and basic services. While shelter deprivations are most apparent in slums, they *do not necessarily* denote levels of urban poverty and are only a subset of a wider range of urban poverty experiences. However, the huge gap between official urban poverty figures and the proportion of people living in slums does raise questions about the validity of the methodologies used to measure urban poverty, with some experts claiming that it has been grossly underestimated. For instance, official figures show that only 9.7 per cent (or 14.6 per cent, depending on the survey used) of Phnom Penh's population lived below the poverty line in 1999, yet 2001 data shows that an estimated 40 per cent of the Cambodian capital's population lived in informal settlements, or slums. Similar gaps between official figures for urban poverty levels and slum estimates in other cities suggest that poverty in cities is still being viewed through the income lens that does not take into account the living conditions of city dwellers and their needs for goods and services that are specific to urban areas. While not all people who live in slums necessarily live below the poverty line, the huge discrepancy between these two figures suggests that the poverty line used in cities is not realistic.

Until a universal knowledge base is developed to produce information on other forms of urban poverty, the thrust of UN-HABITAT's discussion remains focused on the knowledge and data it has gathered and generated on shelter deprivations, which tend to be concentrated in slums. This is being done bearing in mind that "slum dwellers" neither represent all individuals who live in poverty in cities, nor are all slum dwellers, as a heterogeneous group, "living below the poverty line". Nonetheless, slums are a good starting point – but not the ultimate or definitive point – for describing urban poverty and capturing the scale and depth of shelter deprivations in cities.

Sources: World Bank 2000/2001; Asian Coalition for Housing Rights 2001.

Various dimensions of urban poverty

Inadequate and often unstable income, which impacts people's ability to pay for non-food items, such as transport, housing and school fees.

Poor quality, hazardous, overcrowded, and often insecure housing

Inadequate provision of basic services (piped water, sanitation, drainage, roads, footpaths, etc.) which increases the health burden and often the work burden.

Inadequate, unstable or risky asset base (non-material and material) including lack of assets that can help low-income groups cope with fluctuating prices or incomes, such as lack of access to land or credit facilities.

Inadequate public infrastructure, such as schools and hospitals.

Limited or no safety nets to ensure basic consumption can be maintained when incomes fall and which can be easily accessed when basic necessities are no longer affordable, such as public housing and free medical services.

Inadequate protection of rights through the operation of the law, including regulations and procedures regarding civil and political rights, occupational health and safety, pollution control, environmental health, protection from violence and forced evictions and, protection from discrimination and exploitation.

Voicelessness and powerlessness within non-responsive political systems and bureaucratic structures, leading to little or no possibility of receiving entitlements to goods and services; of organizing, making demands and getting a fair response; and of receiving support for developing initiatives. Also, no means of ensuring accountability from aid agencies, NGOs, public agencies and private utilities, and of being able to participate in the definition and implementation of urban poverty programmes.

Adapted from Satterthwaite 2004.

Defining and monitoring slums: Seeing beyond the stereotypes

Until recently, empirical evidence regarding living conditions in the world's slums was not available in a universally comparable format. Slums were the "invisible" parts of cities – neither reflected in official data or maps, nor recognized by authorities. Speculative analysis suggested that people living in slums were experiencing a continuous deterioration of their living environments, yet figures were often inaccurate or contradictory within and among countries. Sometimes, figures overestimated or underestimated the reality of the situation, based on subjective concepts about the nature of slums or the motivation of researchers or political entities involved in estimation. The production of data was thus not the result of a reliable monitoring process.

Global monitoring has also faced a political obstacle over the term "slum", which has often been deemed derogatory by urban planners, city authorities and slum dwellers themselves. The preferred terms, including "informal settlement", "squatter settlement" and "unplanned neighbourhood", have been used interchangeably with "slum", but have not heretofore been linked to specific indicators regularly reported on by governments and stakeholder organizations. The difficulty in developing a workable measurement strategy and the lack of reporting on slums illustrates that they are conceptually complex and methodologically elusive.

Different cultures and countries define the physical and social attributes of slums differently. UN-HABITAT acknowledges this diversity and the fact that slums take many different forms and names. Bearing this in mind, in 2002, UN-HABITAT, the United Nations Statistical Division and the joint UN-HABITAT/World Bank Cities Alliance gathered together a group of experts to define slums and propose a way to measure them. The resulting definition and methodology represent a compromise between theoretical and methodological considerations. The agreed-upon definition is simple, operational and pragmatic: it can be easily understood and adapted by governments and other partners; it offers clear, measurable indicators, provided as a proxy to capture some of the essential attributes of slums; and it uses household-level data that is collected on a regular basis by governments, development agencies and non-governmental organizations, which is accessible and available in most parts of the world.

Shelter deprivation indicators

Slum dwellers often live in difficult social and economic conditions that manifest different forms of deprivation – physical, social, economic and political. Four out of five of the slum definition indicators measure physical expressions of slum conditions: lack of water, lack of sanitation, overcrowding, and non-durable housing structures. These indicators focus attention on the circumstances that surround slum life, depicting deficiencies and casting poverty as an attribute of the environments in which slum dwellers live.

The fifth indicator – security of tenure – has to do with legality, which is not as easy to measure or monitor, as the status of slum dwellers often depends on *de facto* or *de jure* rights – or lack of them. This indicator has special relevance for measuring the denial and violation of housing rights, as well as the progressive fulfillment of these rights. There currently exists no mechanism to monitor secure tenure as part of Millennium Development Goal 7, target 11, as household-level data on property entitlement, evictions, ownership, and other indicators of secure tenure is not uniformly available through mainstream systems of data collection, such as censuses and household surveys. In this Report, UN-HABITAT points to trends that suggest levels and severity of insecure tenure around the world, but these figures are based on secondary sources that may or may not reflect the reality on the ground.

Using the first four slum definition indicators, it has been possible to estimate the prevalence and magnitude of slums and to calculate projections in most countries of the world using existing household surveys and censuses, including Demographic and Health Surveys and UNICEF Multiple Indicator Cluster Surveys, conducted between 1990 and 2001. These indicators are considered "shelter deprivation indicators" by UN-HABITAT and its partners.

The physical and visible manifestations of housing that lacks basic services, space and security take many forms, resulting in diverse types of slums. Not all are as easily distinguishable or visible as the shanty towns cramped together on the periphery of cities such as Mumbai, Nairobi or Cape Town. In some places, slums are less visible to the eye: dwellings may look durable or permanent from the outside, but living conditions inside the dwelling may portray another picture. For instance, many multi-storey public housing projects at the periphery of urban cores or old, dilapidated buildings in inner cities could qualify as slums if they have been neglected or ill-serviced for significant periods of time, as would many workers' hostels or dormitories. Such places typically do not look like slums, but if their residents experience some form of shelter deprivation or insecure tenure, then, according to the UN-HABITAT definition of a slum household, these residents qualify as slum dwellers.

Rural migrants, Mumbai JOERG BOETHLING / STILL PICTURES

Woman-headed households in cities

Although "the proportion of woman-headed households" is not among the indicators used to monitor progress on Millennium Development Goal 3 on promoting gender equality and empowering women, there is a general belief that woman-headed households deserve special attention as they fall under the category of the poorest households. This belief usually translates into misconceptions about women living in slums, as it is assumed that low incomes and single motherhood go together.

However, UN-HABITAT data and analyses have shown that no clear pattern emerges on the marital status of women living in slums or about their household responsibilities. The situation varies from country to country. In some African countries, such as Kenya, Malawi and Rwanda, woman-headed households are mostly found in rural areas, whereas in Burkina Faso, Chad, Central African Republic, Egypt and Tanzania, the majority of woman-headed households are found in slums, as opposed to non-slum and rural areas. In a significant number of countries, including Cameroon, Côte d'Ivoire, Ethiopia, Ghana and Morocco, the majority of woman-headed households are found within non-slum areas of cities.

In Latin America, most women heading households live in cities. With the exception of Haiti, a majority of these households are located in non-slum areas of cities. The overall share of urban households headed by women in Haiti is quite high (50 per cent), compared with 38 per cent in rural areas.

In Asia, an overview of selected countries reveals two findings: firstly, with the exception of Yemen, women heading households tend to settle in urban areas as opposed to rural areas; and secondly, with the exception of Indonesia, the proportion of women heading households in urban areas is larger in non-slum areas than in slum areas. Among countries of the Commonwealth of Independent States (CIS), such as Kazakhstan and Uzbekistan, the proportion of women heading households in cities is exceptionally high, 50 per cent.

What these statistics indicate is that the prevalence of woman-headed households, whether in rural, slum or non-slum areas is quite high. With the exception of Africa, the share of such households is also much greater in cities.

Woman and child in Luanda, Angola EDUARDO LÓPEZ MORENO

FIGURE 1.2.2 PROPORTION OF WOMAN-HEADED HOUSEHOLDS IN SELECTED COUNTRIES

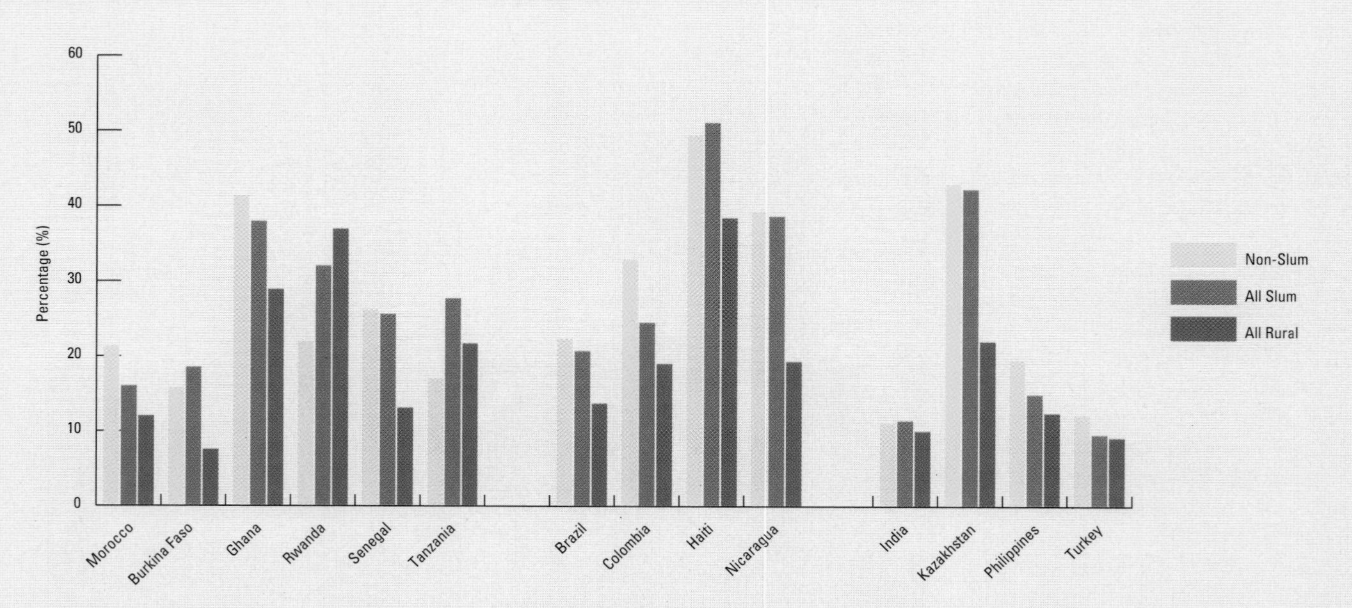

Source: UN-HABITAT 2005, Urban Indicators Programme, Phase III.
Note: Computed from Demographic and Health Surveys (DHS) data (1996-2004).

Age pyramids for slum and non-slum populations in Brazil and South Africa

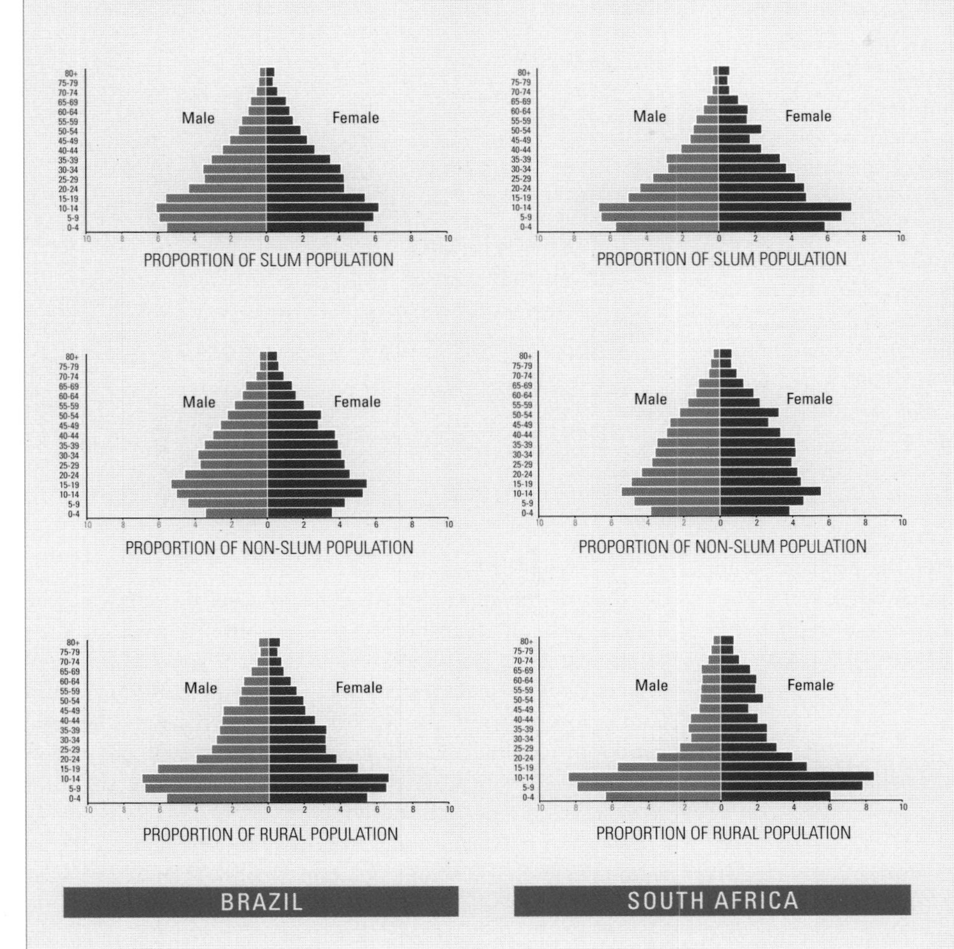

The average slum dweller is younger and dies sooner than the average non-slum dweller.

UN-HABITAT urban indicators provide an overview of the distribution of the 0-4, 5-14 and 15-24 age groups in some of the countries that are part of the organization's global sample of cities. This distribution goes beyond the traditional dichotomy of urban-rural, and presents a new breakdown into four categories: urban, rural, urban slum and urban non-slum.

The above graphs of two middle-income countries – Brazil and South Africa – show very conventional age pyramids for their respective urban and rural populations. The number of children (0-14), relative to the population as a whole is higher. This is typical of an expansive pyramid that has a wider base indicating that a large proportion of the population is young, which is also an indicator of high fertility rates. The rural pyramids of the two countries show that the proportion of people of the most productive age group (20 to 54) shrinks. This may be attributed to rural-to-urban migration because of the greater existence of employment opportunities in cities, which attract people in this age bracket. In countries with lower incomes, the proportion of rural people in this age bracket shrinks further among males as more men migrate to cities than women, particularly in sub-Saharan Africa. Yet, when the pyramids of these two countries are compared with the age group of the slum population, the findings are noteworthy: instead of shrinking, similar age brackets (20 to 54) start to grow, which probably reflects immigration from rural areas. This is particularly clear in the age pyramid for South Africa. The age distribution of Brazil and South Africa for the urban slum and non-slum population shows a significant demographic shift: the age bracket from 0 to 14 years for slum dwellers is as much as 20 per cent larger than the same age distribution of the non-slum population. Brazil exhibits a disproportionately large prevalence of females under the age of 19 years. This youth bulge is increasingly seen as a major development challenge, since in the coming 5 years this group will enter young adulthood and will be demanding housing and other basic services as they marry, have children and form new households.

When data is disaggregated between urban slum and urban non-slum, interpreting age pyramids at simple glance becomes more complicated. While comparing the above graphs of the urban slum and non-slum population, the most striking finding is that the age pyramid of the slum population has a clear expansive pattern, similar to a conventional pyramid of a developing country. On the other hand, the pyramid of the urban non-slum population, while still maintaining a generally expansive pattern, tends to be relatively constricted. This shows a trend toward reduction in the younger age categories, which is typical of more developed countries.

Intra-city inequalities are clearly reflected in the age-sex distribution of the slum and non-slum populations in Brazil and South Africa. The two age pyramids serve to identify two clear trends: an *expansive* broad base in the slum population structure, indicating a high proportion of children, a rapid rate of population growth, and a low proportion of older people; and *stable growth* in the non-slum population structure that suggests a reduction of fertility rates and higher life expectancy, respectively.

Source: UN-HABITAT 2005, Urban Indicators Programme, Phase III.

Throughout Asia, slums are primarily the consequence of only one shelter deprivation, reflecting fewer problems with infrastructure and housing policy than in sub-Saharan Africa. Therefore, governments or regional entities could tackle the slum problem with simple sectoral interventions to improve the living conditions of slum dwellers and their access to secure tenure. For instance, in many sub-regions, one of the main shelter deprivations identified is lack of access to adequate sanitation; by addressing this shelter deprivation, countries can drastically reduce the number of slum dwellings.

Asia's rapid economic growth over the last decade has not made a significant impact on the elimination or reduction of the urban inequalities on several fronts, and there are indications that disadvantaged groups remain disadvantaged, even when economies improve.[10] In fact, despite the region's newly industrializing countries, particularly China and India, which rank among the world's fastest-growing economies, gross domestic product (GDP) rates have risen much more quickly than national poverty rates have fallen, prompting the president of the Asian Development Bank to refer to the scale of deprivation in the region's cities as "daunting".[11] It is interesting to note that despite the sheer size of the problem, slum dwellers in Asia have not received as much attention as they warrant, although a recent Asia-Pacific Economic Cooperation summit did emphasize the need to address urban poverty as a priority issue.[12]

Slum trends in Latin America and the Caribbean

Latin America and the Caribbean has almost completed its urban transition; urbanization rates are stabilizing and slum growth rates in the region are slowing down. The region's share of slum dwellers is 134 million people, less than the total number of slum dwellers in just one country in Southern Asia – India. In the 1960s and 1970s, **Latin America's** experience with slum growth and slum prevalence was comparable to that of the current situation in sub-Saharan Africa. This trend was reversed in the late 1980s and early 1990s, when the process of "re-democratization" resulted in the adoption of progressive policies aimed at promoting more inclusive governance and reducing inequalities. Notwithstanding these positive developments, there is no room for complacency on the part of policymakers. Extreme urban inequalities persist throughout the region, and a considerably large share of slum dwellers live on the edge of destitution. Although it is too early to associate crime and violence levels with the number of shelter deprivations in slum communities, many researchers are beginning to make a link between inequality and violence in the region's larger cities.[13]

Argentina, Brazil and Mexico – the region's three largest countries – will be influential in reducing the proportion of slum dwellers in the region by 2020. Brazil and Mexico have already achieved a remarkable reduction in the share of slum dwellers within urban areas, while in absolute numbers, the increase was minor in both countries. Slum growth rates are very low in both Brazil and Mexico, at 0.34 per cent and 0.49 per cent per year, respectively. By 2020, the combined slum population of Brazil and Mexico will have increased by only 4 million, totalling 71 million, if declining growth continues. Argentina, however, is experiencing faster rates of slum growth, at 2.21 per cent per year.

Latin America and the Caribbean has almost completed its urban transition; urbanization rates are stabilizing and slum growth rates are slowing down.

Low slum prevalence does not apply to all countries in the region. In both Haiti and Nicaragua, more than 80 per cent of the urban population lives in slums; in Bolivia, Guatemala and Peru, slums host two-thirds of the urban population. Bolivia, Guatemala, Nicaragua and Peru are all experiencing high urban and slum growth rates, and slums with more than one shelter deprivation are prevalent.

In the **Caribbean** country of Haiti, urban poverty levels and slum prevalence go hand-in-hand. The slum growth rate in Haiti, 3.63 per cent per year, approximates slum growth patterns in sub-Saharan Africa. Only one in four households has a proper kitchen, and only 23 per cent of households have access to improved water supply. Access to improved sanitation is low, with only 16 per cent of all households, rich and poor, having toilet facilities within the home. A substantial proportion of households in the capital Port-au-Prince, 20 per cent, have only a pit latrine in the immediate vicinity. A large proportion of the city's garbage – 65 per cent – goes uncollected. Many of the city's families are at risk of being left behind in an emergency, as 40 per cent of the dwellings do not have a road leading to them and are not accessible by fire trucks or ambulances.[14] However, the region as a whole stands out as one that has in recent years aggressively pursued and implemented inclusive urban governance and slum upgrading policies aimed at the most vulnerable populations.

MAP 3 URBAN POPULATION AND SLUM PROPORTION IN COUNTRIES OF LATIN AMERICA AND THE CARIBBEAN, 2001

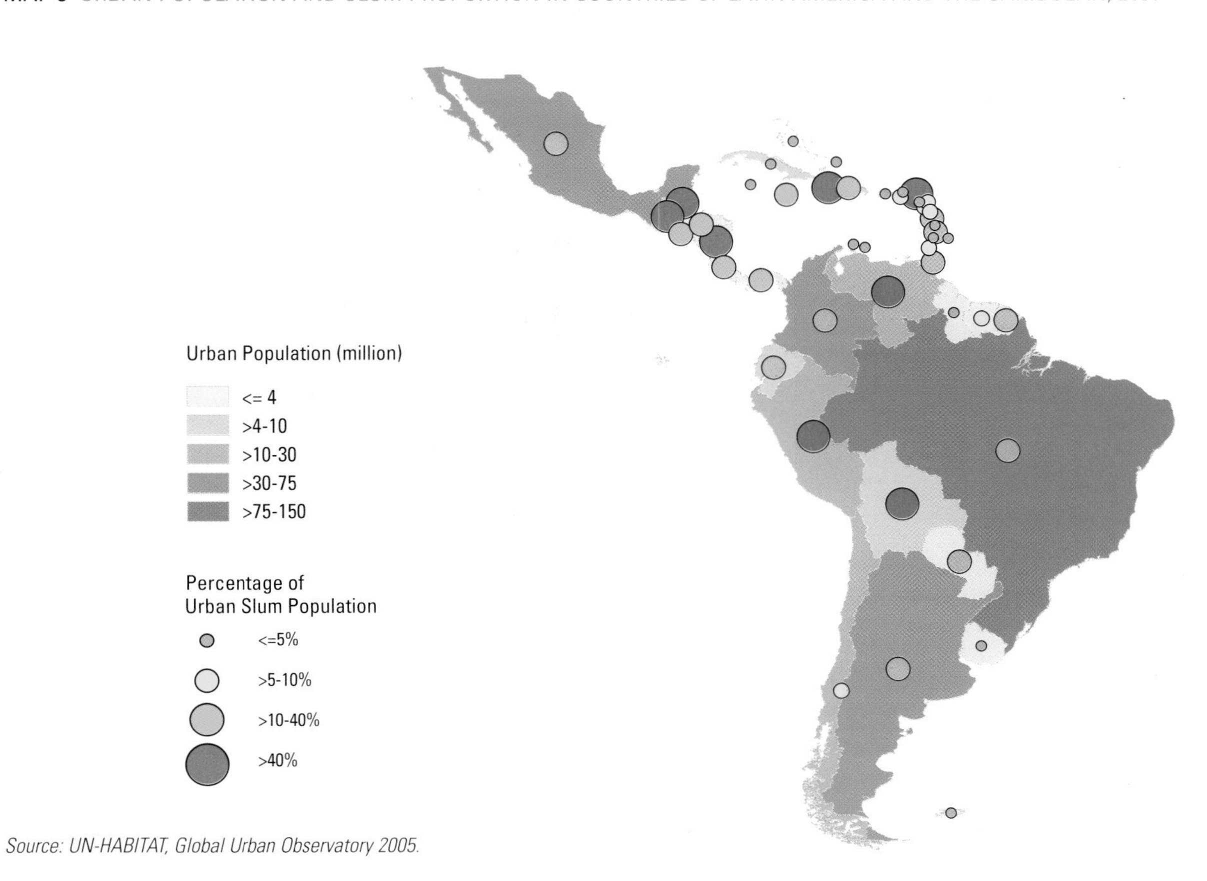

Source: UN-HABITAT, Global Urban Observatory 2005.

Degrees of shelter deprivation: A regional analysis

Not all slums are homogeneous and not all slum dwellers suffer from the same degree of deprivation. The degree of deprivation depends on how many of the five conditions that define slums (poor access to improved water, poor access to sanitation, non-durable housing, insufficient living area and insecure tenure) are prevalent within a slum household. UN-HABITAT analyses show that sub-Saharan Africa's slums are the most deprived; over 80 per cent of the region's slum households have one or two shelter deprivations, but almost half suffer from *at least* two shelter deprivations. Approximately one-fifth of slum households live in extremely poor conditions, lacking more than three basic shelter needs.

Generally, the lack of sanitation and water in the region's slums is compounded by insufficient living space for families and inadequate, makeshift housing. One major reason families have limited access to water and sanitation is that municipal authorities often refuse to extend essential services to their unplanned neighbourhoods. But the burden does not stop there. When lack of sanitation is coupled with lack of water, or temporary or overcrowded housing, the disease and labour burden, especially on women and children, is even more intense. This means that slum dwellers living under such hardship conditions have to cope with survival issues for a considerable part of each day. Many slum dwellers also lack secure tenure, which makes their housing even more precarious. Slum formation and growth is a complex problem to which African governments must commit multiple sectoral interventions and investments to lift their citizens out of poverty.

Slum households with the most shelter deprivations are highly visible in most African cities, as many are clustered within geographically contiguous high-density neighbourhoods, either within or on the outskirts of cities. The concentration of slum households is highest in Ethiopia, followed closely by Burkina

Favelas in Rio de Janeiro JOHN MAIER

The most common deprivation experienced by urban households in sub-Saharan Africa is lack of access to improved sanitation; 45 per cent of the urban population suffers from this deprivation, while 27 per cent suffers from overcrowding.

Faso, Chad and Mali. In general, if a neighbourhood reflects slum characteristics, so do most of the individual households within that neighbourhood. This rule of thumb, however, does not apply to all countries within the region. In Cameroon, Ghana, Guinea, Namibia, and Zimbabwe, most slum households are individual dwellings in different neighbourhoods; some also exist within serviced, middle- and high-income areas.

While sub-Saharan Africa's cities suffer from the most severe shelter deprivations, cities in Northern Africa have managed to reduce the severity of slum conditions markedly; a vast majority of slum households – 89 per cent – suffer from only one shelter deprivation. Simple, low-cost interventions in increasing access to improved sanitation for instance, are all that are needed to help the countries in Northern Africa create "cities without slums" – a goal they have been trying to reach since long before the Millennium Declaration was adopted in 2000.

Urban households in Southern Asia suffer from similar levels of deprivation as those in sub-Saharan Africa. Nearly one in five households lack two basic shelter needs. Among slum households, one in three families has to cope with the lack of two essential services. Nonetheless, unlike sub-Saharan Africa, very few slum households in Southern Asia suffer from three or more shelter deprivations, despite the huge magnitude of slums within the sub-region, lack of sanitation and overcrowding are the most common deprivations experienced by urban populations in the region. These figures may not reflect the reality on the ground, however, as UN-HABITAT and official data sources do not capture the shelter deprivations experienced by pavement dwellers or street families, who are not normally categorized as "households" in censuses and surveys in countries such as India.

Western Asian cities are similar to, albeit somewhat worse off than, Northern African cities. Among slums, nearly one in four lack more than two indicators of adequate shelter. The problem in Western Asia is compounded by a volatile political situation, which has contributed to an influx of refugees and internally displaced persons (IDPs) to cities. In contrast, the majority of slum households in South-Eastern Asia – 74 per cent – suffer from only one shelter deprivation.

In Latin American cities, neither the magnitude of slums nor the degree of severity is as daunting as in other regions. However, the proportion of slum households that suffer from at least one shelter deprivation is quite high: 66 per cent.

TABLE 1.2.3 & FIGURE 1.2.3 PROPORTION OF SLUM HOUSEHOLDS IN DEVELOPING REGIONS BY NUMBER OF SHELTER DEPRIVATIONS, 2001

Region	Deprivations One	Two	Three	Four
	Percentage			
Northern Africa	89	11	0	0
Sub-Saharan Africa	49	33	15	3
Latin America and the Caribbean	66	25	8	1
Eastern Asia	-	-	-	-
Southern Asia	66	29	5	0
South-Eastern Asia	74	20	5	1
Western Asia	77	16	6	1

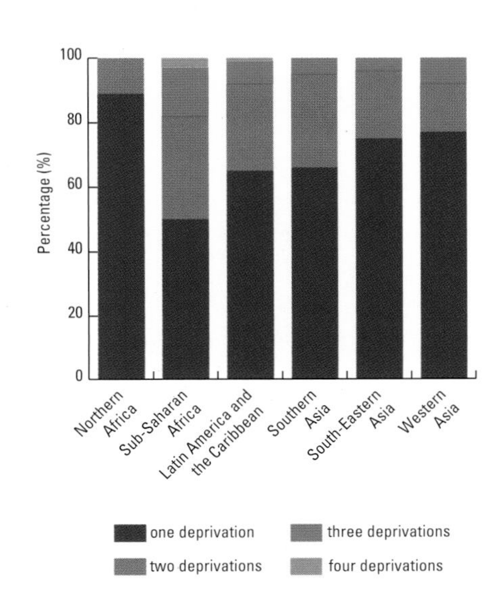

Source: UN-HABITAT 2005, Urban Indicators Programme, Phase III.

■ The prospect of reaching the slum target

Forecasts reveal that the magnitude of slums will continue to increase if the trends dominant between 1990 and 2001 are projected into the future. In light of recent evidence, even if governments collectively manage to improve the lives of 100 million slum dwellers by 2020 – as per the Millennium Development Goals and targets – this achievement will be insignificant in relation to creating "cities without slums",[15] a stated objective of the Millennium Declaration. Assuming that the leaders who developed the slum target were aiming to address a major development issue, policymakers should adjust the benchmark to reflect the reality of slums of today and tomorrow.

In view of the existing slum situation around the world, UN-HABITAT constructed three scenarios to aid planners and policymakers who have a stake in improving the lives of slum dwellers. The worst-case scenario (*Scenario 1*) assumes that the rate of slum growth between 1990 and 2001 will remain the same in all five-year periods between 2000 and 2020 – that is, slums will continue to grow. The second scenario (*Scenario 2*) assumes that there will be 100 million fewer slum dwellers in 2020 than in 1990, which means the target will be met, but in an environment in which the annual growth rate of slums exceeds the rate at which they are being improved. The best-case scenario (*Scenario 3*) assumes that the proportion of slum dwellers in 1990 will be reduced by half, in alignment with most of the other Millennium Development Goals and targets.

The result of the projection for *Scenario 1* suggests that by 2020, there will be nearly 1.4 billion slum dwellers in the world, if present trends of urban and slum growth continue unabated into the future, and if governments do not upgrade slums or provide positive alternatives to new slum formation. Under *Scenario 2,* if the lives of 100 million slum dwellers are improved, 1.3 billion people around the world will continue to live in slum conditions. Scenario 2 reveals that slum growth rates over the next 15 years will be highest in sub-Saharan Africa. While the international community would have achieved the very modest target set out in the Millennium Development Goals, it would have made no significant impact on reducing the proportion of people living in slum conditions. *Scenario 3* is based on the assumption that in the next 15 years, international and national stakeholders will have adopted and implemented policies to prevent slum formation and reduce by half the number of slum dwellers. If these policies were effective, the share of slum dwellers would decline from 31 per cent of the urban population in 1990 to 15 per cent in 2020. This would amount to improvement in the lives of at least 700 million people.

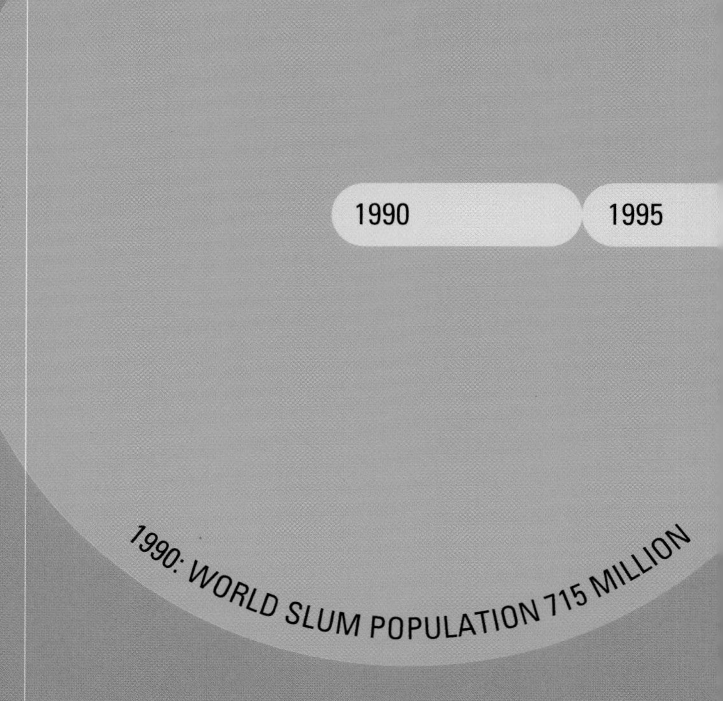

FIGURE 1.2.4 MAGNITUDE OF SLUMS IN 2020 UNDER THREE SCENARIOS

Scenario 1
This assumes present trends of urban and slum growth continue unabated into the future.

Scenario 2
Improve the lives of 100 million slum dwellers by 2020 i.e. achieve MIllennium Development Goal 7, target 11.

Scenario 3
Reduce proportion of slum dwellers from 31 per cent of the global urban population in 1990 to 15 per cent in 2020. This would reduce by half the proportion of people living in slums.

In view of the general lack of international and national commitment to achieving target 11 in most countries, the best-case scenario seems quite elusive. If current trends continue, it is highly likely that in 2020 the slum population will be 1.4 billion. This has serious implications in terms of policies for improving lives of slum dwellers, as well as the situation of slum dwellers *vis-à-vis* the poverty, health, education, and employment targets stated in the other Millennium Development Goals. The growth of slums in the world's cities, which will host the majority of the world's population after 2007, should therefore be a cause for concern, as they may eventually jeopardize the achievement of all the Millennium Development Goals and targets.

SCENARIO 2: WORLD SLUM POPULATION IN 2020 1.3 BILLION

SCENARIO 1: WORLD SLUM POPULATION IN 2020 1.4 BILLION

FIGURE 1.2.5
THE RISE AND RISE OF SLUMS, 1990-2020

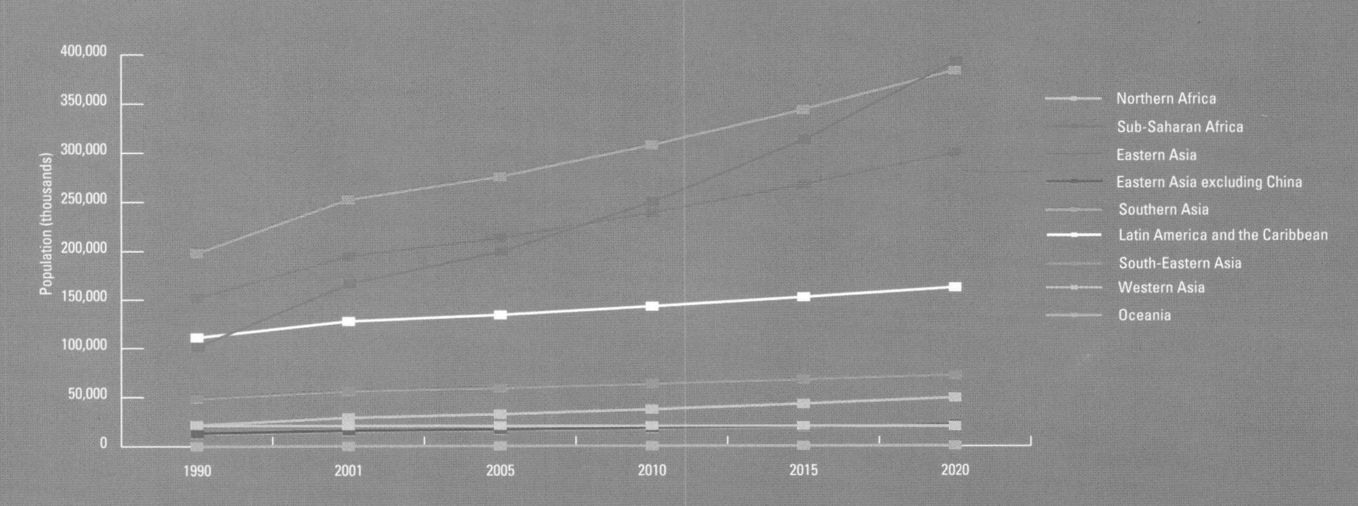

Source: UN-HABITAT, Global Urban Observatory 2005.

SCOTT LECKIE

THE SLUM TARGET IS NOT IN LINE WITH HOUSING RIGHTS

Children of pavement dwellers, Mumbai RASNA WARAH

Reviewing the Millennium Development Goals through a human rights prism, one is forced to ponder whether these Goals were designed to supplant, integrate or ignore human rights concerns of the world's poor. All fair-minded people, of course, would hope for the second of this trilogy of options. Unfortunately, looking closely at the Goals, particularly Goal 7, target 11 on improving the lives of slum dwellers, it appears the Goals' flaws, as far as housing rights are concerned, far out-number their benefits.

Three shortcomings stand out as particularly worrying. First, the objective of improving the lives of at least 100 million slum dwellers by 2020 is almost obscene in its conscious exclusion of a huge majority of the world's urban and rural poor. In a world of almost one billion slum dwellers, to speak of improving the lives of less than 10 per cent of the world's poorest citizens and rights-holders underscores just how far the Goals stray from the language, sentiments and vision of human rights. Which 100 million slum dwellers are we actually talking about? Who will choose those whose lives will be improved? What say will they have in the matter? Which 900 million or more slum dwellers and homeless citizens will fall through the cracks? Who will inform them of their plight of having been so deliberately barred from the Millennium Development Goals? And what of the additional 400 million slum dwellers that UN-HABITAT projects will be in need of improved housing by 2030? Does the world truly accept that perhaps some 1.3 billion people will call slums home some fifteen years from now, and that this is somehow an acceptable future for the planet's urban poor?

Second, the Millennium Development Goals fail to address some of the most pressing housing rights concerns affecting the world's slum dwellers, as if these, too, were somehow not part of the poverty trap facing growing numbers of people. The global forced evictions epidemic, decimated budgets and reductions in public expenditure on housing for the poor, spiraling house, land and property prices reaching bubble-like proportions in many countries, illegal land grabs, entrenched discrimination against women, the disabled and the elderly, ethnic cleansing, the demolition of homes during war and so many other core housing rights themes are all remarkably absent from the Goals. Not only would addressing these poverty-expanding processes have been logical given their impact on hundreds of millions of dwellers throughout the world, but by focusing on and forcefully discouraging these practices, the Goals would have improved the lives of many more millions of slum dwellers than will possibly see improvements based on Goal 7, target 11.

Third, the Goals – as with so many of the agreements emerging from the various global summits during the past decade, whether by decision or by default – all too often end up taking the wind from the sails of the human rights movement, slowing human rights progress and shifting burdens of proof from national governments to the international community, which can never alone transform the human rights dreams of the poor into reality.

The Millennium Development Goals are formulated as if to tease those who – in precisely the same manner as international human rights law – treat issues of poverty not only as development questions, but as rights. By labeling what are, in fact, core human rights principles merely as "goals", these internationally-agreed targets are effectively supporting a creed which sees only half of the human rights equation as actual human rights, somehow relegating the other half to "goals", "aspirations" or "needs" – not enforceable rights held and rightfully expected by all, especially those currently without the protection these rights are meant to provide.

As far as housing rights are concerned, it would be difficult not to conclude that the Millennium Development Goals let governments off the hook; they almost insinuate that a staggeringly large portion of humanity is condemned by circumstance to live in life- and health-threatening conditions, without security, as if to say, "Yes, we care, but as far as slums dwellers go, we know and accept that our care will only reach a few of you." Good luck to the 900 or so million slum dwellers that the Millennium Development Goals forgot about. They will need it. In the coming years, these neglected millions will, as always, gain ground, organize, and support themselves. Whatever rights slum dwellers accrue, or security they can claim, will come from their own energies and sadly, not from a global accord agreed to by governments.

Scott Leckie is the Executive Director of the Centre on Housing Rights and Evictions (COHRE).

Endnotes

1 Milanovik 2005.
2 Definition used in Concise Oxford English Dictionary. Tenth Edition, 2002.
3 UN-HABITAT slum estimations revised in 2005, based on coordination of definitions with the WHO/UNICEF Joint Monitoring Programme for Water Supply and Sanitation.
4 Only data for the developing world, collected with household surveys and censuses, are reported here, as UN-HABITAT's estimates for the developed world are based on secondary sources, including reports, printed censuses and modeling.
5 UN-HABITAT 2004b.
6 Ibid.
7 The Orangi Project, which started as a pilot project within a township in Karachi in 1965, has been touted as one of the most successful demonstration projects in the world and has since been replicated in seven cities in Pakistan.
8 Dialogue at Second World Urban Forum, Barcelona, September 2004.
9 In China, people's status as residents is determined by their ability to acquire residency permits for a particular city, town or village. An independent study conducted by the Fafo Institute for Applied International Studies in Beijing in 2000 attests to the fact that the population that comes either from other cities or villages or from the rural areas of Beijing is not considered part of the total city population. Since they are not regarded as dwellers of the city, they are not considered slum dwellers of the city either. However, the study indicates that those who belong to settlements outside Beijing's urban areas constituted 16 per cent of the *de facto* population of Beijing at the time. Most of the slum characteristics that UN-HABITAT uses as criteria are prevalent among this group of people. The share of people originating from outside villages could be even higher, as many migrant workers do not live within established housing, as they do not have permits to live in cities - many live on construction sites. Quite a substantial proportion of these inhabitants suffer from overcrowding; they often live in dormitories, jerry-built houses or workers' huts, shared apartment units or simple apartment units with shared facilities.
10 Forum on the Global South 2003.
11 Asian Development Bank President Haruhiko Kuroda quoted in *Newsweek*, 21 November 2005.
12 The Asia-Pacific Economic Cooperation (APEC) summit held in Busan in November 2005.
13 For instance, the 2005 United Nations report entitled *The Inequality Predicament* shows a strong correlation between inequality in Latin American cities and levels of homicide.
14 Fafo Institute for Applied International Studies 2003.
15 UN Millennium Declaration 2000, paragraph 19.

1.3 How Well is Your Country Performing on the Slum Target? A Global Scorecard

A snapshot of country performance

Developing countries have pursued a range of policies and practices to deal with the deficiencies in the provision of basic services, housing, health, and education for the urban poor and slum dwellers over the course of the last thirty years. The international community, including United Nations agencies, multilateral funding agencies and bilateral donors, have, through successive strategies, arguably played a pivotal role in transforming government attitudes and policy responses to slums. The policy options and interventions by both the international community and governments have changed over time: neighbourhood-level self-help solutions to housing and *in situ* upgrading during the late 1970s and 1980s; getting the "enabling environment" right and improving urban management in the 1990s; and scaling-up of slum upgrading through national and citywide programmes since the end of 1990s.

During this period, such remedies have brought hope to many of the world's urban poor. Initiatives such as the Kampong Improvement Programme in Indonesia, the *Favela Bairro* programme in Rio de Janeiro, Brazil, and the Million Houses Programme in Sri Lanka have significantly improved the lives of the urban poor. For example, the programme in Indonesia managed to reach 15 million people over its 30-year history working with some 300 local governments in the provision of water, sanitation, shelter and roads.[1]

And yet for much of the rest of the developing world, policy reforms or interventions have not been enough or simply failed to materialize. Despite their good intentions, some governments and donors have struggled to cope with overwhelming demographic pressures, massive backlogs in basic services and housing provision, and growing environmental degradation and unemployment, while other governments continue to ignore the issue of slums in official policy circles. When remedies fail to reach people they are meant to serve, it is the poor who fall back on their own capacity and resilience to make a home and living for themselves in the city.

However, it is still not very clear where actions to upgrade slums or prevent their formation have collectively made the biggest difference in improving lives of the urban poor or where they have failed to address the problem of slums. While for the past three decades, researchers and practitioners have produced mostly anecdotal or qualitative evidence to evaluate progress and failure of countries and cities in slum improvement, what has been missing is a much more systematic, rigorous attempt to compare performance among and within

Laying new water pipes in Mathare Valley, Nairobi FRIEDRICH STARK/STILL PICTURES

Millennium Development Goal 7, target 11:
By 2020, to have achieved a significant improvement in the lives of at least 100 million slum dwellers

countries using statistical data and internationally-agreed definitions and methods.

But things are changing. The movement that is now challenging countries and the international community to live up to the promise of the Millennium Development Goals is helping to trigger a new way of measuring the performance of countries in improving slums and meeting the other Goals and targets in urban areas. International agencies are indeed seizing the opportunity to produce more and better statistics with which to monitor the Goals.[2] UN-HABITAT has devised, through extensive consultation with its partners, a new methodology for measuring slums and has subsequently produced estimates for the numbers of slum dwellers at global, regional, national and city levels.[3] The results have been used in this part of the Report to construct a *global scorecard* showing the varying performance of over 100 countries in improving the lives of slum dwellers and reducing slum growth rates.

Results at a glance

Countries "on track" – those starting to make urban poverty history

Among all developing countries, Thailand has seen the sharpest decline in slum growth rates. In 1990, there were almost 2 million people living in slum conditions; by 2005, this figure had been slashed to just 119,000. This dramatic decrease is attributed largely to the government's long-standing commitment to implementing programmes to improve the housing conditions of the urban poor. Egypt, Georgia, Sri Lanka and Tunisia are also among the "on track" countries registering falling slum growth rates since 1990, and are seeing significant reductions in the number of people living in slums. Egypt succeeded in reducing the number of slum dwellers by 3 million from 1990 to 2005. Tunisia has succeeded in more than halving the number of slum dwellers in the same period to approximately 190,000. However, these countries are among a meagre 14 countries out of the over 100 analysed that made it to the "on track" category.

Countries in the "stabilizing" category – those starting to put the brake on slum growth rates, but need to monitor and make sure they don't slip back

Brazil, Indonesia, Iran, Mexico, South Africa and Turkey appear to have made fairly good progress in basic service provision to the urban poor, which is reflected in low to almost stable slum growth rates over the last 14 years. Yet, these countries still have large numbers of people lacking adequate housing and basic services – 52 million in Brazil, 22 million in Indonesia, 15 million in Mexico and 8 million in South Africa. These countries are on the right path but clearly have some way to go in order to make the kind of reductions registered by the "on track" countries.

Warning signals light up in "at risk" and "off track" countries

The worrying trend is that most developing countries have failed to make much headway in reducing slum growth rates. Of particular concern are those countries that experienced substantial slum growth rates (ranging between 4 per cent and 6 per cent annually) and high incidence of slums in the last 15 years. This combination has had lethal effects. For instance, the slum population in Tanzania has more than doubled in the last 15 years, from 5.6 million in 1990 to 14 million in 2005. Likewise in Bangladesh, Nigeria and Sudan, the numbers of slum dwellers grew from 19 million, 24 million and 5.7 million in 1990 to 36 million, 46 million and 12 million in 2005, respectively. This "off track" group needs to take drastic action now to improve existing slum conditions and prevent future slum formation; otherwise the numbers of slum residents will continue to rocket upwards. Other countries, including Argentina, China, India and Morocco, are doing slightly better than this group in terms of managing slum growth rates at roughly 2 per cent a year, but they are still in the "at risk" category of countries, as the proportion of people living in slum conditions is relatively high (over 30 per cent) and they still need to revisit existing policies and improve performance. Over 70 per cent of the countries analysed fell under these two bottom-end performance categories.

Track record of regions

Moving from a country to a regional portrait, the scorecard shows countries in *sub-Saharan Africa* struggled above all to cope with the rising numbers of slum dwellers – 34 out of the 50 countries in the "off track" group are in this part of the world, including, Kenya, Lesotho and Mali. These countries not only experienced some of the highest slum growth rates, but also tended to have a large proportion of their total urban population living in slums. South Africa is the only country in the region that made it into the higher "stabilizing" category, recording an almost negligible annual slum growth rate.

Southern Asia, as a whole, also demonstrated a poor track record. In particular, Bangladesh, Nepal and Pakistan continued to lag behind, while India performed somewhat better in managing overall slum prevalence, with a slum growth rate of 1.7 per cent. As a notable exception, Sri Lanka, following decades of social investment, ranks among the top performers not only in the region but also worldwide; with an annual decline of 3.7 per cent, the total number of slum dwellers stands at half a million, down from nearly 900,000 in 1990.

Scorecard Methodology

The chart presents the first findings of the global scorecard on slums. Countries are judged to be doing well if they have managed to reduce or reverse slum growth rates and if, at the same time, they have succeeded in keeping the proportion of slum dwellers at relatively low levels.

The cut off points for the **annual slum growth rates** were: 0%, > 0-2%, > 2-4%, > 4%. These growth rate categories were combined with the **proportion of slums** in the discussed regions. As the incidence of slums is very different in these regions, different categories for the slum proportions were chosen to characterise these different conditions. For Africa and Asia, where the incidence of slums is generally higher than in the other regions, the cut-off points were the same (30%, >30-60%, >60%). For Latin America and the Caribbean, the cut off points were 10, >10-40, 40% and for the Commonwealth of Independent States the categories are 20, >20-50, >50.

As a result of this procedure the countries could be grouped within the four performance categories. Covering over 100 countries in the developing world, the scorecard measures trends in country performance between 1990 and 2005. Countries are grouped into the following four performance categories:

ON TRACK

Countries experiencing rapid, sustained decline in slum growth rates in urban areas and/or those with low slum prevalence.

STABILIZING

Countries starting to stabilize or reverse slum growth rates but which need to monitor progress to ensure sustained reductions.

AT RISK

Countries experiencing moderate to high slum growth rates but also having moderate incidence of slums that require remedial policies to reverse growth in numbers of slum dwellers.

OFF TRACK

Countries with already high slum proportions, facing rapid, sustained slum growth rates and which require immediate, urgent action to slow down or reverse slum trends.

ON TRACK

Income group	Country	Slum Annual growth rate	% slum 1990	% slum 2005	Slum Pop. 1990 (000s)	Slum Pop. 2005 (000s)
UPPER MIDDLE INCOME*	Puerto Rico	1.6	2.0	2.0	50	63
	Uruguay	-10.3	6.9	1.3	191	41
LOWER MIDDLE INCOME*	Cuba	0.7	2.0	2.0	156	174
	Egypt	-1.6	57.5	34.9	14,087	11,015
	Sri Lanka	-3.7	24.8	10.9	899	515
	Thailand	-18.8	19.5	0.9	1,998	119
	Tunisia	-5.4	9.0	2.7	425	188
LOW INCOME*	Georgia	-7.2	18.4	6.4	558	189

*WORLD BANK INCOME GROUP

MAPS 4, 5, 6 A LOOK AT COUNTRY PERFORMANCE IN DEVELOPING REGIONS, 2006

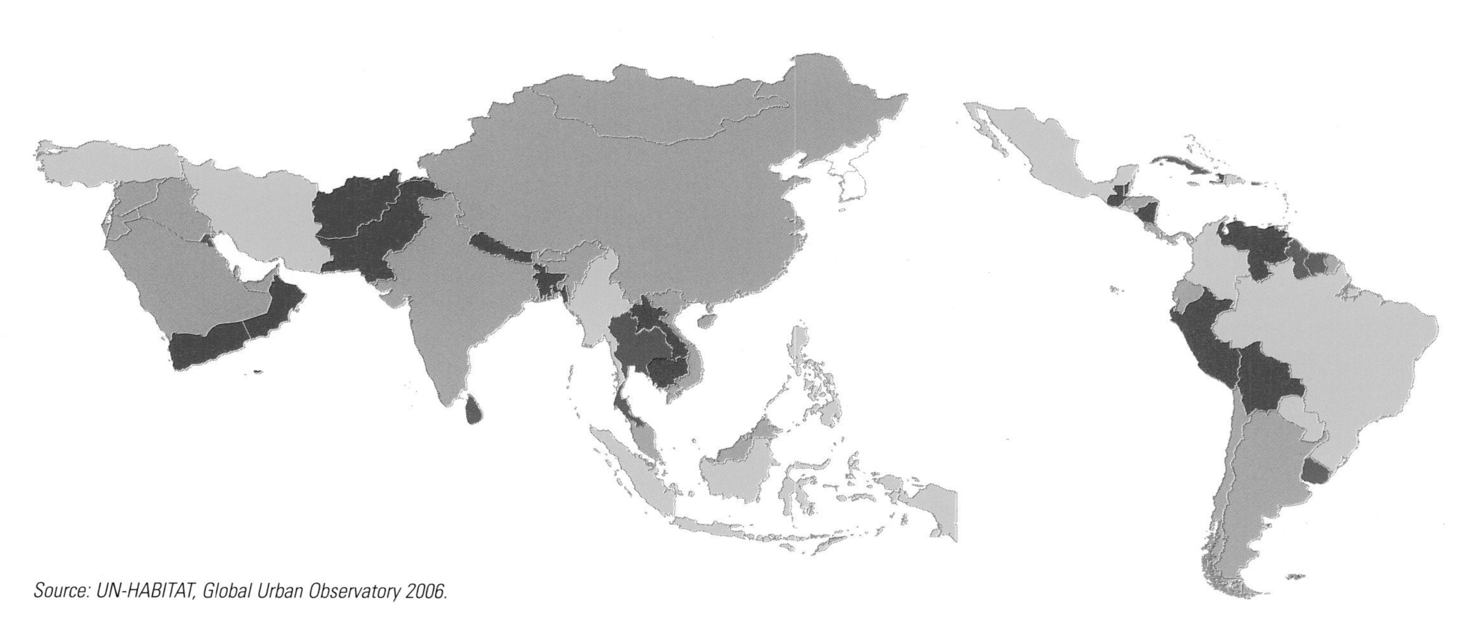

Source: UN-HABITAT, Global Urban Observatory 2006.

STABILIZING

	Slum Annual growth rate	% slum 1990	% slum 2005	Slum Pop. 1990 (000s)	Slum Pop. 2005 (000s)
Brazil	0.3	45.0	34.0	49,806	52,374
Mexico	0.5	23.1	18.5	13,923	14,983
Trinidad and Tobago	0.5	34.7	31.0	292	317
Colombia	1.1	26.0	20.5	6,239	7,381
Dominican Republic	-0.9	56.4	32.4	2,327	2,038
El Salvador	1.9	44.7	32.3	1,126	1,495
Iran, Islamic Rep. of	1.6	51.9	41.6	17,094	21,763
Paraguay	0.5	36.8	21.7	756	812
Philippines	1.9	54.9	40.7	16,346	21,792
South Africa	0.2	46.2	29.4	8,207	8,439
Turkey	0.0	23.3	16.3	7,997	8,016
Kazakhstan	-0.6	29.7	29.7	2,835	2,605
Indonesia	1.4	32.2	20.5	17,964	22,049
Myanmar	1.3	31.1	24.9	3,105	3,794
Republic of Moldova	-1.3	31.0	31.0	634	522

AT RISK

	Slum Annual growth rate	% slum 1990	% slum 2005	Slum Pop. 1990 (000s)	Slum Pop. 2005 (000s)
Argentina	2.2	30.5	34.1	8,597	11,978
Botswana	3.7	59.2	61.3	311	540
Costa Rica	4.3	11.9	13.1	195	372
Lebanon	3.1	50.0	50.0	1,142	1,811
Panama	2.2	30.8	30.8	397	552
Saudi Arabia	3.8	19.8	19.8	2,385	4,196
Algeria	3.0	11.8	11.8	1,508	2,370
China	2.3	43.6	35.9	137,929	195,682
Ecuador	2.5	28.1	24.8	1,588	2,317
Honduras	2.4	24.0	16.3	488	703
Iraq	2.5	56.7	56.7	6,825	9,992
Jamaica	3.5	29.2	38.4	356	604
Jordan	4.3	16.5	15.4	388	741
Morocco	2.0	37.4	31.1	4,457	6,054
Namibia	2.9	42.3	36.4	155	239
Syrian Arab Republic	3.2	10.4	10.4	629	1,012
Bhutan	1.2	70.0	37.3	61	73
Congo	3.6	51.9	48.7	5,366	9,227
India	1.7	60.8	53.6	131,174	169,671
Mongolia	0.7	68.5	63.6	866	969
Viet Nam	1.2	60.5	43.4	8,100	9,632

OFF TRACK

	Slum Annual growth rate	% slum 1990	% slum 2005	Slum Pop. 1990 (000s)	Slum Pop. 2005 (000s)
Gabon	5.9	56.1	70.4	357	872
Oman	5.4	60.5	60.5	671	1,506
Venezuela	2.5	40.7	40.7	6,664	9,642
Bolivia	2.3	70.0	58.4	2,555	3,597
Guatemala	2.5	65.8	60.4	2,192	3,186
Peru	3.4	60.4	71.1	8,979	14,862
Afghanistan	6.4	98.5	98.5	2,458	6,375
Angola	5.3	83.1	83.1	2,193	4,839
Bangladesh	4.3	87.3	83.8	18,988	36,079
Benin	5.3	80.3	84.8	1,288	2,870
Burkina Faso	4.0	80.9	74.9	987	1,791
Burundi	2.7	83.3	59.8	294	438
Cambodia	6.1	71.7	72.4	870	2,162
Cameroon	5.0	62.1	68.9	2,906	6,197
Central African Rep.	3.1	94.0	91.8	1,038	1,646
Chad	4.3	99.3	99.1	1,218	2,308
Congo, Dem Rep. of the	5.2	84.5	92.2	1,050	2,276
Côte d'Ivoire	6.0	50.5	75.6	2,532	6,203
Eritrea	3.6	69.9	69.9	342	590
Ethiopia	4.8	99.0	99.5	5,984	12,315
Gambia	5.4	67.0	67.0	155	348
Ghana	1.8	80.4	66.0	4,083	5,372
Guinea	3.4	79.6	69.8	1,145	1,918
Guinea-Bissau	5.2	93.4	93.5	210	456
Haiti	3.6	84.9	86.0	1,728	2,976
Kenya	5.9	70.4	70.9	3,985	9,620
Lao People's Dem Republic	4.7	66.1	66.1	422	850
Lesotho	6.3	49.8	59.8	168	434
Liberia	2.0	70.2	51.2	632	853
Madagascar	5.3	90.9	93.7	2,562	5,696
Malawi	3.9	94.6	89.9	1,033	1,860
Mali	4.9	94.1	92.9	1,968	4,083
Mauritania	5.6	94.3	94.3	827	1,915
Mozambique	6.9	94.5	93.9	2,722	7,710
Nepal	4.8	96.9	90.9	1,574	3,213
Nicaragua	3.4	80.7	81.0	1,638	2,730
Niger	5.9	96.0	96.3	1,191	2,882
Nigeria	5.0	80.0	71.9	24,096	46,272
Pakistan	2.7	78.7	71.8	26,416	39,722
Rwanda	3.5	82.2	90.1	296	504
Senegal	4.1	77.6	76.0	2,276	4,181
Sierra Leone	3.6	90.9	97.6	1,107	1,895
Somalia	3.6	96.3	97.4	1,670	2,867
Sudan	5.2	86.4	85.5	5,708	12,441
Togo	4.3	80.9	80.5	796	1,510
Tanzania, U. Rep. of	6.2	99.1	89.6	5,601	14,113
Uganda	5.3	93.8	92.7	1,806	4,010
Yemen	5.0	67.5	64.2	1,787	3,803
Zambia	2.9	72.0	74.8	2,284	3,519

FIGURE 1.3.1 A GLOBAL SCORECARD ON SLUMS 2006

Source: UN-HABITAT 2005, Urban Indicators Programme, Phase III.

ON TRACK

STABILIZING

AT RISK

OFF TRACK

NO OR NOT RELIABLE DATA

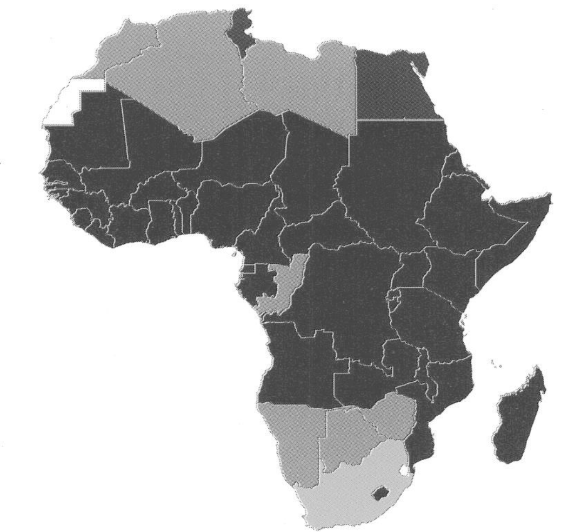

Businesswoman in Tunisia VINCENT KITIO

Egypt and Tunisia are the "high-flyers" in Northern Africa with a good track record of investments in water and sanitation, as well as informal settlements upgrading.

Eastern Asia demonstrates even more mixed results, but averages a better overall performance than Southern Asia. While Thailand appears to be very much in a league of its own, Indonesia and the Philippines have succeeded in keeping slum growth rates low. China, despite impressive policy interventions, recorded a relatively large slum population. Preliminary UN-HABITAT analysis attributes this to high prevalence of poor sanitation in cities; a WHO/UNICEF assessment in 2000 showed that 33.8 per cent of the urban population lacked access to improved sanitation. Cambodia struggled to stay ahead of demographic and poverty curves during the 1990s and experienced a high annual slum growth rate of 6.1 per cent, resulting in a more than doubling of the slum population from 870,000 in 1990 to over 2 million in 2005.

In largely Arabic-speaking *Northern Africa* and *Western Asia*, the results have been mixed: while the former performed well on reducing slum growth rates, the latter had higher slum growth rates than those of the rest of Asia. Slum growth in many Western Asian countries is not necessarily confined to lower-income countries, but also extends to middle-income and high-income countries in the region, such as Iraq, Jordan and Saudi Arabia, which experienced a drop in economic performance in the 1990s and witnessed sustained growth in the number of slum dwellers (3 per cent to 4 per cent a year), from 6.8 million, 390,000 and 2.4 million in 1990 to 10 million, 740,000 and 4.2 million in 2005, respectively. Mauritania and Yemen also performed poorly, recording annual slum growth rates of 5 per cent or more. In contrast, Egypt and Tunisia are the "high-flyers" in Northern Africa with a good track record of investments in water and sanitation, as well as informal settlements upgrading, during the course of the 1990s that resulted in a significant decline in the number of slum dwellers, from 14 million and 425,000 in 1990 to 11 million and 188,000 in 2005, respectively.

The best-performing developing region appears to be *Latin America and the Caribbean*. The majority of countries mainly fall under the "at risk" and "stabilizing" groups. The top performers, according to the scorecard, are Cuba and Uruguay. On the other hand, urgent attention is needed to deal with overwhelming demand for services in countries such as Bolivia, Guatemala, Haiti, Nicaragua, Peru and Venezuela. However, the majority of countries in the region, including Brazil, Colombia, Costa Rica, Honduras and Mexico, appear to be coping more effectively and have seen relatively low to moderate increases in the numbers and proportions of slum populations.

Understanding the dynamics of improved performance

Six years after the Millennium Declaration committed governments to strive for "cities without slums", the results of the global scorecard suggest that only a handful of countries in the developing world have made real progress in providing better public services and housing to slum dwellers. "On track" countries, where slum growth rates have actually fallen, include Egypt, Sri Lanka, Thailand and Tunisia. In these countries, governments and other stakeholders have taken on the responsibility of providing decent housing, water and sanitation to slum dwellers – and they have shown how to get the job done. These countries also provide a beacon of hope and inspiration to their neighbours. However, the worrying trend is that most developing countries, particularly in sub-Saharan Africa, failed to make much headway in reducing or reversing the slum growth rates. A detailed explanation as to why some countries succeed in reducing slum growth and preventing slum formation, while others struggle to deal with growing poverty and inequality in their cities is provided in Part Four of this Report. The analysis draws on recent data compiled and generated by UN-HABITAT to compare country performance and national policy reforms in slum upgrading and prevention.[4]

Based on the scorecard and policy analyses carried out by UN-HABITAT, we now have a better understanding of what drives a country's performance in improving the living conditions of slum dwellers. We know that ***political commitment and long-term government policies*** are essential ingredients. The experience of some of the best-performing countries shows that there is nothing like the commitment of top political leadership to give clarity of purpose, direction and a sense of urgency in tackling head-on the growth of slums – it has often proven to be the surest way of committing actions and resources to the problem. In Morocco, when the King declared slum upgrading as one of his top four priorities, more than 40 per cent of the budget of the national development plan went to upgrading. Brazil's urban pro-poor policies were given a boost when the President set up a housing fund of $1.6 billion for house building and *favela* upgrading. And in Tunisia, scaled-up upgrading has been a core business of the government for three decades and is consistently included in its successive five-year national development plans.

"On-track" countries have also effectively hit the most critical policy levers, carrying out reforms and scaling-up of slum upgrading programmes that have led directly to improvement in the lives of slum dwellers. We know that some of the most ***progressive sectoral reforms*** that drive better performance focus on: improving secure land tenure; setting up a proper land regulation system; providing affordable and accessible housing; and improving the efficiency and coverage of water supply and sanitation services.

Some of the better performing countries also score highly on ***local governance***, accelerating effective decentralization policies, municipal reforms and broad-based participation in planning. With new-found mandates and powers and increasing pressure to respond to voters' demands, municipalities are paying more attention to improving slums and are getting more involved in slum upgrading. The analyses also show that, in many cases, there are no hard-and-fast rules regarding which policy interventions or governance structures work best. For instance, some of the best performing countries, including Egypt, Russia and Tunisia, have succeeded in reversing slum growth rates under very ***centralized systems of governance***, while others, such as Colombia and India, have been unable to reverse slum growth despite highly decentralized governance structures.

The evidence also points to the high performers improving the ***financial sustainability*** of local governments, and finding innovative ways of raising domestic investment. We also know that the most successful countries face up to a double challenge – that of ***scaling up improvements*** that will reach the large numbers of people who are living in slums today, as well as ***preventing future slums***. This experience is something that other struggling nations and cities could adapt, but they may not find it easy in conditions where economic development, good local governance and political leadership are weak.

Using the scorecard: Measuring, managing and motivating change

The scorecard presented in this Report is only the first step towards a better understanding of which countries are on track to meet target 11 and why they appear to be more successful than others. It does not pretend to provide definitive answers to many of the questions indicated above. There is still a long way to go to further develop the scorecard approach, especially in terms of refining the outcome performance measures for slum growth, producing more complete information about policies for countries, and setting up a more rigorous way of rating and comparing policy performance.

Political commitment to slum prevention and upgrading is essential to reducing slum growth.

Cairo VINCENT KITIO

Slum Housing in Manila, Phillipines A. APPELBE/UNEP/STILL PICTURES

The scorecard should be seen as more than just a measuring yardstick for measuring progress; rather, governments, civil society and the international community have an opportunity to use the results to better manage their performance. By comparing the performance outcomes of different slum policies among countries, governments and international development agencies, stakeholders can use the results to improve policy and resource allocation decisions, drive performance improvements in service delivery for the urban poor, and communicate to the public how well or otherwise progress is being made towards target 11 and the other Millennium Development Goals and targets in cities. More specifically, the scorecard could be used to:

- **Identify best practices in slum upgrading and prevention** using reliable statistical information. The scorecard may help point toward success stories in improving the lives of slum dwellers, and provide a basis for analysing what factors led to better performance. Other countries could adapt and implement the most effective practices in order to achieve similar results.

- **Motivate countries to improve their performance.** One of the most critical barriers to effective service delivery in the public sector is the lack of competition – comparing and reporting on performance can help demonstrate how well agencies are doing compared with others.[5] The scorecard could foster a competitive spirit among countries, particularly those at a similar level of development, motivating progressive countries to make further, continuous improvements and to alert lagging countries to take urgent, remedial action.

- **Strengthen the accountability of governments.** Ultimately, the performance information contained in the scorecard, when made publicly available in countries, could help to make governments more accountable to the urban poor. While governments do not have the primary responsibility of providing basic services, a responsibility that is often shared with non-governmental organizations, the private sector and the urban poor themselves, such a scorecard can lead to wider public debate and create pressure for change.

- **Help the international community to target aid more effectively.** The scorecard could provide donors, multilateral funding agencies and the United Nations system with a tool for comparing how well countries are performing on target 11. Results could be used to identify where problems exist and to set priorities for resource allocation according to level of need. The scorecard could also provide a baseline against which future changes can be tracked. If there is evidence of sustained lack of effective policies to bring down slum growth, the international community, including donors, could use the scorecard to decide if making any further investments in slum upgrading or prevention to non-performing countries is worthwhile.

The real energy behind this tracking system should come, not from the donors, but from the countries that fall under the spotlight of the scorecard. It is their policies, their political leaders and their future prospects that are under scrutiny. The issue is whether the results will grab the attention of politicians, policymakers and the general public to produce sufficient pressure on governments for change. Arguably, indicators and statistics make headlines if they are about issues that people really care about and feel that they, or their leaders, have the power to change for the better. It is too early to say if target 11 can trigger such fundamental changes in attitude – but it is clear that the momentum behind this movement is growing.

Progressive sectoral reforms, such as improving land tenure and regularization, providing affordable housing and improving coverage of water and sanitation, are key to slum prevention.

Endnotes

1. MIT website, "Upgrading urban communities": http://web.mit.edu/urbanupgrading.
2. Moreno 2005.
3. Refer to UN-HABITAT 2003c.
4. The scorecard simply shows where and by how much there has been progress or otherwise in reducing the numbers of people living in slums. In order to investigate the reasons behind good and bad performance, the results were used to trigger an in-depth analysis of the policy environment in 23 of the over 100 countries broadly representative of the four performance categories. The 23 countries analysed were: Cuba, Egypt, Sri Lanka, Thailand, and Tunisia in the "on-track" category; Brazil, Colombia, Indonesia, Mexico, Philippines, and South Africa in the "stabilizing" category; Chile, Haiti, India, and Morocco in the "at-risk" category; and Afghanistan, Bangladesh, Burkina Faso, Ethiopia, Ghana, Liberia, Senegal, and Tanzania in the "off track" category.
5. Morley, et al. 2001.

1.4 The Struggle to Achieve the Millennium Development Goals will be Won or Lost in Cities

Singapore ©PHIL DATE. IMAGE FROM BIGSTOCKPHOTO.COM

THE GOOD NEWS

Cities drive national economies

The global fight against poverty[1] – encapsulated in the Millennium Development Goals[2] – is heavily dependent on how well cities perform. The link between urbanization and socio-economic development cannot be disputed. Cities make countries rich. Countries that are highly urbanized have higher incomes, more stable economies, stronger institutions and are better able to withstand the volatility of the global economy than those with less urbanized populations.[3] The experiences of developed and developing countries also indicate that urbanization levels are closely related to levels of income and performance on human development indicators.

Cities around the world are playing an ever-increasing role in creating wealth, enhancing social development, attracting investment and harnessing both human and technical resources for achieving unprecedented gains in productivity and competitiveness. As countries develop, urban settlements account for a larger share of national income. In both developed and developing countries, cities generate a disproportionate share of gross domestic product (GDP)[4] and provide huge opportunities for investment and employment.

Urban-based economic activities account for up to 55 per cent of gross national product (GNP)[5] in low-income countries, 73 per cent in middle-income countries and 85 per cent in high-income countries.[6] In the United States, for example, some

FIGURE 1.4.1 HUMAN DEVELOPMENT IS CLOSELY RELATED TO LEVELS OF URBANIZATION

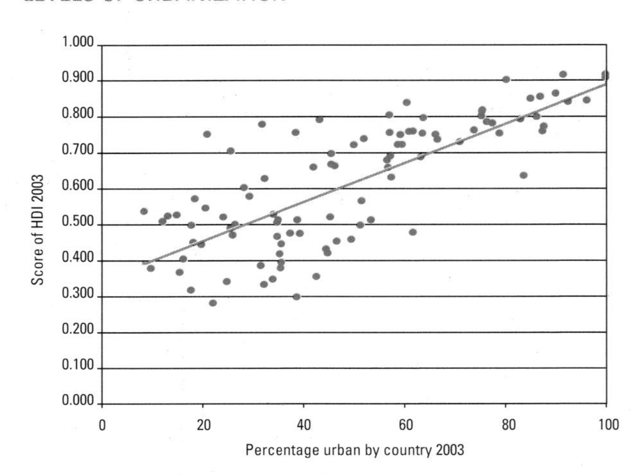

Sources: United Nations, World Urbanization Prospects: The 2003 Revision, UNDP, Human Development Report 2005.
Note: Developing countries with populations of more than one million are plotted. The human development index (HDI) is a summary measure of human development that measures the average achievement in a country in three basic dimensions: a long and healthy life, as measured by life expectancy at birth; knowledge, as measured by the adult literacy rate and the combined primary, secondary and tertiary gross enrolment ratio; and a decent standard of living, as measured by GDP per capita (PPP US$).

cities outpace even some countries in economic output. If the five largest cities in the United States – New York, Los Angeles, Chicago, Boston and Philadelphia – were treated as a single country, it would rank as the fourth largest economy in the world.[7] This trend is also evident in the developing world: São Paulo, Brazil's largest city, and Bangkok, the capital of Thailand, both host just over 10 per cent of the total population of their respective countries, but both account for more than 40 per cent of their countries' GDP. Cities also generate a disproportionate amount of revenue for governments; the residents of India's commercial capital Mumbai, for instance, pay almost 40 per cent of the nation's taxes.[8]

Goods and services are generally produced more efficiently in densely populated areas that provide access to supportive services, transport and communication links, a pool of labour with appropriate skills, and a critical mass of consumers – all attractive qualities associated with cities. In the new, increasingly knowledge-based global economy, cities are particularly efficient producers. Improved economic and social infrastructure, together with economies of scale and agglomeration benefits associated with large urban centres, allow businesses and enterprises in cities to flourish. The concentration of economic activity in cities makes them prime generators of non-agricultural employment in both the formal and informal sectors. While the formal sector accounts for a much larger share of urban employment in industrialized countries, the informal sector employs a significant proportion of the non-agricultural labour force in developing regions (up to 80 per cent in sub-Saharan Africa and more than 60 per cent in Asia and Latin America).

Cities are also engines of rural development. They provide many opportunities for investment, which not only support urban development but also contribute to rural development in an environment of strong urban-rural linkages. Improved infrastructure between rural areas and cities increases rural productivity and enhances rural residents' access to education, healthcare, markets, credit, information and other services. On the other hand, enhanced urban-rural linkages benefit cities through increased rural demand for urban goods and services and added value derived from agricultural produce. Increased productivity and competitiveness also fuels the urbanization process: all over the world there are examples of sleepy fishing villages becoming thriving ports, barren outposts becoming major trading centres and railway depots or harbours becoming capital cities. Urban transformations often translate into positive performance on human development indicators and reduced poverty in both rural and urban areas. Put together, all of these factors provide an apt environment for the attainment of the Millennium Development Goals and targets.

However, the relative absence of infrastructure, such as roads, water supply, communication facilities, and adequate housing in small- and medium-sized cities – which are currently absorbing most of the world's urban population growth – makes these cities less competitive at the national, regional and global levels. In many countries, a disproportionate amount of public investment, especially investment in infrastructure, goes to the larger cities, particularly national capitals. Attempts to "decentralize" economic activities to secondary cities are unlikely to be successful unless the decentralization is supported by pro-poor investment in infrastructure and public services, and by the financial and institutional strengthening of local authorities.[9]

Countries that are highly urbanized have higher incomes and more stable economies than those with less urbanized populations.

Contrary to popular perception, infrastructure investments in urban areas are not only cost-effective but also environmentally sound. The concentration of population and enterprises in urban areas greatly reduces the unit cost of piped water, sewers, drains, roads, electricity, garbage collection, transport, health care, and schools. However, the cost-effectiveness of infrastructure investment is greatly reduced when these investments are made too late. For instance, when informal settlements or slums are allowed to proliferate, it becomes more difficult and more expensive to install infrastructure and services because no prior provision was made for the settlement's development.[10] Moreover, population densities and the spatial configuration of slums often do not allow for the development of roads, sewerage systems and other facilities that may be easier to install in less dense and better-planned areas.

The Millennium Development Goals provide an apt framework for linking the wealth of cities with increased opportunity and improved quality of life for their poorest residents. In many countries, however, prosperity has not benefited urban residents equally. Mounting evidence suggests that economic growth in itself cannot reduce poverty or increase opportunities if it is not accompanied by equitable polices that allow low-income or disadvantaged groups to benefit from that growth. Recent World Bank reports show that the best policies for poverty reduction involve more redistribution of influence, advantage and subsidies away from wealthier, more powerful groups to those that are disadvantaged.[11] Countries that have attempted to address inequality by investing in the health, housing and education of their most vulnerable populations tend to perform better on all human development indicators, including GDP.[12] Countries such as Brazil, Cuba, Egypt, South Africa, Sri Lanka, Thailand and Tunisia, for instance, have performed relatively well on many human development indicators and have managed to contain or reduce slum growth because of a political commitment – backed by resources – to invest in the urban poor. Inclusive and visionary urban planning and governance that includes slum upgrading and prevention, combined with pro-poor urban development policies that expand and improve opportunities for employment are, therefore, key ingredients for sustainable urban development; these are also key ingredients for the achievement of the Millennium Development Goals in cities.

THE BAD NEWS
The locus of poverty is shifting to cities

Despite the enormous potential of cities to reduce poverty and to achieve the Millennium Development Goals, recent evidence shows that the wealth generated by cities does not automatically lead to poverty reduction; on the contrary, intra-city inequalities are on the rise, particularly in the cities of Africa and Latin America.

In fact, urbanization in many developing countries, particularly in sub-Saharan Africa, has not been accompanied by economic growth, industrialization or even by development *per se.*[13] On the contrary, the population of some African cities has grown *despite* poor economic growth; the region as a whole has the highest urban growth rate in the world, at 4.58 per cent per year. This phenomenon, combined with inequitable distribution of resources and anti-poor policies, has led to rising urban poverty, which impedes the sustainability of cities and impacts their economic viability. In many parts of the region, high rates of urban population growth, high prevalence of unskilled labour and the HIV/AIDS pandemic are further undermining poverty reduction efforts in cities. Even in Asia's economically successful and rapidly industrializing countries, such as China and India, urban poverty remains a persistent problem as national GDP rates have risen much more quickly than national poverty rates have fallen.[14] The economic growth models used by governments and local authorities have widened not only disparities between rural and urban populations, but also inequalities between high- and low-income populations within cities.

Poverty is already becoming a severe, pervasive and largely unacknowledged feature of urban life. Poverty is shifting to urban areas and growing in magnitude. World Bank estimates indicate that while rural areas are currently home to a majority of the world's poor, by 2035, cities will become the predominant sites of poverty.[15] But in Africa, the proportion of people living in poverty in urban areas (43 per cent) is catching up

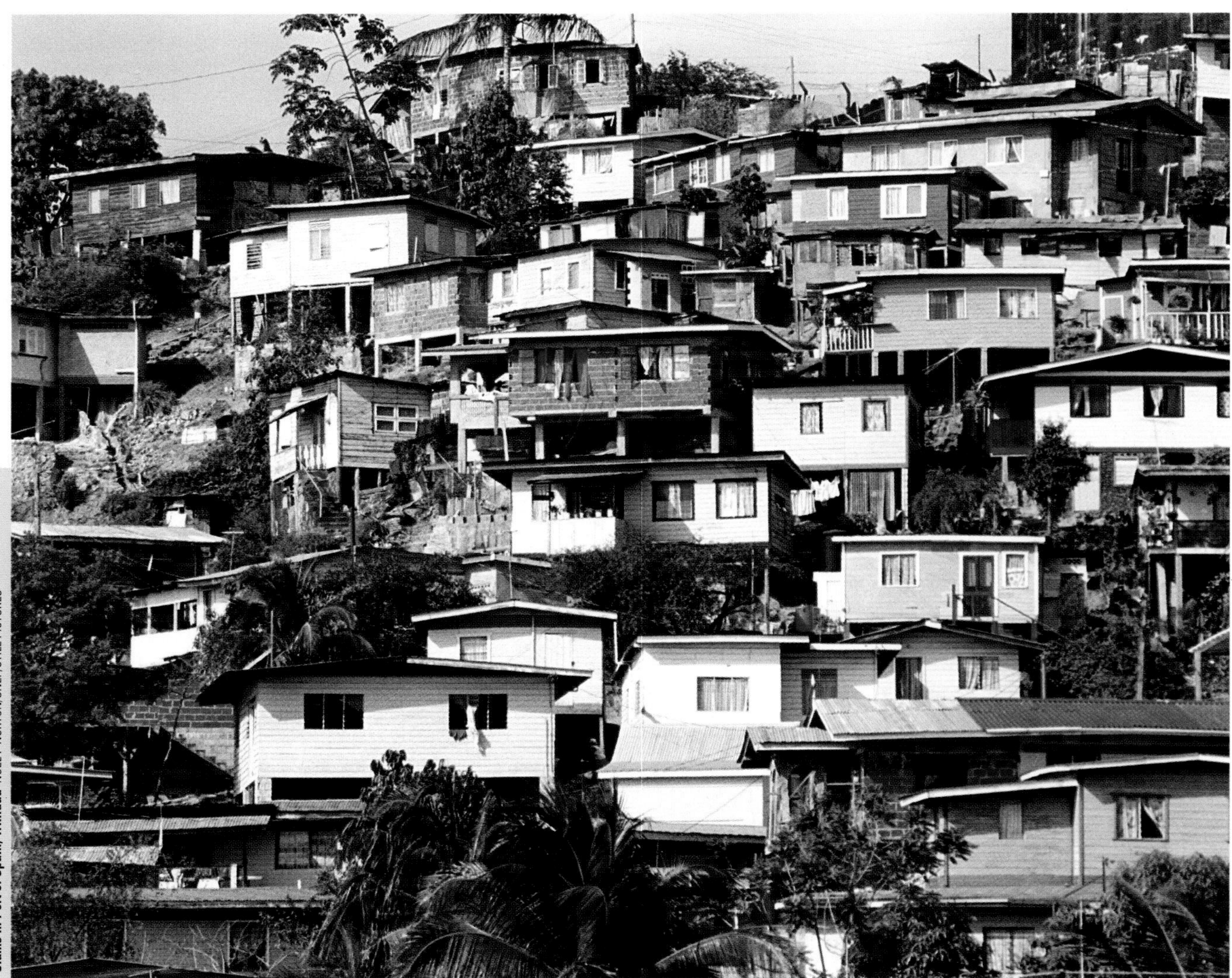

Slums in Port of Spain, Trinidad NOEL P NORTON/UNEP/STILL PICTURES

much faster with the proportion of people living in poverty in rural areas (59 per cent).[16] Sub-Saharan African countries have some of the world's highest levels of urban poverty, extending to more than 50 per cent of the urban population in the poorest countries, including Chad, Niger and Sierra Leone. In other countries – notably Nigeria – urban and rural poverty percentages are almost equal.[17] In Latin America, the most urbanized region in the developing world, there are more poor people living in cities than in rural areas. In 1999, only 77 million of the region's 211 million poor lived in rural areas, while the remaining 134 million lived in urban areas. Proportionally, however, far more of those living in rural areas than in urban areas were poor: 64 per cent of the rural population lived in poverty, as opposed to 34 per cent of the urban population; levels of deprivation are also more extreme in rural areas than in urban areas.[18] The picture is quite different in the Caribbean countries, where urban poverty levels already exceed rural poverty levels.[19] Relatively low levels of urban poverty exist in countries of Northern Africa and Western Asia, where urban poverty levels are near or below 20 per cent; the highest prevalence of urban poverty in Asia is in India, at 30 per cent.

UN-HABITAT analyses have further shown that people living in slums – where a large proportion, but not all, of the urban poor live – have worse health outcomes and are more likely to be affected by child mortality and acute respiratory illnesses than their non-slum counterparts. They are also more likely to live in or near hazardous locations with few basic services, making them more vulnerable to natural disasters such as floods, and saddling them with heavy health and social burdens, which ultimately affect their productivity.

Despite the existence of increasingly large pockets of deprivation within cities, many governments continue to assume that poverty is mainly a rural phenomenon and that those who live in or move to cities escape the worst consequences of this scourge, including hunger, illiteracy and disease. A prevalent view among governments and the international development community is that urban poverty is a transient phenomenon of rural-to-urban migration and will disappear as cities develop, thus absorbing the poor into the mainstream of urban society. This view is reflected in most national poverty reduction strategies, which remain rural-focused, and in international donor assistance to cities, which continues to be modest in scale and impact,[20] with the result that both national and international interventions during the last two decades have had the net effect of increasing poverty, exclusion and inequality in cities.[21]

The concept of cities as islands of privilege and opportunity is supported by national and international statistics on health, education and income, which generally reflect better outcomes in urban areas. What these statistics fail to reveal are the severe inequalities within cities, and the various dimensions of urban poverty that are not captured by income-based indicators, including political exclusion and poor quality, hazardous and insecure housing.

FIGURE 1.4.2 THE ANNUAL GROWTH RATE OF CITIES AND SLUMS

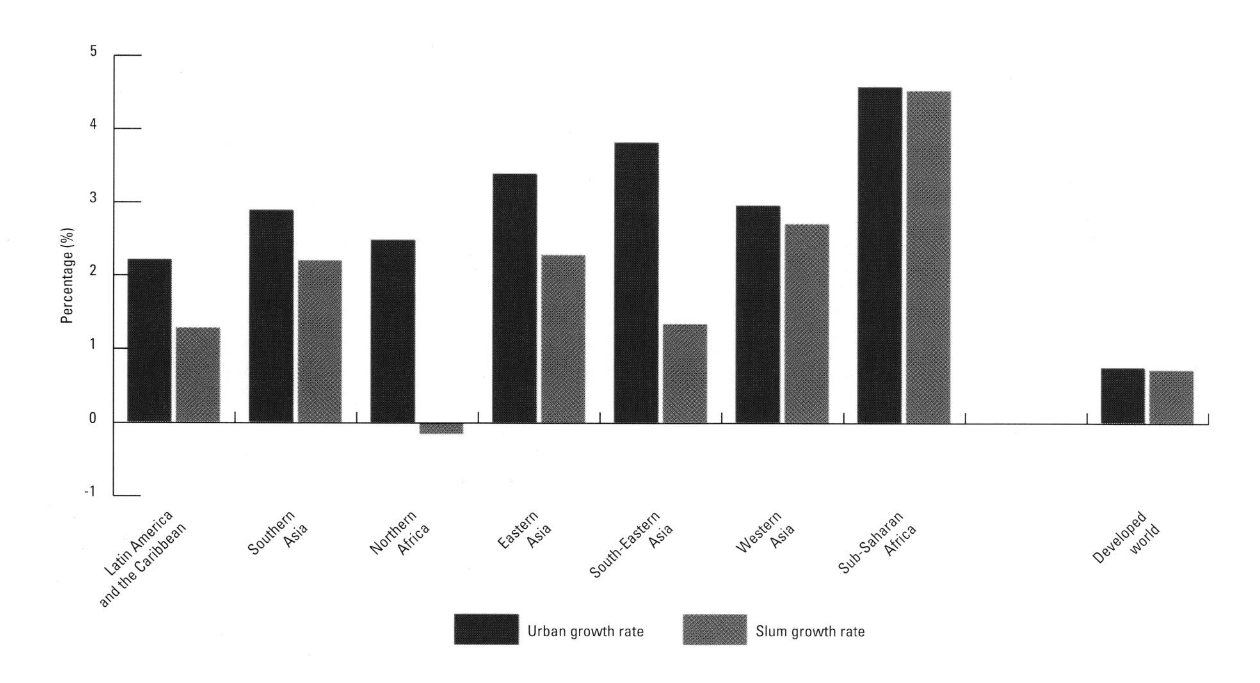

Source: UN-HABITAT 2005, Global Urban Observatory, Urban Indicators Programme, Phase III.

Poverty is becoming a severe, pervasive and largely unacknowledged feature of urban life.

The high cost of non-food items, such as transport, health, education and water in cities, coupled with poor living conditions, including inadequate housing and poor access to basic services, impact the ability of the urban poor to rise out of poverty. If the definition of poverty is broadened to include the social assets available to the poor and their vulnerability to stress and shocks, including evictions, crime, disease, environmental disasters and unpredictable employment markets, then it is likely that the proportion of poor people living in cities is much higher than current estimates. Poverty reduction interventions, including efforts to achieve the Millennium Development Goals, therefore, have to be focused on cities of the developing world, which are absorbing a significant number of the world's poor and where slums are growing at unprecedented rates.[22]

By 2007, half the world's population will be urban, and in the next two decades, more than 95 per cent of the population growth in the world's poorest regions will occur in cities. Urban growth rates are particularly high in the least developed countries, at almost 5 per cent per year. This shift in population implies that the major development challenges – and the struggle to achieve the Millennium Development Goals and targets – will have to be focused on cities of the developing world, where an increasingly large proportion of the world's poor will live.

Achieving the slum target

Millennium Development Goal 7, target 11 on improving the lives of at least 100 million slum dwellers by 2020 forces the international community to address one specific aspect of urban poverty that until now was neither captured in national statistics nor reflected in urban data. The slum target is a recognition by the international community that slums cannot be considered an unfortunate by-product of urbanization; rather, by ignoring the plight of slum dwellers, governments are inadvertently adopting urbanization models that are neither sustainable nor acceptable.

Slums currently house one out of every three urban dwellers. Although slums[23] represent a physical manifestation of urban poverty and do not capture the myriad facets of urban poverty that have little to do with housing or basic services, their rapid growth in the last 50 years, particularly in Asia and Africa, indicates that they are home to a large number of the world's urban poor who are not benefiting from the wealth and opportunities generated by the cities in which they live.

Slums are becoming the norm rather than the exception in the poorest cities of the world. In sub-Saharan Africa, where more than 70 per cent of urban residents live in slums, many slum dwellers are unable to escape the material deprivation, illiteracy and disease that are normally associated with impoverished rural areas. Asia's cities, which host almost 60 per cent of the world's slum population, are becoming sites of severe environmental degradation and pollution, which are impacting recent economic gains. And despite progressive legislation and improved governance structures in recent years, Latin America's cities remain the most unequal in the world.

The sheer scale of the problem warrants attention. Nearly one billion people around the world are currently living in urban slums that lack basic services and adequate housing, and their numbers will increase if no remedial action is taken. While the target of improving the lives of at least 100 million slum dwellers may be achieved by 2020, particularly in countries that have put in place slum upgrading and prevention policies, the scale of the problem may worsen: UN-HABITAT estimates indicate that if governments continue with business as usual, then an additional 400 million people will be drawn into the misery of slum life as the global slum population reaches 1.4 billion in 2020.

However, slums do not simply ensnare impoverished urban dwellers; they also act as intermediate urban spaces, situated between destitution and opportunity – key places of transition that can help or hurt individuals, depending on the actions of governments, the private sector, civil society and slum dwellers themselves. They can also provide upward mobility to urban dwellers and become sites of immense economic opportunity, culture and innovation – the hallmarks of successful cities.

Despite the substandard living conditions prevalent within them, slums can also represent a kind of opportunity for the urban poor. In most regions, slums are "stepping stones" out of rural poverty, as slum dwellers have more urban-based employment opportunities than villagers and have better access to publicly financed services and infrastructure. Indeed, most slum dwellers choose to remain in cities because of the perceived and actual social and economic benefits they provide. Slums, some have argued, are an integral and natural part of economic growth and industrialization and should be considered marks of "success" in urban areas. They reflect a dynamic and diverse labour market and offer affordable housing to those who cannot, or will not, pay more for accommodation in the city.[24]

Cities and the slums within them also offer governments an "opportunity" or entry point to tackle some of the world's most pressing challenges, including extreme poverty, under-five mortality, HIV/AIDS, environmental degradation, and gender inequality. The sheer concentration of people living in cities and slums means that any investment is likely to reap greater benefits per capita. On the other hand, the economies of scale offered by high density slums also make them ideal targets for interventions aimed at achieving the Millennium

Fruit vendor in Shanghai, China ©AMARK STEPHAN. IMAGE FROM BIGSTOCKPHOTO.COM

Development Goals and targets. Improving the living conditions of slum dwellers, by improving housing, tenure security and access to water and sanitation, will automatically have a positive impact on the attainment of most of the Goals and targets. In regions where slum dwellers do not suffer from multiple shelter deprivations, interventions and investments in just one sector can dramatically reduce the numbers of people living under slum conditions.

UN-HABITAT is convinced that the slum target cannot be achieved in isolation. On the contrary, it is becoming increasingly clear that the failure of the slum target will jeopardize the achievement of all the other Goals and targets; conversely, achieving the other Goals and targets in slums will make the achievement of the slum target more likely.[25]

The struggle to achieve the Millennium Development Goals has to be waged in slums, not at the expense of rural areas, but alongside them. Ultimately, as the world becomes more urban, the battle to achieve the Goals will be won – or lost – in the *zopadpattis* of Mumbai, the *bidonvilles* of Abidjan, the *chawls* of Ahmedabad, the *villas miseria* of Buenos Aires, the *favelas* of Rio de Janeiro, the *barrios ilegales* of Quito, the *shammasas* of Khartoum, the *iskwaters* of Manila, the *chereka betes* of Addis Ababa, the *aashwai'is* of Cairo, the *corticos* of São Paulo, the *colonias populares* of Mexico City, the *vijiji* of Nairobi, the *gecekondus* of Ankara, the *hoods* of Los Angeles, the *museques* of Luanda and the *katchi abadis* of Karachi.

Endnotes

1 The campaign to eradicate poverty began in earnest in 2000 when world leaders pledged to halve it by 2015 at the UN Millennium Summit, held in New York in September 2000.

2 The Millennium Development Goals and targets are derived from the Millennium Declaration adopted by the world's governments at the Millennium Summit in 2000.

3 World Bank 2000.

4 GDP refers to the total value of goods and services produced within a country by both nationals of the country and foreigners.

5 GNP refers to the total value of goods and services produced by nationals of the country.

6 World Bank 2000.

7 U.S. Conference of Mayors 2004.

8 Mehta 2004.

9 UN-HABITAT/DFID 2002.

10 Hardoy, et al. 2001.

11 This point has been extensively argued in the World Bank's *World Development Report 2006*, which shows that the best policies for poverty reduction involve more redistribution of influence, advantage or subsidies away from dominant groups.

12 This finding, for instance, is reflected in UNDP's *Human Development Report 2005*.

13 Davis 2004.

14 This fact was highlighted at the 2005 Asia-Pacific Economic Cooperation Summit in Busan.

15 Ravallion 2001.

16 United Nations 2005a.

17 World Bank 2002 estimates.

18 United Nations 2005a.

19 Inter-American Development Bank 2004.

20 Cohen 2004. Author estimates that total urban assistance to developing countries from 1970 to 2000 was just $2 billion a year.

21 UN-HABITAT 2003a.

22 For a detailed analysis of this phenomenon, see Chapter 1.2.

23 See Chapter 1.2 for the UN-HABITAT definition of "slums".

24 Babar Mumtaz makes a case for why "Cities Need Slums" in *Habitat Debate*, Vol 7. No. 3, September 2001. He argues that insisting on a "city without slums", especially when no alternative housing has been developed, can mean even more hardship for the very group that is essential to urban development: the rural migrant.

25 Moreno 2005.

Cities, Slums and the Millennium Development Goals

Goal 1: Eradicate extreme poverty and hunger

Halve, between 1990 and 2015, the proportion of people whose income is less than $1 a day
Halve, between 1990 and 2015, the proportion of people who suffer from hunger

- Cities act as catalysts for poverty reduction: they generate the wealth and the economic opportunities needed to make the achievement of the Millennium Development Goals possible.

- Urbanization levels are closely related to levels of income and better performance on social indicators, including health and literacy. The achievement of the Millennium Development Goals is, therefore, more likely in cities. Conversely, urban economic growth provides the basis on which cities can contribute to the achievement of the Goals, particularly in the area of poverty reduction.

- Cities and slums are often the "first step" out of rural poverty. The rural poor move to cities, where there are more employment opportunities and better access to services such as health care and education.

- The locus of poverty is moving to cities. In the next two decades, more than 95 per cent of the population growth in the world's poorest regions will occur in urban areas, with the result that cities will become the predominant sites of poverty in coming years.

- Malnutrition, hunger and disease are becoming more prevalent in slums, particularly in developing countries. Because hunger experienced in cities is directly related to income (rather than agricultural productivity), the urban poor are much more vulnerable to income-dependant hunger than their rural counterparts.

Goal 2: Achieve universal primary education

Ensure that, by 2015, children everywhere, boys and girls alike, will be able to complete a full course of primary schooling

- Cities are nodes of education and learning, a key contributing factor in rural-to-urban migration. Urbanization has been associated with economic and social progress, the promotion of literacy and education.

- Educational facilities are generally more advanced and accessible in cities but the cost of other items (transport, housing, food) is higher in cities than in rural settings, which impacts the ability of the poorest households to send children to school. Slum children, particularly girls, are more at risk of dropping out of school than children living in non-slum urban areas.

- In many slums, overcrowding and inadequate or non-existent toilet facilities and other amenities further impact the quality of education that children receive. In some countries, adolescent girls drop out of school because of insufficient toilet facilities in slum schools, or because of family responsibilities, such as taking care of siblings while parents are at work.

Goal 3: Promote gender equality and empower women

Eliminate gender disparity in primary and secondary education, preferably by 2005, and in all levels of education no later than 2015

- Cities offer women social mobility, which has a positive impact on gender equality and has helped reverse some socially prescribed roles. In many countries, urban women have more access to land and property than their rural counterparts as they are not constrained by discriminatory customary laws. In general, urbanization has had a positive impact on women's access to resources and enlarged their decision-making roles.

- With the exception of Africa, the share of woman-headed households is greater in urban areas than in rural areas. This trend has socio-economic implications, which can translate into deepening poverty among urban women in some countries.

- In situations of limited resources and urban impoverishment, women and girls are the first to be sacrificed when families have to make difficult choices about sending their children to school.

- Slum life forces many women and girls to engage in sexually risky behaviour, making them more vulnerable to HIV/AIDS and other sexually transmitted diseases. HIV prevalence among urban women in sub-Saharan Africa is already is much higher than among rural women.

- Poor access to water and sanitation places an enormous labour and health burden on women living in slums, who are not only charged with ensuring that their families have water, but who also suffer disproportionately from the health and environmental hazards associated with poor sanitation. In slums, where there are few or no toilets, many women are forced to defecate in the cover of darkness, which renders them more vulnerable to sexual and physical assault.

Goal 4: Reduce under-five mortality

Reduce by two-thirds, between 1990 and 2015, the under-five mortality rate

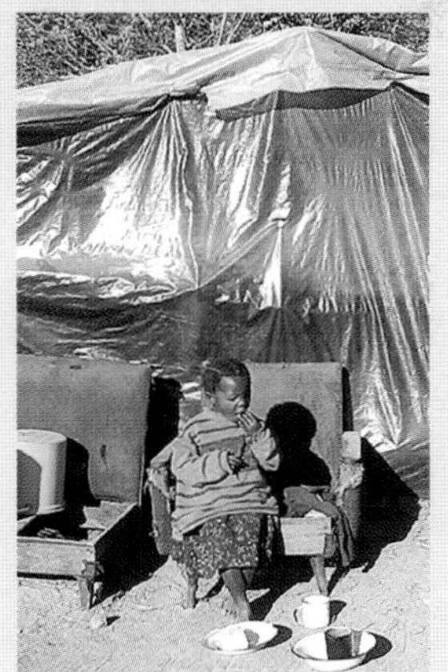

- Better access to health facilities in cities means that children born in urban areas have a better chance of surviving than their rural counterparts, who may not live near facilities that offer services such as immunization and post-natal care.

- Empirical evidence suggests that sectoral interventions in water, sanitation and housing have positive outcomes in the reduction of child mortality rates.

- Under-five mortality rates are higher in slums than in non-slum urban areas. High child mortality rates in slums are not so much related to whether or not children are immunized; rather, they have more to do environmental factors, such as overcrowding, indoor air pollution, poor wastewater treatment and lack of drainage, sewerage and sanitation facilities. The use of solid fuels, combined with overcrowding and poor ventilation, in slum households increases the chances of children contracting acute respiratory illnesses, such as pneumonia. Many slums are also located in or near hazardous or toxic sites, which expose children to additional environmental and health hazards.

- Access to more health care facilities in urban areas does not automatically lead to reduced mortality rates in slums. Parents struggling to pay for food, school fees and transport costs may be unwilling or unable to pay for the health care of their children, which has an impact on child mortality rates.

Goal 5: Improve maternal health

Reduce by three-quarters, between 1990 and 2015, the maternal mortality ratio

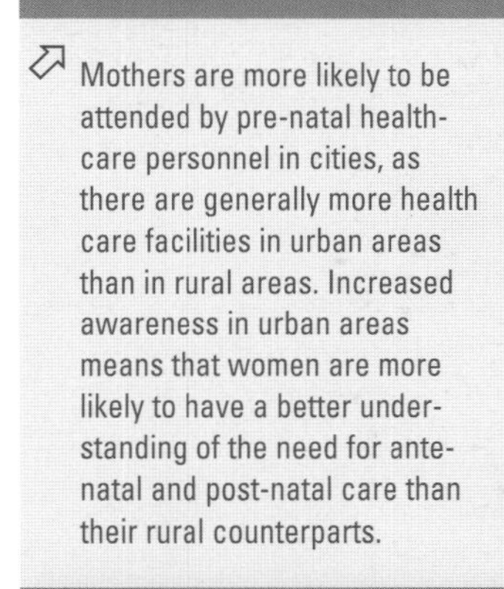

- Mothers are more likely to be attended by pre-natal health-care personnel in cities, as there are generally more health care facilities in urban areas than in rural areas. Increased awareness in urban areas means that women are more likely to have a better understanding of the need for ante-natal and post-natal care than their rural counterparts.

- Many women living in slums cannot afford the relatively more expensive delivery and post-natal health services in urban areas, and are, therefore, less likely to seek these services, especially when a choice has to be made between paying for these services and buying food or meeting other household expenses.

- Poor urban women who supplement their incomes by engaging in sexually risky behaviour expose themselves to a variety of sexually transmitted diseases that have a negative impact on maternal health. Numerous studies have shown high rates of HIV infection among urban women; in sub-Saharan Africa, HIV prevalence is highest among women living in urban areas, and particularly high among women living in slums.

Goal 6: Combat diseases including HIV/AIDS, malaria and other diseases

Have halted by 2015 and begun to reverse the spread of HIV/AIDS
Have halted by 2015 and begun to reverse the incidence of malaria and other major diseases

- Access to information and health care facilities in cities is higher than in rural areas. Increased awareness has led to the prevention of malaria and other major diseases and widened people's choices in the treatment and prevention of the HIV/AIDS epidemic.

- Slum upgrading and prevention policies that incorporate voluntary testing and counselling facilities have helped reduce HIV prevalence in urban areas.

- Increased awareness about prevention in urban areas has not had the desired effect of reducing HIV prevalence in cities. In fact, trends suggest that HIV prevalence is much greater in urban areas than in rural areas, and is also higher among urban women than among rural women.

- The HIV-AIDS pandemic may shave off up to 2 per cent of annual economic growth in the worst affected countries. This has a direct impact on urban economic growth and in some countries severely undermines poverty alleviation efforts.

- The situation of extreme deprivation in cities, particularly in slums, encourages residents to engage in risky sexual behaviour for economic survival. Slum residents often start sexual intercourse younger, have more sexual partners, and are less likely than other city residents to know of or adopt preventive measures against contracting HIV/AIDS.

- HIV/AIDS has contributed to the growing problems of AIDS orphans; many of these orphans become street children caught in the poverty trap of hunger, malnutrition, disease and illiteracy.

- Slums are characterized by overcrowding and poor ventilation, the leading contributors to the rise in tuberculosis cases worldwide. Studies have shown that HIV-related tuberculosis is becoming an increasingly urban phenomenon, particularly in slums.

Goal 7: Ensure Environmental sustainability

Integrate the principles of sustainable development into country policies and programmes and reverse the loss of environmental resources
Halve, by 2015, the proportion of people without access to safe drinking water and basic sanitation
By 2020, to have achieved a significant improvement in the lives of at least 100 million slum dwellers (Goal 7, Target 11)

Air Pollution ©ALISTAIR SCOTT. IMAGE FROM BIGSTOCKPHOTO.COM

Sustainable urban development

- Sustainable urbanization policies have been incorporated in many city and national plans and are contributing to reversing the impact of environmental degradation and pollution. Cities with sound and sustainable land, air and water management policies have also managed to reduce soil erosion, improve air and water quality and protect biodiversity within cities and in their hinterlands. Some cities in the developed world are unilaterally reducing greenhouse gas emissions and other pollutants as part of their respective governments' commitment to adhere to the Kyoto Protocol.

- Cities concentrate production and population, which gives them obvious advantages over rural settlements or dispersed populations. For example, the concentration of populations in urban areas greatly reduces the unit cost of piped water, sewers, drains and roads. The use of environmentally friendly energy sources and transport can reduce these costs even further.

Slums

- Slums provide an important entry point for the achievement of all the Millennium Development Goals in cities; the sheer concentration of people living in slums make them ideal targets for interventions aimed at reducing poverty, reducing child mortality and HIV/AIDS, improving literacy and promoting environmental sustainability in urban areas.

Sustainable urban development

- Urbanization can bring about irreversible changes in production and consumption of water, energy and land. Both developed and developing countries are witnessing rapid urban sprawl with direct consequences for the surrounding hinterland.

- Air pollution is concentrated in cities. The concentration of industrial emissions and increased motorized transport in cities is severely eroding their environmental sustainability and is affecting the health of urban populations. Acute respiratory illnesses associated with poor air quality and poor housing conditions are impacting the human and economic productivity of cities, particularly in Asia; it is estimated that the health costs from pollution reduce gross domestic product (GDP) by some 2 per cent in developing countries.

Slums

- The rate at which slums are growing exceeds the rate at which they are being improved. This severely impacts the achievement of Goal 7, target 11: by 2020, to have achieved a significant improvement in the lives of at least 100 million slum dwellers. Some 200 million more slum dwellers have been added to the world's urban population since 2000; if current trends continue, by 2020, there will be 400 million more people drawn into the misery of slum life and the global slum population will reach 1.4 billion.

- Slum dwellers are more likely to live in hazardous or toxic locations, which are more prone to natural disasters, such as floods, and which pose severe health risks, not just to slum dwellers but to city dwellers in general.

- Although access to water and sanitation is generally better in urban areas than in rural areas globally, the consequences of poor access in cities are more severe. Many slum dwellers have no choice but to use water sources, such as rivers, to bathe and wash clothes. Poor sanitation in some cities has also led to large sections of the population defecating in the open. This contributes to contamination of water and land resources within cities, and is a cause of many of the water-borne diseases prevalent in slums.

- Indoor air pollution caused by the use of solid fuels is prevalent in slums and is a leading cause of respiratory illnesses in urban areas, particularly among women and children.

Goal 8: Develop a global partnership for development

Address the special needs of the least developed countries and small island developing States
Develop further an open, rule-based, predicable, non-discriminatory trading and financial system
Deal comprehensively with developing countries' debt
In cooperation with developing countries, develop and implement strategies for decent and productive work for youth
In cooperation with pharmaceutical companies, provide access to affordable and essential drug in developing countries
In cooperation with the private sector, make available the benefits of new technologies, especially information and communications

Tokyo NATSUO ITO

Partnerships for development

In the last decade, the international community has become much more aware of the problems, challenges and opportunities of urbanization. City-to-city cooperation and decentralization policies are becoming more common as governments at the local and national levels cooperate to make cities more liveable. Some national governments are for the first time addressing urban poverty in their national poverty reduction strategies and programmes. Many local governments are also recognizing the need to plan, manage and govern their cities better, which has improved the lives of many urban residents worldwide. As part of new structures of governance, where cities are being given more authority to manage their affairs, cities are playing a more prominent role in developing partnerships with central governments, regional organizations and development partners to promote sustainable urbanization within cities, countries and regions.

Some multilateral agencies and regional development banks are recognizing the need to intervene and invest in urban areas as part of their development assistance programmes. In Latin America, for instance, the portfolio of loans for urban development have grown in both volume and complexity.

Partnerships for development

Increased awareness of the potential and challenges of cities has not led to a commensurate increase in international development assistance or to greater allocation of national finances to urban poverty reduction. Most development aid is focused on eradicating extreme poverty and improving the living conditions of rural populations. Many governments have not identified "urban poverty" as an area of intervention in their development plans.

Although investments have been made in various sectors, such as health and infrastructure development, they are often not targeted specifically at slums.

Part Two

The State of the World's Slums

Slums are the emerging human settlements of the 21st century. In order to monitor the state of the world's slums, this Part presents a detailed analysis of the five indicators that reflect conditions that characterize slums. These indicators, known as "shelter deprivations", are: lack of durable housing; lack of sufficient living area; lack of access to improved water; lack of access to improved sanitation; and lack of secure tenure. Information on the five indicators is analysed at global, regional, national and city levels. This Part also presents an analysis of the degrees of shelter deprivation in some selected countries and regions.

Manila TINA ELAINE RUSTE/UNEP/STILL PICTURES

RCL
PIYA BHUM

2.1 Neither Brick nor Mortar: Non-Durable Housing in Cities

Durability is one of the least understood attributes of a nation's housing stock, and life expectancy of a house is a neglected indicator.[1] In many parts of the world, housing durability assessments are not systematic, and when conducted, they do not generate data that is comparable with data in other areas. Estimates suggest that worldwide, 18 per cent of all urban housing units (some 125 million units) are non-permanent structures, and 25 per cent (175 million units) do not conform to urban building codes or regulations.[2] These figures, however, could be highly underestimated as global data on durability is based primarily on permanence of individual structures, not on location or compliance with building codes. Mainstream reporting mechanisms are not designed to capture data on unsafe or hazardous location of housing, but compelling reasons exist for collecting such data, as it is estimated that at least three or four in every 10 non-permanent houses in cities in developing countries are located in dangerous areas that are prone to floods, landslides and other natural disasters.[3] Other unsafe locations include living on garbage dumpsites or in highly polluted areas. Non-durable dwellings located on hazardous locations are particularly at risk when natural disasters strike, as they are least able to withstand the destruction caused by flood, earthquakes or hurricanes. However, people living in extreme poverty in cities often have little choice but to take shelter in ramshackle structures haphazardly constructed in the most polluted, dangerous areas.

Informal settlements in the developing world typically do not meet local building codes and other regulations for urban development. Disregard of building codes has clear implications for the durability and safety of housing. The destruction of residential buildings wrought by recent earthquakes in cities – such as those in Turkey in 1999, India in 2001 and Iran in 2003 – confirmed the importance of enforcing building codes and regulations, as the structures that collapsed were found to have violated building norms and technical requirements deemed necessary in earthquake-prone regions. Poor enforcement of building codes was also the primary reason cited for the collapse of a multi-storey building under construction in Nairobi, Kenya, in January 2006, which claimed 14 lives and injured several others.[4] The urban poor suffer the greatest loss when natural disasters strike, in part because their housing is built to lower standards than housing for higher-income residents.

In the developed regions, it is estimated that almost all housing is in compliance with codes and regulations. The Canadian government, for instance, reports that just a fraction of the country's housing stock – 5 to 10 per cent – is non-compliant, comprising older construction that has not been upgraded or add-on structures that have not been brought up to the standard of the rest of the dwelling.[5] In some European cities, a small amount of unauthorized construction of dwelling units takes place in absence of building permits and in violation of zoning rules. Such houses are illegal in the technical sense, but in most cases, they respect current building standards. There are, however, a small number of houses that do not comply with building codes, some of which are regarded as unhealthy dwellings. In Naples, such buildings are called "bassos"; in Barcelona, "illegal pensions" are a problem; and in Paris, many low-income families live in illegal boarding houses and other forms of sub-divided apartments.[6]

It is estimated that at least three or four in every 10 non-permanent houses in cities in developing countries are located in dangerous areas that are prone to floods, landslides and other natural disasters.

In Europe, North America and other developed regions, housing durability has undergone a fair amount of analysis, particularly with regard to building materials, maintenance and construction methods and systems. Issues such as affordability, accessibility, financial costs, and quality of the dwelling are relatively well researched. Studies also include measures to improve housing durability, not only for economic reasons, but also for disaster mitigation and vulnerability reduction.[7]

The vast majority of housing in developed countries is in decent condition. In Canada in 2001, for example, just one in 12 homes (8.2 per cent) were in need of major repairs, and nearly two-thirds of homes needed only regular maintenance.[8] In the United States, approximately 2 per cent of occupied units had severe physical problems with plumbing, heating, electricity, public areas, or maintenance in 1999.[9] This ratio is similar to other developed countries that have a housing stock comprised almost entirely of permanent buildings, even in the lowest-income parts of the city. In some European cities, a few unique exceptions exist, such as the Roma encampments built with temporary or "non-permanent" materials.

An interesting pattern emerges from the data about the housing stock in developed countries, however, which reveals that the units that are in the worst condition and require the most repairs often house members of ethnic minorities or immigrants.[10] Many such houses are overcrowded and dilapidated

Buenos Aires MARK EDWARDS/STILL PICTURES

Affordability, accessibility and durable houses

Durability – the longevity of residential structures – is directly associated with housing accessibility and affordability.

In the *formal housing market* in developing countries, mortgage institutions, such as commercial or trading banks, restrict loans to builders who use materials that are "acceptable" in the market because they are considered "durable" (e.g. bricks, stones, concrete, and the like). The use of durable materials decreases the bank's risk over the period of the mortgage loan. This housing policy excludes local building materials such as wood, bamboo or other innovations that arise out of the use of indigenous materials; such materials are inexpensive, but are subject to rapid deterioration and subsequent depreciation. Financial institutions, therefore, have adhered to standards and building codes in a way that limits access to affordability schemes for the working poor.

In the *informal housing sector,* slum dwellers recognize the trade-offs between durability and affordability. They understand that a durable house has a clear positive economic benefit, is safer and can be healthier for its occupants than one made of temporary materials. However, putting up a house that lasts requires a high initial investment, even though it could cost less in the long run. "Building to last" in terms of using stronger and more durable materials that would reduce maintenance and replacement costs remains out of the financial reach of most slum dwellers. Poor families requiring housing right away are more likely to opt for temporary solutions than wait until they can afford something more permanent. These solutions increase their vulnerability, as the units may be structurally unstable or located in hazardous areas.

Lack of affordable land and the absence of affordable self-build housing schemes leave impoverished urban residents susceptible to risks and harms that could be alleviated by improved urban planning and the development of more affordable building materials that are durable and easy to use by self-builders.

London MJS

dwellings located in low-income neighbourhoods, built in the 1970s or earlier as part of government housing development projects, in older parts of central cities and first-ring suburbs. Many are substandard units that are contaminated and lacking light, air and open space. Others are poorly built, poorly maintained and isolated, often situated in inaccessible or unhealthy locations, such as along motorways and industrial wastelands. Even if the number of these housing units is statistically insignificant, they typify some of the exclusion patterns and forms of physical and social decay prevalent among impoverished minorities in the developed world. Government agencies have also observed that rental units often have twice as many durability problems as owner-occupied units.[11]

A close relationship clearly exists among durability, affordability and accessibility in housing markets and building patterns around the world. These linkages are very often neglected and thus are not properly analyzed for policy purposes.

UN-HABITAT data and analysis of housing durability in developing countries

To estimate the prevalence of slums around the world using data collected between 1990 and 2001, UN-HABITAT included a measurement of housing durability as one of its five indicators of slum households. In principle, the estimation procedure considered the nature of the roof, wall and floor materials of dwellings. Data on all three was easily obtainable for developed countries, but in the developing world, estimations were made considering only the nature of the floor material. Roof and wall materials were considered inappropriate variables for the durability indicator, as information on them is collected in few countries.

The research revealed that in 2003, 94 per cent of the world's housing units in urban areas were considered permanent based on the "floor criterion", meaning that most of the floor in each dwelling was constructed or covered with permanent materials and was not simply earthen. Using this criterion alone masks

MAP 7 PROPORTION OF HOUSEHOLDS WITHOUT FINISHED MAIN FLOOR MATERIAL IN SELECTED CITIES, 2003

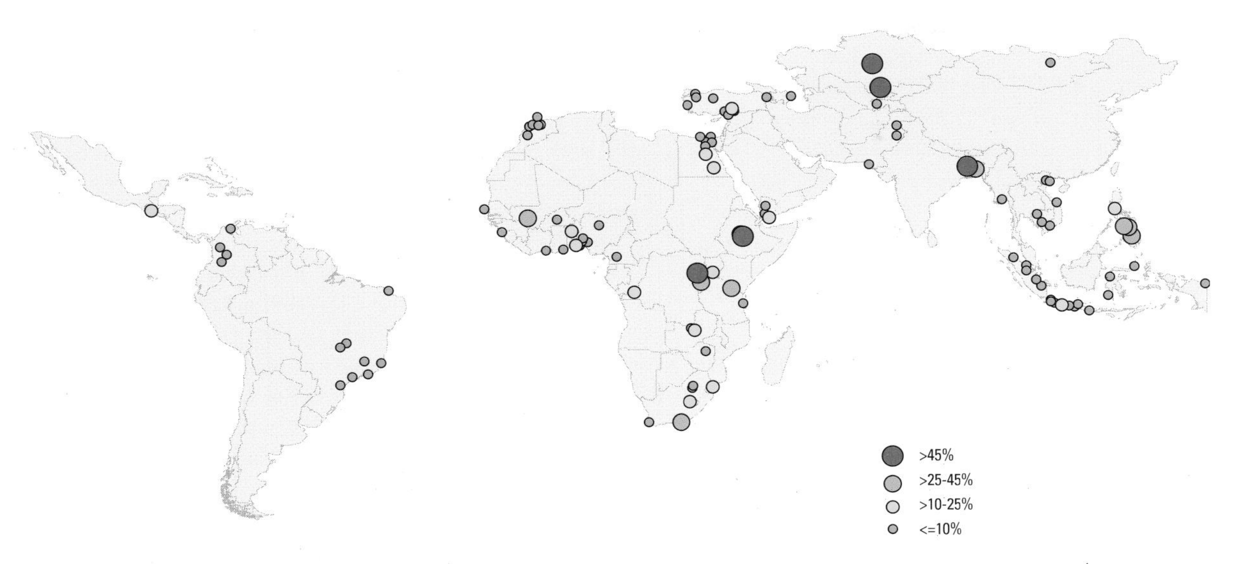

Source: UN-HABITAT 2005, Global Urban Observatory, Urban Indicators Programme, Phase III.

potential problems with other key structural materials of dwellings. For example, if housing durability estimations include quality of roof and wall materials, the figure for many countries would drop. (See table.) For instance, in Bolivia, when only floor material is considered, 83.8 per cent of the urban population is counted as living in durable housing, but when wall and roof materials are considered, this figure drops to 27.7 per cent. In Nicaragua, when floor and roof materials are combined to determine durability, only 9 per cent of the urban population qualifies as living in a durable home.

In order to produce more accurate data on housing durability, a statistical analysis was conducted in the countries where information is available for the three main physical structure variables – floor, walls and roof – at the urban level. The results provide a more realistic image of housing durability, and two examples aptly illustrate this point. In Indonesia, the percentage of durable housing in urban areas in 2002 was 69.8 per cent when the three components of the dwelling were considered, whereas when only the floor criterion was used, 83.7 per cent of houses were deemed durable.[12] In Benin, 80 per cent of houses qualified as permanent and durable in 2001, considering only the floor criterion; however, when materials for the three elements of the house were taken into account, housing durability dropped to 60 per cent.[15] UN-HABITAT results are consistent with data produced by governmental sources that assess quality of housing combining the three variables. India, for instance, reported that 73 per cent of urban households lived in *pucca* or permanent houses in 1991,[13] and Sri Lanka reported that permanent structures with brick walls, tiled roofs, and cement floors constituted 70 per cent of houses in urban areas in the early 1990s.[14]

This data only presents the national and urban aggregates, which are useful for monitoring urban poverty and sustainability at city, national and regional levels. Further research is therefore needed to monitor spatial inequalities within cities where segregated urban social structures persist.[16] UN-HABITAT's urban inequities study conducted in Addis Ababa, Ethiopia, for instance, highlights the disparity between slum dwellers and other urban groups with regard to adequate housing. While more than half (51.6 per cent) of slum households used natural, non-permanent, materials for the floors of their dwellings, 58.4 per cent of non-slum households used polished cement, a permanent material. Also, a significantly higher percentage of

TABLE 2.1.1 HOUSING DURABILITY, BASED ON FLOOR, ROOF AND WALL MATERIALS, IN THE URBAN AREAS OF 13 SELECTED COUNTRIES, 2001

	Proportion of households with durable housing based on three building materials			
Country	Floor	Wall	Roof	All Three
Benin	80.2	61.6	88.3	60.1
Central African Republic	26.2	9.1	52.5	7.7
Chad	15.3	5.3	52.9	4.7
Togo	94.4	72.6	88.7	66.2
Uganda	68.6	58.5	91.8	53.7
Bolivia	83.8	52.0	41.5	27.7
Brazil	89.2	95.2	98.9	86.1
Dominican Republic	95.7	92.5	98.6	88.6
Guatemala	77.4	66.3	96.7	60.8
Nicaragua	65.3	61.2	21.9	9.0
Peru	66.5	58.4	84.8	47.7
Bangladesh	53.2	64.3	27.3	26.4
Indonesia	83.7	72.6	93.9	69.8
	69.2	**59.2**	**72.1**	**46.8**

Source: UN-HABITAT, Global Urban Observatory, 2005.

slum dwellers (74.6 per cent) used traditional (non-permanent) materials to construct their walls, compared to non-slum dwellers (58.7 per cent). While UN-HABITAT recognizes that slums are not always geographically contiguous, in most cities, slum households are clustered to some extent, indicating specific areas of cities in which housing conditions are distinctly worse than in other areas. It should therefore be emphasized that an understanding of the spatial patterns of inequality is fundamental to formulating area-based policies to address lack of housing durability and other deprivations.[17]

A global and regional overview of housing durability

UN-HABITAT estimates indicate that in 2003, 133 million people living in cities of developing regions lived in housing that lacked finished floor materials. In the developing world, Asia had the largest proportion (73 per cent) of urban dwellers living in non-permanent housing. Over 50 per cent of this population lived in Southern Asia, followed by South-Eastern Asia (11 per cent). Stark contrasts exist between Northern Africa and sub-Saharan Africa. Northern Africa hosts only 1 per cent of the developing world's urban population with non-durable housing, while sub-Saharan Africa hosts 20 per cent.

Housing durability in Africa

Over 10 per cent of the urban population in sub-Saharan Africa lives in non-durable housing. Important progress in increasing housing durability has been registered in the intermediate cities of Assyut, Aswan and Beni Suef in Egypt. These cities have also made progress on other shelter indicators, leading to an overall decline in the number of slum households. The city of Porto Novo, Benin – an important political capital and host of an annual international festival – has also shown significant improvements on housing durability and other indicators that are attributed to increased government investments in housing. Kigali, the capital city of Rwanda, has experienced a significant improvement in housing durability and on all of the slum indicators. Despite the conflict that ravaged the country in 1994, Rwanda has made steady progress in economic recovery and has made impressive sectoral improvements in education and health. The city of Dar es Salaam, Tanzania, on the other hand, showed an improvement in housing durability but did not have similar progress on other shelter indicators.

Urban-level aggregates conceal intra-city disparities. In cities such as Luanda, Angola, and Arusha, Tanzania, the population living in durable houses is approximately 50 per cent.[18] It is likely that housing deficits in these countries will increase, as their slum growth rate – on average, 5.4 per cent – is higher than the rest of the sub-region, where the growth of slums is already staggeringly high, at 4.5 per cent. In other cities such as Port Elizabeth in South Africa, lack of durable housing is surprisingly high if compared with other urban agglomerations in the country. Only 58 per cent of Port Elizabeth residents have housing considered durable, the lowest in the country.

In Northern Africa, housing durability is less of a concern, since more than 99 per cent of the total urban population lived in durable houses in 2003.

TABLE 2.1.2 HOUSING DURABILITY AMONG URBAN POPULATIONS IN DEVELOPING REGIONS, 2003

	Access to finished main floor materials, 2003 (%)	Urban population in 2003 (thousands)	Urban population lacking finished main floor materials (thousands)	Distribution of urban population lacking finished main floor materials in developing world (%)
Northern Africa	98.3	77,910	1,363	1.0
Sub-Saharan Africa	89.1	251,166	27,416	20.6
Latin America and the Caribbean	98.2	417,229	7,630	5.7
Eastern Asia	98.4	564,871	9,271	7.0
Southern Asia	84.8	448,738	68,415	51.4
South-Eastern Asia	93.6	228,636	14,650	11.0
Western Asia	96.4	124,370	4,480	3.4
Total			**133,226**	

FIGURE 2.1.1 DISTRIBUTION OF THE URBAN POPULATION LACKING FINISHED FLOOR MATERIALS IN DEVELOPING REGIONS, 2003

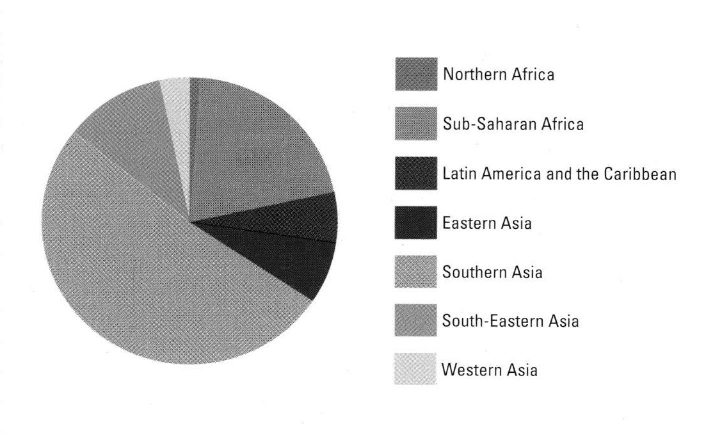

Source: UN-HABITAT (2005), Urban Indicators Programme Phase III and United Nations, World Urbanization Prospects; The 2003 revision.
Note: Access to finished main floor materials was computed from Demographic and Health Surveys (DHS) data.

Defining durability

Old Town in Fez, Morocco. NATSUKO ITO

Presently, global data on housing durability is not possible to collect and analyze because researchers and governments have not agreed upon definitions, specific indicators, classifications, and approaches. What little data has been collected has several methodological weaknesses. For instance, wood is considered durable in developed regions, but not in most developing countries. Other building materials are classified as "rudimentary" (e.g. mud or palm), but in certain cases, they are recorded as "permanent". A material may not be deemed durable in terms of other, more modern building materials, but when combined with skilled construction and regular repair, it could be coded as "durable". Moreover, a problem that arises when measuring the permanency of the structure is that durability manifests itself differently in different cities. In Nairobi, for instance, non-durable houses may be made of a patchwork of tin, cardboard or plastic sheeting, whereas in Mumbai, a temporary house may be made of thatch, bamboo or mud. In Moscow, Tokyo and other developed cities, a non-durable unit is often the equivalent of a dilapidated house made of older materials, or an apartment in a substandard building.

"Durable housing" is generally defined as a "unit that is built on a non-hazardous location and has a structure permanent and adequate enough to protect its inhabitants from the extreme of climate conditions such as rain, heat, cold, and humidity".

However, the permanence of housing is defined by different criteria in different countries. In the Netherlands, the Department for Housing considers the number of houses that do not comply with the local building codes but are still livable. In the United States, information about the condition of housing, including the need for structural repairs and maintenance, is obtained through self-reporting in censuses. Similarly, the Canadian government asks its citizens to assess the condition of their own dwellings as part of its census every five years. Canadians indicate whether their dwellings require any major or minor repairs, with "major repairs" defined as the need for "repair of defective plumbing or electrical wiring, structural repairs to walls, floors or ceilings, etc." In Japan, "durable" housing is any that is not dilapidated, as defined by the government. According to this definition, housing durability can be captured by three main conditions: permanence of structure, non-hazardous location and compliance with local building codes.

In developing countries, the nature of the location and compliance with building codes are rarely used in durability assessments. Whenever a developing country considers the definition of durability in a more comprehensive manner, the measurement is undertaken by integrating two variables: "permanence of structure", under which the state of repair is sometimes also considered; and "hazardous location", which normally encompasses compliance with building codes. However, no information is provided about the description of the area, such as road conditions and pathways, vehicular access roads and other variables of neighbourhood quality, including the type of housing (i.e. detached houses, multi-storey buildings, and the like).

People living in poverty in cities endure a complex array of housing conditions, the range of which is difficult to define and capture for statistical purposes. Recent migrants may utilize temporary materials – sometimes scrap – to build their homes and gradually improve them over time. At one stage of this process, the roof of the house may be made of permanent materials, such as galvanized iron sheets or ceramic tiles, while the rest of the unit is made of temporary materials. This further complicates classifications, since the same house could be considered permanent, semi-permanent or temporary, depending on definitions and measurement criteria. Most of the censuses collected around the world present data on housing quality based on the material used for construction of walls, roofs and floors separately. The Indian government, for instance, classifies a house as *pucca* when the materials used both for the roof and wall are regarded as "permanent" (burnt brick, stones, concrete, tiles, and the like). The government classifies houses as *kutcha* when both materials (roof and wall) correspond to non-permanent (temporary) materials that are replaced frequently (grass, bamboo, leaves, mud, and the like). In the cases where there is a mixture of both permanent and temporary materials, the dwelling is classified as *semi-pucca*. (For definitions refer to the National Census of India 2001.) Moreover, very often the settlements' occupants do not have legal title to their property, allowing them to bypass the request for a building construction permit. Houses built in a non-authorized settlement, even if they are permanent, may be classified as "temporary" until the government grants security of tenure. Since these units are produced by the informal sector, it is difficult to estimate the number of houses built every year and the quality of the structure and materials in terms of durability.

In developed countries, "compliance with building codes" is most often used as a separate variable from the hazardous location of housing. Thus, all inhabitants of the cities of Amsterdam and Den Haag in the sub-sea level Netherlands are classified as living in a disaster-prone area; yet, their housing units can still be considered "durable" if the dwellings are in compliance with the local building codes.

Sources: UN-HABITAT 2003b; Planning Commission of India 2002; Canada Mortgage and Housing Corporation 2005.

Afghanistan RASNA WARAH

Housing durability in Asia

The Asian region is far from homogenous. More than half inhabitants who lack durable housing in the developing world live in Southern Asia, a sub-region that has among the highest prevalence of slums, infant mortality and poor performance on other social indicators.[19] Around one-third of the urban dwellers in Bangladesh, Nepal and Pakistan lack durable housing. One-tenth of the urban population in India, or around 28 million people, are living in non-permanent structures. In Afghanistan, despite some advances, housing challenges remain daunting. Most people in the country live in extreme poverty in cities ravaged by war; housing and physical infrastructure in most Afghan cities need to be rebuilt.[20]

Throughout Asia, advances in housing durability have been less conspicuous than in other developing regions, with some secondary cities in Indonesia (Bitung and Jaya Pura) experiencing important improvements. In 2003, one-fifth of the Asian slum population lacked durable housing – a proportion that may be higher, considering that these estimates do not include information about other construction variables, such as walls and roofs.

Housing durability in Latin America and the Caribbean

In Latin America and the Caribbean as a whole, the number of non-durable houses in urban areas is relatively low. Durability is not a determining factor of slums in the region, since more than 99 per cent of the urban population lived in houses that were considered durable in 2003. Overcrowding remains a much more significant determinant of slum households in the region. However, serious deficits in housing durability have been reported in Guatemala, Nicaragua and Peru.

Apart from Haiti, no reliable data on durability is available for the Caribbean sub-region. Haiti is notorious for having the worst social and health indicators in the whole region but surprisingly, lack of durable housing is not the most prevalent shelter deprivation, as only 10 per cent of the urban population is estimated to live in non-durable houses. In Haiti, slum dwellers suffer most from lack of access to improved water and sanitation – less than 50 per cent of the urban population has access to either – and lack of sufficient living area, or overcrowding, which affects 35 per cent of the urban population.

Expanding the definition of "durability"

There is no doubt that housing durability figures for Africa, Asia and Latin America and the Caribbean are underestimated if the indicator is not widened to include the condition of floor, wall and roof materials combined. Methods to measure lack of durability require further refinement to include more information regarding compliance with building codes, the hazardous location of residential buildings and the condition of individual dwellings. In Brazil, for instance, according to information provided by Munic/IBGE in 2001, *all* municipalities with more than 500,000 inhabitants had a certain number of *favelas* (slums), and most of the *favelas* had houses that were non-permanent. Of the municipalities, 87 per cent had non-authorized subdivisions, called *loteamentos clandestinos*, and 65 per cent showed different forms of inadequate housing, including non-durable structures.[21]

Studies on housing durability should be expanded to take into account other shelter deprivation indicators as well, since non-durable houses are very often associated with dwellings that lack some basic services, such as water and sanitation. In Indonesia, for instance, of the non-permanent housing stock, only 28 per

Havana ©ALEX BRAMWELL. IMAGE FROM BIGSTOCKPHOTO.COM

cent had toilets and 38 per cent had access to an improved water source in 2000, whereas in permanent houses, these ratios were 64 per cent and 75 per cent, respectively.[22] In addition to being inadequate for families' needs, non-durable housing structures are often situated in urban areas that lack adequate infrastructure, especially access roads and bridges. When emergencies happen, residents of such neighbourhoods may find themselves trapped and unable to escape or call for help. In cases when help does arrive, poor access roads and high densities in the settlement may prevent fire engines or ambulances from reaching victims. Physical accessibility is considered an essential part of the durable housing indicator, but data collection systems do not as yet capture this vital information, including data on the surroundings of the household unit.

As part of the United Nations Expert Group on the 2010 World Programme on Population and Housing Censuses, UN-HABITAT has made recommendations that census questionnaires be revised and updated to include questions that address three key housing durability variables:

1. houses in hazardous locations;[23]
2. building codes;[24]
3. and hazard mapping.[25]

The next round of censuses will include relevant questions to obtain data on these variables to further clarify the extent of housing durability.

FIGURE 2.1.2 PROPORTION OF URBAN HOUSEHOLDS WITH FINISHED MAIN FLOOR MATERIALS, BY REGION, 1990 AND 2003

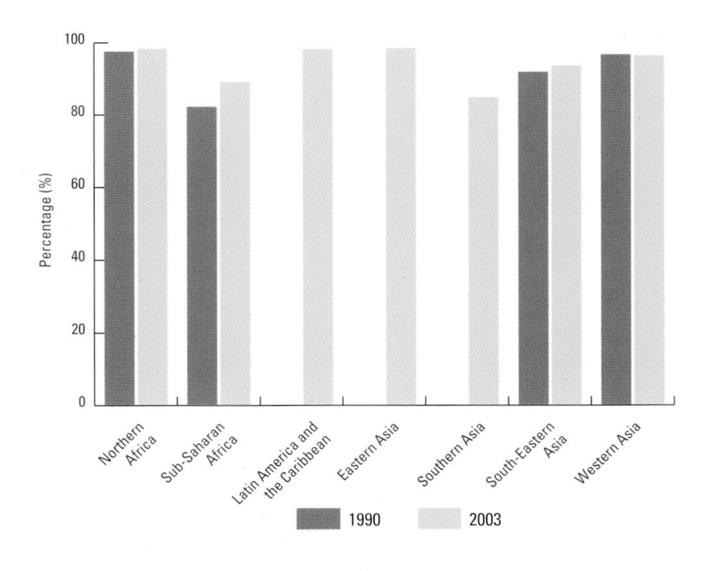

Source: UN-HABITAT Urban Indicators Programme Phase III.
Note: Data for 1990 not available for some regions.

Europe's forgotten Roma community

Roma Settlement in Eastern Europe BARBARA GALASSI

The Roma or Romani (meaning "man" or "people") have also been called Gypsies, Tsigani, Tzigane, Cigano, Zigeuner – labels the Roma themselves consider derogatory. Most Roma identify themselves by their tribes or groups, which include the Kalderash, Machavaya, Lovari, Churari, Romanichal, Gitanoes, Kalo, Sinti, Rudari, Manush, Boyash, Ungaritza, Luri, Bashaldé, Romungro, and Xoraxai.

In 1999, the United Nations Economic and Social Council (ECOSOC) reported that the majority of Roma people in Europe lived in the "most squalid and derelict housing estates" with sanitary facilities that were either extremely poor on non-existent. Since 1999, the situation has not changed much. In fact, according to most reports, it is getting worse. The United Nations, the World Bank, the European Commission and other organizations have produced indicators that show that the standard of living in Roma ghettos in Europe is appallingly low, with life expectancy rates sometimes 20 years less than the majority populations, unemployment rates close to 100 per cent, excessive school drop-out rates, and high incidence of violence. Despite these indicators, such ghettos are not only allowed to remain, but more are being created. In fact, the construction of a wall built around one such community in the Czech town of Usti nad Labem was cited by a recent report as "a single but vivid example of such ghettoization in all its shocking reality".

Roma settlements in Central and Eastern Europe have increased, as have the number of Roma who are forced to live in them. Most affected by the collapse of the Communist economies, evicted because they cannot afford market rents, the Roma often have no other choice but to settle in makeshift housing located on land no one else wants: contaminated industrial properties or garbage sites isolated from the majority population, without public utilities such as clean drinking water, electricity or waste collection.

In Western Europe, slum-like settlements housing Roma communities can be found in Greece, Portugal and Spain. In Italy and in France, it is mainly Roma asylum seekers from Eastern Europe and Asia who end up in camps because of official unwillingness to provide proper housing for refugees. The squalor of these camps was revealed in 2004 when two Roma girls from Romania died in a fire that destroyed the hut in which they slept.

The European Roma Information Office in Brussels has compiled reports showing that the exclusion of the Roma from mainstream life in Europe runs contrary to the norms of European legislation and international human rights conventions. In September 2002, Alvaro Gil-Robles, the Commissioner for Human Rights for the Council of Europe, stated that the Roma in Greece were living under conditions "very remote from what is demanded by respect for human dignity". And in 2003, a World Bank report found that "the Roma inhabit approximately 95 per cent of the *chabolas* (makeshift housing and slums) around larger cities in Spain. Approximately 80 per cent of these houses are smaller than 50 square meters and house more than 4 people … . The lack of sanitation and running water in these areas threatens the health of the inhabitants."

In Romania – which has Europe's largest Roma population of approximately 2 million people (although the name Roma is *not* derived from Romania) – persecution of the Roma has been occurring since the mid-1800s. During World War II, the Nazis also slaughtered between 500,000 and 1,500,000 Roma during the holocaust.

According to Valeriu Nicolae of the European Roma Information Office in Brussels, although more than 20 years have passed since the European Parliament discovered in its 1984 report that the Roma in Europe faced societal and legal discrimination, the 8 to 12 million Roma living in Europe are still considered third-class citizens and there are few, if any, concrete actions targeting the improvement of their living conditions. If the situation continues unabated, it is likely that the Millennium Development Goals will not be achieved among this much-neglected community.

Source: Nicolae 2005.

Endnotes

1 US Department of Housing and Urban Development 2005.

2 UN-HABITAT 2001.

3 Interviews by Eduardo López Moreno with National Directors of Housing of various countries as part of the regional workshops on "Training on Data and Indicators for Monitoring Progress towards the Millennium Development Goals." The countries included, among others, Venezuela, Ethiopia, Indonesia, Angola, and Mexico.

4 Ayieko 2006.

5 In British Columbia, Canada, it is estimated that 20 to 25 per cent of all urban rental housing is in this form, which is 8 to 10 per cent of the total housing stock. Canadian Mortgage and Housing Corporation 2005.

6 Scaramella 2003.

7 National Institute of Standards and Technology 1999.

8 Canada Mortgage and Housing Corporation 2005.

9 US Census Bureau 2000. Data from the 1999 American Housing Survey. In 2003, the American Housing Survey further found that 11 per cent of the total housing stock was considered as having minor technical deficiencies.

10 In Canada, of the housing stock that needed major repairs, 39.2 per cent were houses on First Nations reserves. In the United States, the share of households living in houses with severe problems was more than two times higher for blacks and Hispanic populations than white non-Hispanic populations (3.4 and 3.8 compared to 1.5). US Census Bureau 2000.

11 In the United States, this ratio is 15.7 per cent of renter houses with selected deficiencies versus 7.5 per cent of owner-occupied houses. (US Department of Housing and Urban Development 2005.) In Canada, the ratio is only 9.3 per cent versus 7.4 per cent. (Canada Mortgage and Housing Corporation 2005).

12 However, the Housing Statistics of Indonesia published that in 2000 only 14 per cent of the national urban housing stock was considered non-permanent. Data on floor, roof and walls requires further refinement. Refer to Housing Statistics Indonesia 2000.

13 Planning Commission of India 2002.

14 Observers and researchers doubt that the situation has evolved positively in recent years due to economic uncertainty and additional expenditures on security matters. US Library of Congress 1998.

15 For countries with information on the three components, regression equations using Demographic and Health Survey data show that the percentage of durable housing is strongly correlated with the percentage of houses with durable flooring material in Africa, Asia and Latin America and the Caribbean. For the three regions the regression equations are statistically similar; that is, one regression equation from these three data sets is sufficient to estimate the durable housing from the nature of the floor whatever the region.

16 UN-HABITAT 2004a.

17 Martínez-Martín 2005.

18 UN-HABITAT 2005c.

19 In countries such as Bangladesh, the prevalence of non-durable housing is around 50 per cent. UN-HABITAT 2005c.

20 Economic and Social Commission for Asia and the Pacific (ESCAP) 2003.

21 Presidency of the Republic of Brazil 2004.

22 Housing Statistics Indonesia 2000.

23 The census will gather additional information on dwellings that are considered hazardous, namely: housing located in areas subject to disaster more than once every hundred years (disasters include flooding, earthquakes, volcano, storm surge, landslide, or avalanche); housing not adequately protected against cyclones or bushfires which occur at this frequency; housing settled on garbage mountains or dumps; housing around high industrial-pollution areas; and housing around other high-risk zones, including railroads, airports and energy transmission lines.

24 This will include anti-cyclonic and anti-seismic building standards, which should be based on hazard and vulnerability assessment.

25 Hazard mapping is a simple and effective way of ensuring that hazards are recorded and updated on a regular basis. The maps shall cover the entire city and its boundaries, be available to the public and as recent as possible (less than five years old).

2.2 Not Enough Room: Overcrowding in Urban Households

"This one room is my bedroom, my kitchen and my sitting room."
Jared Odhiambo, a slum dweller in Kibera, Nairobi.[1]

Having only one room for sleeping, eating and socializing may be perfectly fine for a single person on a tight budget, but imagine managing a family of four, five or even more in an area fit for just one. Renting a squalid, overcrowded one-roomed house or apartment is the only way many low-income families around the world can afford shelter at all in urban areas. Family members must adapt small spaces to suit their daily needs, often at the cost of their privacy and health.

Recent studies of connections between housing conditions and rates of illness and child mortality have contributed to the growing realization that good-quality housing conditions are essential to ensuring a healthy, productive population. The risk of disease transmission and multiple infections increases substantially as the number of people crowded into small, poorly ventilated spaces increases. A study on overcrowding in low-income settlements conducted by UN-HABITAT in 1995 confirmed that infectious diseases are likely to thrive in overcrowded and low-income households, owing to lack of ventilation, lack of hygiene and exposure to environmental contaminants.[2] The prevalence of overcrowding in inadequate dwellings has also been linked to increases in negative social behaviours, such as domestic violence and child abuse, and to negative outcomes of education and child development. Children's education may be affected by overcrowding directly, owing to a lack of space to do homework and the disruption of sleep patterns, and indirectly, through absenteeism caused by illness arising in part from overcrowding.[3] Research has also suggested that overcrowding may lead to the eviction of some tenants, since congestion increases the likelihood of property damage and may violate rental agreements.

In cities of the developing world, overcrowding in low-income areas is often related to other forms of social and physical deprivation. It is not just a question of parents sharing a bedroom with their grown-up sons and daughters, or too many people sleeping in the same room; rather, as one inhabitant living in a Nairobi slum put it, overcrowding "takes one's dignity away". Living in crowded quarters intensifies interpersonal contact and the experience of sights, sounds and smells – often for the worse. At the community level, high residential densities can put excessive strains on social services, such as medical clinics, and on schools, natural environments and other resources.

Overcrowding is a manifestation of housing inequality that results from a combination of factors. Insufficient housing stock and lack of affordable housing are perhaps the most prominent factors leading to overcrowding, coupled with market and policy situations unfavorable to low-income residents: inefficient housing markets, inadequate public and private investment in affordable rental units and inappropriate design of available units, among others. From a structural perspective, unemployment, lack of living-wage jobs and the spatial concentration of ethnic minorities and people living in poverty are significant factors.

Global trends in overcrowding

Scholars have argued that overcrowding is a hidden form of homelessness. People without an adequate place to live are often forced to search for accommodation with friends or relatives. This can place stress on the hosts, whether tenants or owners, making accommodation for visitors and kin even more tenuous. Consequently, many people occupy dwellings that exceed local standards of occupancy. In Chile, "*los allegados*" (people living with other families) today represent slightly more than one-fourth of the country's urban population.[4] Chileans are clearly coping with the national housing deficit through co-habitation instead of creating new slums. In Australia, 58 per cent of the homeless population takes shelter with friends or relatives for sometimes six months or longer, surpassing the "chronic" homelessness threshold in that country.[5] In Haiti, as in many other parts of the developing world, many individuals and families "time share" the same house, occupying it in shifts.

In developed countries, overcrowding as a physical housing problem has substantially decreased over time. According to the UN-HABITAT definition (see box), which is based on conditions in developing countries, overcrowding is non-existent or extremely rare in most countries (less than one-half of one per cent of the urban population). In Amsterdam, for example, data collected by the Netherlands Department

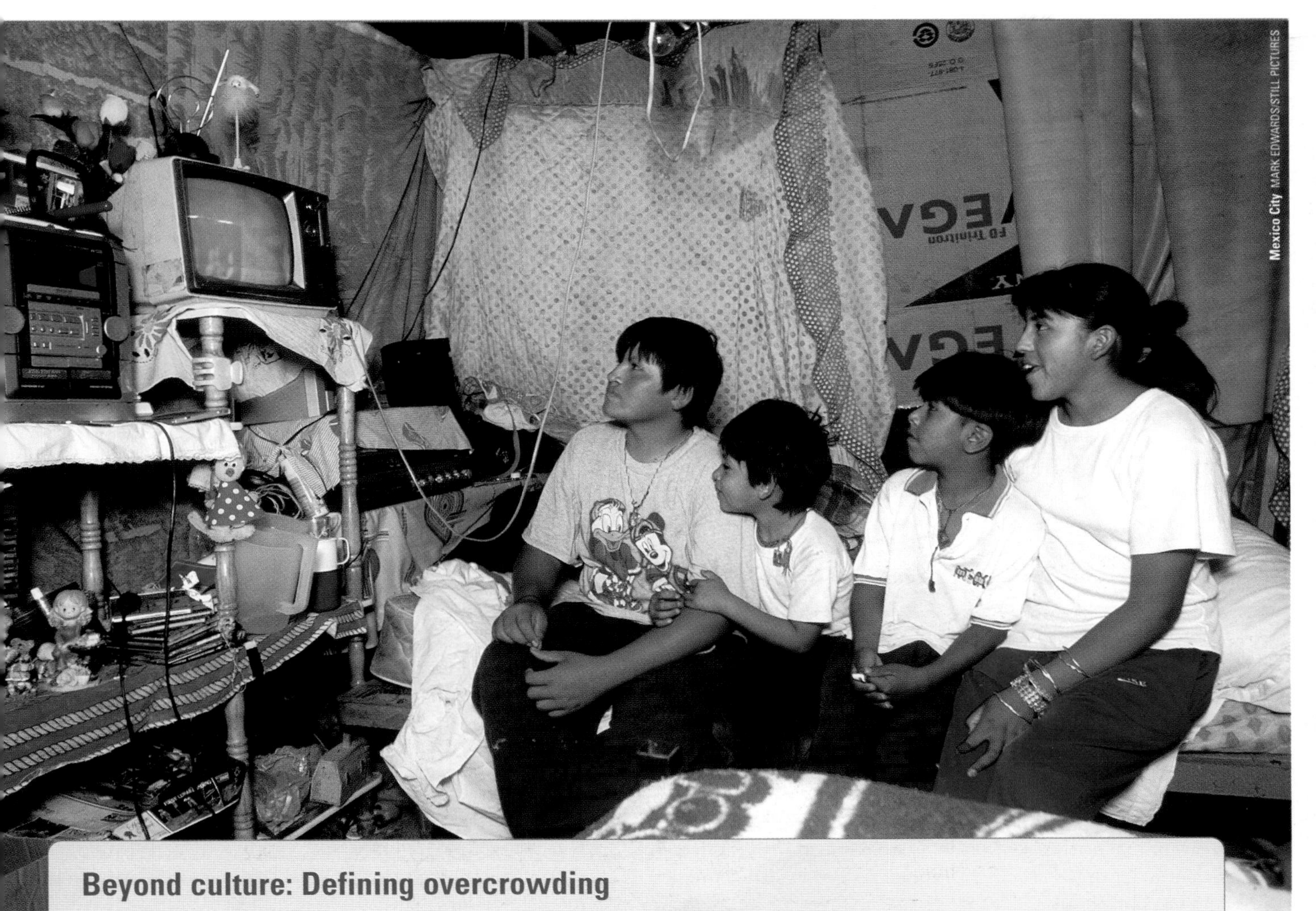

Mexico City MARK EDWARDS/STILL PICTURES

Beyond culture: Defining overcrowding

The perception of overcrowding is subject to cultural definitions and is often a function of standard dwelling unit sizes, family groupings and other cultural norms. In some cultures, and among some ethnic groups, living in close quarters is preferable to living in smaller family groupings, or it is at least tolerated.

Behavioural studies indicate that certain levels of crowding are desirable among some groups. In the United States, for instance, 8 per cent of high-income Asian and Hispanic groups continue to live in houses considered "overcrowded" by American standards, even though they can afford to enlarge their living space or move to a larger house. Non-Hispanic and non-Asian people with incomes comparable to those groups experience overcrowding only half as often. Even though it is difficult to place a value judgment on overcrowding, given its cultural specificity, household surveys suggest that if given the choice, very few people would be willing to share a bedroom with four or five people. Culture, in terms of tolerable crowding levels, has some limits in that sense.

Many people, especially the poor, have few options regarding whether or not to live in crowded spaces. Overcrowding, therefore, can be an important indicator of substandard housing, whereas *sufficient living area* is a key indicator for measuring adequacy of shelter. In this sense, overcrowding can be an objective measure that transcends culture and ethnicity. As one of the shelter deprivation indicators, overcrowding expresses a normative judgment about the degree of crowding, which applies a criterion that defines a particular density as acceptable or unacceptable.

There is no basis in scientific literature for choosing one standard of unacceptable overcrowding over another. Countries define the crowding indicator in different ways. Some developed nations apply the concept of the adult individual's need for a separate bedroom, and any value in excess of 1.0 – any bedroom used by more than one or two adults – represents a measure of crowding. Other countries determine the number of bedrooms a dwelling should have to provide freedom from crowding.

A common standard is defined by the number of people per dwelling, per room or per bedroom, with some countries separating the number of individuals by age group, in which a gender disaggregation is fundamental. Overcrowding can also be defined in terms of the square meters available per person, in which values are determined according to the number of individuals.

UN-HABITAT and its partners developed an operational definition of overcrowding as one of the slum-related indicators: the "*proportion of households with more than two persons per room*". This definition was developed considering that reduced space and high concentration of people in the dwelling is often associated with certain health risks, so may be correlated with slum conditions. After observing the statistical distribution of more than two persons per room throughout the world, UN-HABITAT revised its definition to *three persons per room.* As part of the UN-HABITAT monitoring exercise, the indicator is described in the positive as "sufficient living area".

Sources: Myers & Baer 1996; www.stats.govt.nz; UN-HABITAT 2002b.

for Housing indicates that less than 1 per cent of the city's households have more than three persons per room. However, according to the Amsterdam housing standards criteria (one person per room), 26.2 per cent of houses are overcrowded.[6] In Canada, only 0.014 per cent of households reported having more than three persons per room. Nonetheless, if a much higher standard is applied as per the country's own definition, 6.3 per cent of urban households were below the standard in 2003.[7] In many developed cities, housing overcrowding (as per national standards) is correlated with the prevalence of ethnic minorities. In London, the ward with the highest proportion of households with more than one person per room is Wembley Central, a ward that also has the highest proportion of residents of Indian origin.[8] In Australia, overcrowding is approximately nine times more prevalent among the aboriginal population than it is among non-indigenous people.[9]

People living in cities in developing countries, however, still experience high levels of overcrowding. According to UN-HABITAT urban indicators, approximately 20 per cent of the world's urban population was living in inadequate dwellings, in terms of sufficient living area, in 2003. However, the extent and nature of this phenomenon varies among regions. In 2003, two-thirds of the developing world's urban population without sufficient living area resided in Asia, half of this group lived in Southern Asia (156 million people). Africa ranked second with 75 million people living in overcrowded conditions. Over one-tenth of the developing world's urban population without sufficient living area (49 million people) resided in Latin America and the Caribbean.

Apartments in Phnom Penh ©ERIK DEGRAAF. IMAGE FROM BIGSTOCKPHOTO.COM

TABLE 2.2.1 SUFFICIENT LIVING AREA COVERAGE AMONG URBAN POPULATION BY REGION, 2003

	Access to sufficient living area, 2003 (%)	Urban population in 2003 (thousands)	Population lacking sufficient living area (thousands)	Distribution of urban population lacking sufficient living area in developing world (%)
Northern Africa	90.5	77,910	7,429	1.9
Sub-Saharan Africa	73.1	251,166	67,629	16.8
Latin America and the Caribbean	88.2	417,229	49,176	12.2
Eastern Asia	91.5	564,871	47,813	11.9
Southern Asia	65.0	448,738	156,849	39.1
South-Eastern Asia	73.1	228,636	61,448	15.3
Western Asia	91.1	124,370	11,111	2.8
Total			**401,456**	

FIGURE 2.2.1 DISTRIBUTION OF URBAN POPULATION LACKING SUFFICIENT LIVING AREA, 2003

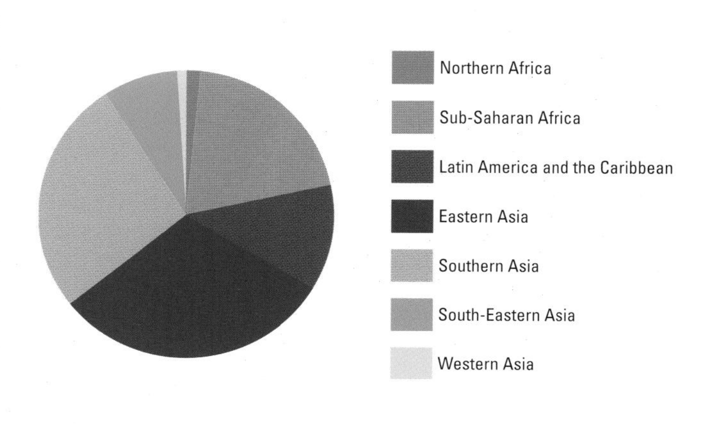

Source: UN-HABITAT (2006), Urban Indicators Programme Phase III and United Nations, World Urbanization Prospects; The 2003 revision.
Note: Access to sufficient living area was computed from Demographic and Health Surveys (DHS) data.

MAP 8 PROPORTION OF HOUSEHOLDS LACKING SUFFICIENT LIVING AREA IN SELECTED CITIES, 2003

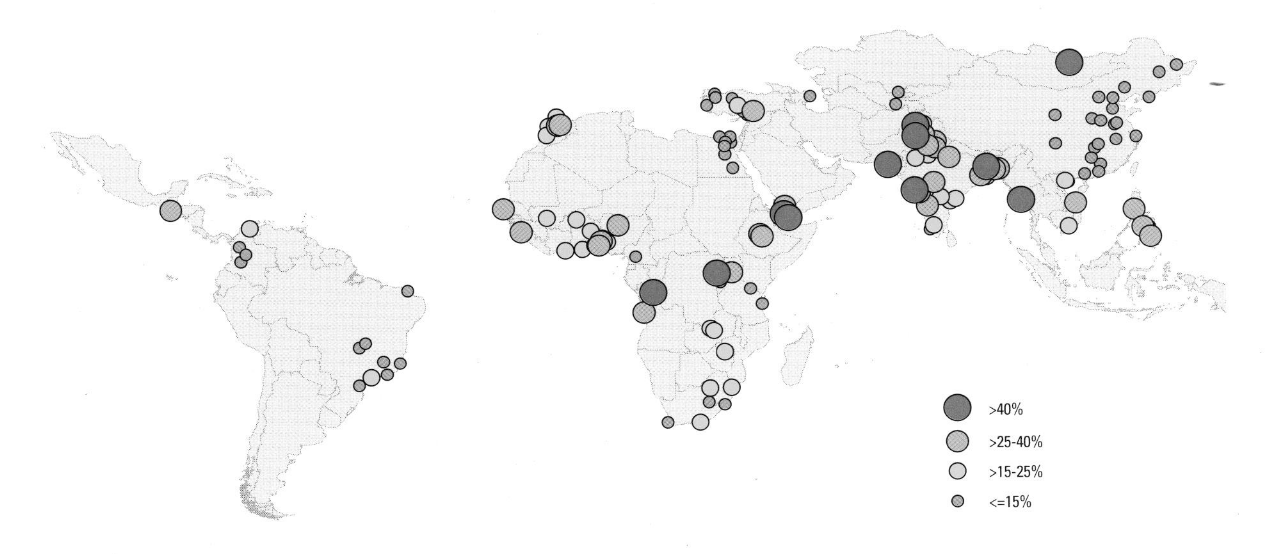

Source: UN-HABITAT 2005, Global Urban Observatory, Urban Indicators Programme, Phase III.

Overcrowding data by region

Overcrowding in Africa

About a quarter of sub-Saharan Africa's urban population lives in overcrowded houses. Residential overcrowding is more prevalent in some cities than in others, including Addis Ababa, Kampala and Luanda, where more than 40 per cent of the urban population lives in housing that does not have sufficient living areas. Similar deficits are found in the Nigerian cities of Lagos and Ibadan. Residents of Ibadan have experienced the sharpest decline in sufficient living area in the world: from 95 per cent of the urban population in 1990 to 70 per cent in 2003. It is likely that this phenomenon was precipitated by the disproportionate increase in the property market value during the 1990s, as a result of the construction of the Ibadan-Lagos expressway that encouraged many workers in Lagos to move to Ibadan, where accommodation is cheaper. This influx raised the housing demand and increased prices, simultaneously forcing the urban poor to find rooms in the cheapest areas of Ibadan – the inner city and peripheral slums – and increasing overcrowding rates.[11]

In *Northern Africa*, the incidence of overcrowding is 10 per cent of the urban population. However, some countries, notably Egypt, have dramatically reduced overcrowding in the last two decades. The cities of Cairo, Alexandria, Port Said and Suez increased the proportion of their inhabitants with sufficient living area from 70 per cent in 1990 to 95 per cent in 2003. In the Moroccan cities of Casablanca and Rabat, the percentage of people with sufficient living space rose from 69 per cent and 79 per cent, respectively, in 1998, to 79 per cent and 87 per cent, respectively, in 2003.

Overcrowding in Asia

In *Asia*, sufficient living area is poorly reported, so determining the actual incidence of overcrowding for the region is difficult. However, some trends are beginning to emerge. Southern Asia has the highest prevalence of overcrowding in the developing world, with a third of its urban population residing in houses that lack sufficient living area, followed by South-Eastern Asia where over a quarter of the urban population lives in overcrowded housing. For Eastern Asia, not enough information was available for analysis, and levels and trends were estimated from those observed in South-Eastern Asia and Western Asia. Differentials across sub-regions should therefore be examined with caution.

An analysis of sufficient living area in this region shows that most of the cities and countries are facing growing trends toward overcrowding. The few exceptions include the cities of Manila in the Philippines, and the cities of Istanbul, Ankara and Adana in Turkey, where urban residents slightly increased their living area.

Overcrowding rates are high in various Asian cities, namely in Yangon, in Myanmar, Dhaka and Rajshahi in Bangladesh, Karachi, Faisalabad and Islamabad in Pakistan, and Ulan Bator in Mongolia, where around 40 per cent of the urban population lived in overcrowded dwellings in 2003. With the exception of Myanmar, which has the lowest proportion of slums among the least developed countries in the world (26 per cent), the other nations are characterized by a high prevalence of urban dwellers living in slum conditions: Bangladesh, at 85 per cent; Pakistan, at 74 per cent; and Mongolia, at 65 per cent. In other countries in which the proportion of slum dwellers is high, such as Nepal

(56 per cent) and India (55 per cent), the incidence of overcrowding is also relatively high – one-third and half of the urban population, respectively.[12]

Overcrowding in Latin America and the Caribbean

In *Latin America and the Caribbean*, information on sufficient living area is lacking for most of the Caribbean countries and many South American nations. Estimations were made using data from countries that represent just over 50 per cent of the entire region's population. However, estimates indicate that whereas the region has made significant progress in improving slums, overcrowding affects over 10 per cent of the urban population.

The highest levels of overcrowding in the region are found in Central America, particularly in Guatemala and Nicaragua (30 and 38 per cent, respectively). Both countries had a high prevalence of slum households in their cities in 2001, at more than 60 per cent, and among the highest rates of slum growth in the region between 1990 and 2001 – 2.4 per cent and 3.4 per cent per year, respectively – more than twice the average slum growth rate of the rest of the region (1.3 per cent). In South America, overcrowding rates were higher in Bolivia and Peru, where around one-third of the urban populations were deprived of sufficient living space in 2003. This is not surprising, considering that slums remain a major challenge in both countries, with more than two-thirds of their urban populations living in slum conditions

Overcrowding in the developed world

Studies carried out in *developed countries* confirm that overcrowding affects some specific populations more than others. Robust research evidence corroborates the fact that tenants are more likely live in overcrowded units than homeowners.[13] In the United States, for instance, overcrowding is approximately twice as prevalent among tenants as among owners. Likewise, households made up of young occupants are more likely to be overcrowded than households comprised of older adults, and higher rates of overcrowding are found among recent immigrants than other residents. Hispanic and Asian communities account for 8.3 per cent of all households in the United States, but they represent 46.6 per cent of all overcrowded households.[14]

In cases in which it has been possible to adjust for confounding variables, such as tenure status, income and age, the conclusions are clear: overcrowding is one of the most significant ways in which urban poverty expresses itself. Statistical analysis conducted by UN-HABITAT shows that overcrowding does not appear as an isolated variable, but it is very often combined with other slum dimensions, such as lack of safe drinking water and improved sanitation.

FIGURE 2.2.2 PROPORTION OF URBAN HOUSEHOLDS WITH SUFFICIENT LIVING AREA

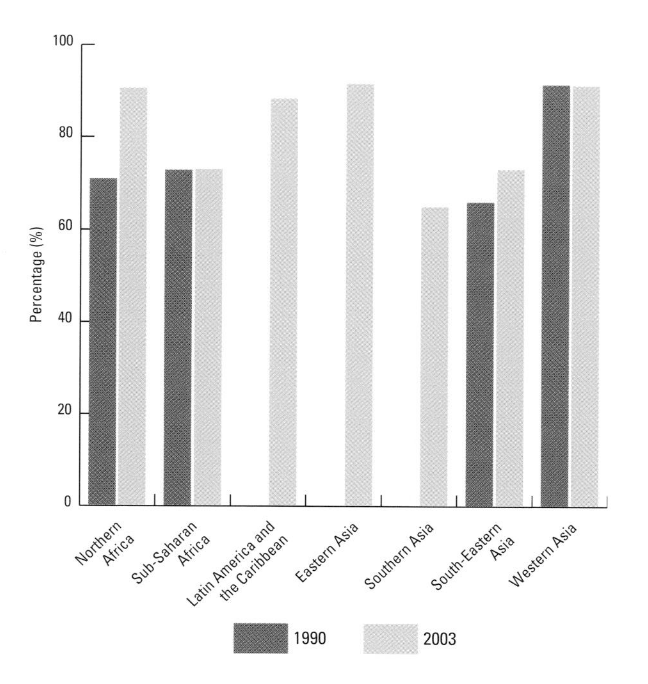

Source: UN-HABITAT Urban Indicators Programme Phase III.
Note: Data for 1990 not available for some regions.

FIGURE 2.2.3 PROPORTION OF URBAN POPULATION WITH ACCESS TO SUFFICIENT LIVING AREA IN SELECTED CITIES, 2003

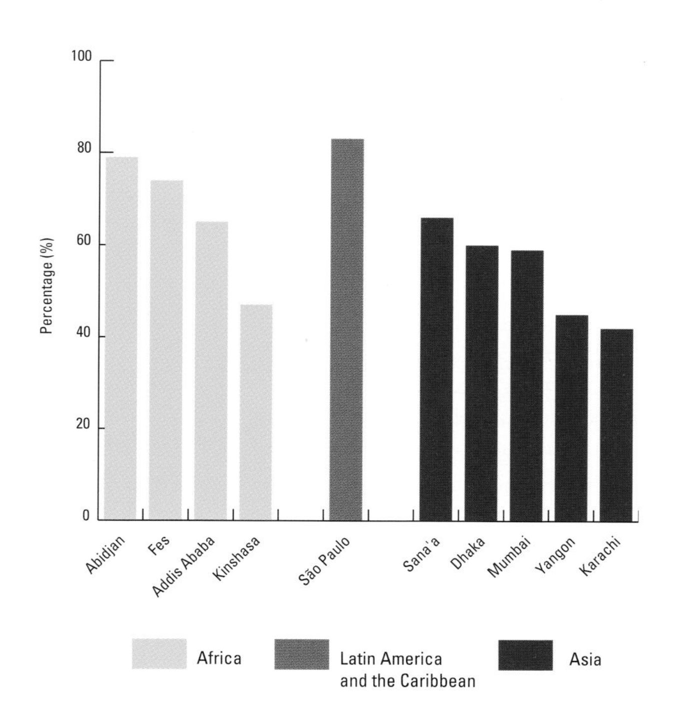

Source: UN-HABITAT 2005, Urban Indicators Programme Phase III.

Kolkata, India RON GILING/STILL PICTURES

Endnotes

1. Quoted in Phombeah 2005.
2. UN-HABITAT 1995.
3. United Kingdom 2005.
4. Ministerio de Vivienda y Urbanismo 2004. "*Allegados*" in Chile may be a hidden form of slum-dwellers. If that is the case, national data should be revised.
5. Australian Bureau of Statistics 2001.
6. Netherlands Department for Housing 2005.
7. Canada Mortgage and Housing Corporation 2005.
8. London Research Centre 2005.
9. Australian Bureau of Statistics 2004. In some regions, such as Warburton, overcrowding stood at 50 per cent for indigenous households and 2 per cent for non-indigenous households.
10. In sub-Saharan Africa, information was obtained from countries representing more than 80 per cent of the region. In Latin America and the Caribbean, data was derived from countries representing slightly more than 50 per cent of the region. Lack of information was observed on this indicator in some sub-regions of Asia; for example, in Eastern Asia none of the countries reported information on sufficient living area, and in other sub-regions, this information was obtained from only a few countries.
11. Fouchard 2003.
12. Other countries that are characterized by having a high prevalence of slums are Afghanistan (98 per cent) and Cambodia (72 per cent); no information currently exists on sufficient living area for these countries.
13. See, for instance, Myers & Baer 1996; and Ellaway & Macintyre 1998.
14. Myers & Baer 1996.

2.3 Safe Drinking Water in Cities

United Nations statistics on safe drinking water provision throughout the world indicate a slight improvement in recent years: between 1990 and 2002, approximately 1.1 billion people gained access to an improved source of drinking water, an increase in global coverage from 77 per cent to 83 per cent.[1] Access to safe drinking water is, however, unevenly distributed around the globe. Significant disparities exist on several levels. First, safe drinking water is unevenly distributed between the *urban* and *rural* populations of the world: 95 per cent of the world's urban dwellers have access, but only 72 per cent of the world's rural population has access.[2] Second, *per capita water consumption* levels vary widely between rich and poor nations, with the former consuming 10 times more drinking water than the latter: 500 to 800 litres per day, compared with 60 to 150 litres per day.[3] Third, asymmetries exist in *water access* and *water management* between high-income countries experiencing low population growth and low-income countries facing rapid population growth and water scarcity problems. And fourth, there are extreme differences in the *quantity* and *quality of water* that rich and poor households can obtain within the same city in different parts of the world.

Some of the disparities in drinking water provision are clear, but others – particularly intra-city differences in access and consumption – are less evident, as they are often disguised by aggregated urban data that averages out quantity and quality of water among those having access to safe water and those who are frequently deprived. This results in a single, and misleading, estimate of access to safe drinking water in a city.

Current United Nations statistics, which use aggregated data, confidently report that 95 per cent of the world's urban population has "improved" water provision (see box). At least 12 countries with low or middle income levels even report that they have 100 per cent coverage.[4] More than 15 countries that perform poorly on a number of health indicators linked to living conditions also report that their national water coverage is above the world's urban average (95 per cent)[5]; and 44 countries (18 in Africa, 12 in Asia and 14 in Latin America and the Caribbean), each with slum populations representing at least one-fourth of their total urban populations, report that water coverage is almost universal – as high as 90 per cent.[6]

Yet, millions of people living in these countries suffer from waterborne diseases, indicating that they do not have adequate access to safe drinking water as officially reported. In some countries, people suffering from waterborne diseases occupy a high proportion of hospital beds – as in India, where 65 per cent of hospital patients are being treated for water-related illnesses at any given time.[7] Many people also spend large proportions of their incomes on the treatment of water-related diseases. In sub-Saharan Africa, for instance, people living in poverty spend at least one-third of their incomes on treatment of water-borne and water-related diseases, such as diarrhoea and malaria.[8] The facts suggest that the drinking water available to people living in poverty is frequently inadequate or contaminated.

Indonesia PETRUS IYAY SAPUTRA/UNEP/STILL PICTURES

Although most statistics reflect better drinking water coverage in urban areas than in rural areas, various surveys show that in many cities, the quantity and quality of water available to low-income residents falls short of acceptable standards. Hundreds of millions of people who supposedly have access to water only have access to communal pipes with intermittent water supply shared by hundreds, or even thousands, of people. In the lowest-income areas, people also pay more for water than their wealthier neighbours. UN-HABITAT urban indicators show that in 2003 in Jakarta, for example, only 29 per cent of the households were connected to piped water in their dwellings; 19 per cent used private wells as a second source of water; and 7 per cent purchased water from vendors, who charged several times more than the official price.[9]

FIGURE 2.3.1 WATER EXPENSE AS A FRACTION OF HOUSEHOLD INCOME IN ADDIS ABABA, 2003

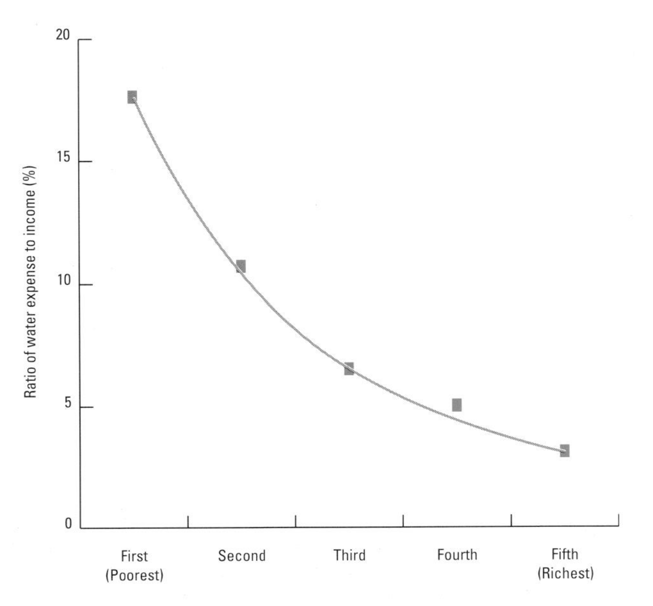

Source: UN-HABITAT, Addis Ababa Urban Inequity Survey 2003.
Note: Based on wealth of households, households divided equally into five groups, First (Poorest), Second, Third, Fourth and Fifth (Richest).

Drinking water supply is often intermittent even in neighbourhoods where people have water pipes and taps inside their houses. In the coastal city of Mombasa in Kenya, a study found that very few parts of the city had continuous water supply; on average, water was available for less than 3 hours per day, and in some parts of the city, pipes had not been functional for several years.[10] In other instances, such as in Dhaka, Bangladesh, slum dwellers complain that the agency that provides water supply and sewerage only makes functional water connections available to land owners, not tenants in slums.[11]

Water is, therefore, not as safe as it appears according to statistics, nor is it supplied in high enough quantities to all households, thus compelling many families to find more water from other sources that are less reliable. Differing definitions, technical measurement problems and political unwillingness to report accurate data prevent the existing water access issues among people in cities from being captured for statistical purposes. The need to acknowledge differential access and the problems associated with it is urgent: nearly half of the urban populations in Africa, Asia, and Latin America suffer from one or more of the communicable diseases associated with inadequate water and sanitation provision.[12] Too often, the myriad issues associated with lack of safe drinking water go unrecognized by the development community. Clearly, however, lack of safe water kills. The

"Improved" or "Adequate" access to water: Definitions and issues of measurement

As part of its strategy to monitor slum improvement, UN-HABITAT included inadequate access to safe drinking water as one of five shelter deprivation indicators. Lack of access to improved water facilities and lack of access to improved sanitation are coincident with the other indicators of slum households – housing durability, overcrowding and secure tenure – and together account for the identification of most slum households.

To measure access to safe drinking water, UN-HABITAT adopted the definition of "improved" water supply developed by the WHO/UNICEF Joint Monitoring Programme for Water and Sanitation. The definition is widely accepted and is backed by a long history of data collection in developing countries.

Definition: *The water should be affordable and at sufficient quantity that is available without excessive effort and time.*

Indicator: *The proportion of households with access to improved water supply with:*

- *Household connection*
- *Public standpipe shared by a maximum of two households*
- *Borehole well*
- *Protected spring*
- *Rainwater collection*

At least 20 liters per person should be available within an acceptable timespan (defined locally). This is a variation from the original WHO/UNICEF definition, which indicates that water should be available within one kilometer of the residence; the definition was changed to accommodate data collection in high-density urban areas. Some data collectors count all residents with a water source within 200 meters of their home as having access to water, but, as UN-HABITAT has found in further research, that having a tap within 200 meters of a dwelling in a rural settlement with 200 people using it is not the same as having a public tap within 200 meters of a dwelling in an urban squatter settlement with 5,000 people using it.

The WHO/UNICEF definition was designed to measure water access in rural contexts and does not necessarily provide a suitable definition for research in urban areas. Urban settlements have particular needs for water that are distinct from rural areas, yet it is still common to refer to "improved" and "adequate" access to water interchangeably in both urban and rural settlements. These terms cannot capture the full extent of two different realities: rural and urban.

"Improved" water provision is often no more than a public tap shared by several hundred people with an intermittent supply of water. Definitions of services and access vary not only among the different types of surveys undertaken, but also over time. It is therefore sometimes difficult to compare surveys undertaken even within the same country. In addition, people often use more than one water source, and it is difficult to ascertain the quality, accessibility, reliability, and cost of each, and whether its use is a problem.

United Nations estimates do not as yet consider the quality and affordability of water services. UN-HABITAT is in the process of refining survey questions on access to improved water through the application of Urban Inequities Surveys, within the framework of the Monitoring Urban Inequities Programme. The new questions will appear on household surveys regularly administered by UN-HABITAT's partner agencies.

Gender issues in safe water access, child access to facilities and the number of households using the same facility are other important issues to consider in urban water research. UN-HABITAT is now working with the WHO/UNICEF Joint Monitoring Programme to harmonize and standardize indicators and methodologies, in order to match definitions and methods of measurement to produce data comparable across countries and cities.

root of this unrelenting catastrophe is well known[13]: people are not getting sufficient quantity and quality of water that is affordable and available without having to invest excessive effort and time.

UN-HABITAT data on urban indicators collected in 2003 provides the distribution of households by major source of drinking water at the national, urban agglomeration and slum/non-slum levels. Analysis of the data at each level reveals that getting water from a tap is a luxury enjoyed by only two-third of the world's urban population. In 2003, 62 per cent of all city dwellers had access to piped water, 46 per cent of whom had water piped into the dwelling and 16 per cent of whom had a water tap in the yard or plot. Public taps serviced 10.4 per cent of the urban residents, and 8 per cent had access to manually pumped water or protected wells.[14]

UN-HABITAT data reveals important inter-regional differences in the way urban residents in the developing world gain access to safe water. In 2003, Latin America and the Caribbean has the highest proportion of urban households dependent on piped water sources (89.3 per cent) and sub-Saharan Africa has the lowest (38.3 per cent). In Africa, 20 per cent of the urban population cites public water taps as a primary source of drinking water – twice the world's average – while in Latin America and the Caribbean, only 2 per cent of residents depend on public taps. Asia has the highest proportion of people using manually pumped water (12 per cent); in Africa, 7 per cent rely on manual pumps, and in Latin America and the Caribbean, 5 per cent use them.

Global trends

The world has made important progress in increasing access to safe drinking water. Despite this increase, however, the total number of people who gained access to improved water sources remained stable at approximately 17 per cent, owing to global population growth.[15]

Data collected by the WHO/UNICEF Joint Monitoring Programme provides additional information on access of the world's urban population to safe drinking water. The share of the urban population with access to improved water sources remained stable from 1990 to 2002 at 95 per cent. The Joint Monitoring Programme counts access to all types of improved water sources that are protected from external contamination, including piped household connections, public standpipes, boreholes, hand-dug wells, springs, and rainwater collection.[16] However, the number of people without access to improved water will double between 1990 and 2010, increasing from 108 million to 215 million.

Widespread inter-regional differences in water access exist, though these asymmetries began to conform around higher coverage in the 1990s, particularly in urban areas. An outline of water access trends in the major regions of the world follows.

MAP 9 PROPORTION OF HOUSEHOLDS WITHOUT ACCESS TO IMPROVED WATER IN SELECTED CITIES, 2003

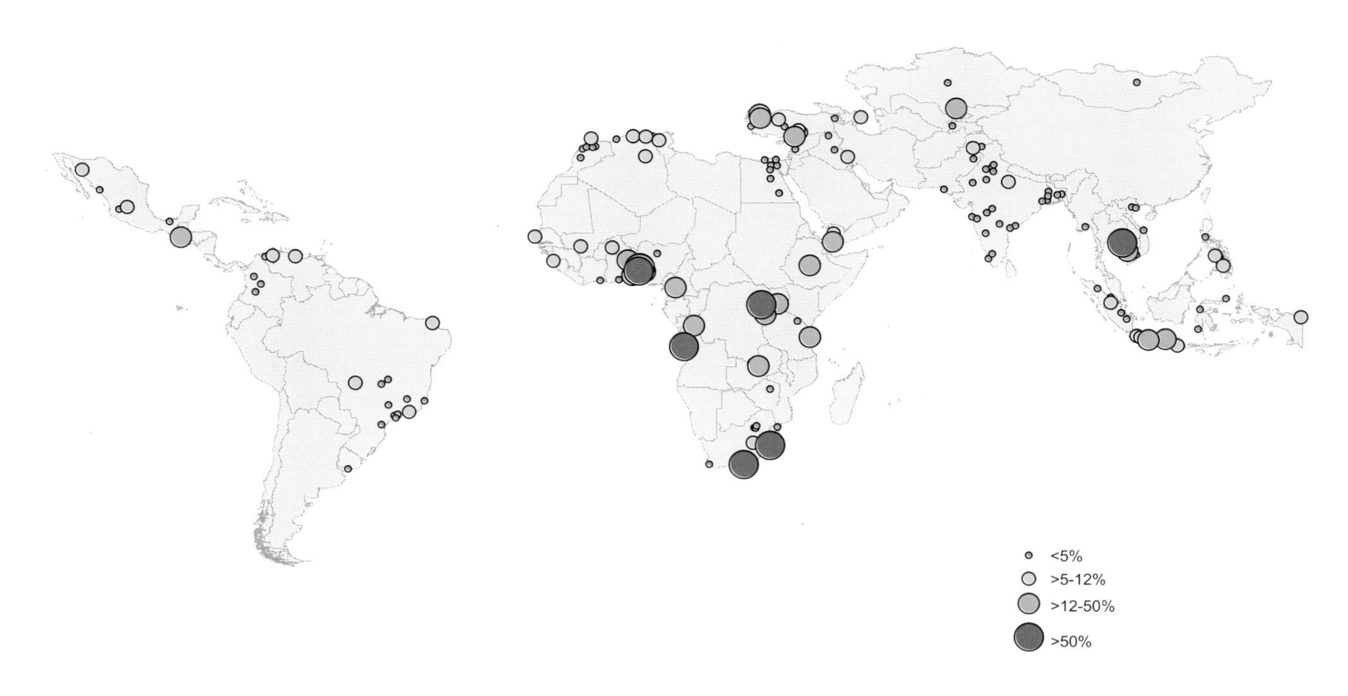

Source: UN-HABITAT 2005, Global Urban Observatory, Urban Indicators Programme, Phase III.

TABLE 2.3.1 IMPROVED DRINKING WATER COVERAGE AMONG URBAN POPULATION BY REGION, 2003

	Access to safe water source, 2003 (%)	Urban Population 2003 (thousands)	Population lacking safe water (thousands)	Distribution of urban population lacking safe water in developing world (%)
Northern Africa	94.9	77,910	3,960	2.4
Sub-Saharan Africa	82.0	251,166	45,210	27.6
Latin America and the Caribbean	95.2	417,229	20,166	12.3
Eastern Asia	92.5	564,871	42,365	25.9
Southern Asia	94.3	448,738	25,428	15.5
South-Eastern Asia	91.0	228,636	20,577	12.6
Western Asia	95.1	124,370	6,115	3.7
Total			**163,822**	

FIGURE 2.3.2 DISTRIBUTION OF URBAN POPULATION LACKING IMPROVED DRINKING WATER BY REGION, 2003

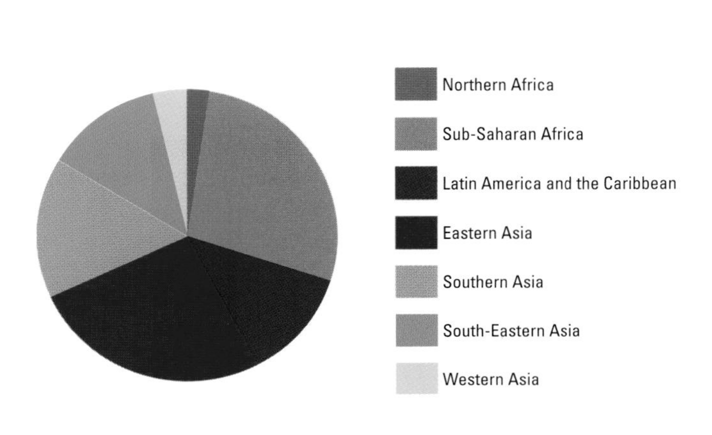

Source: UN-HABITAT 2005, Urban Indicators Programme Phase III and United Nations, World Urbanization Prospects; The 2003 revision.
Note: Access to safe water was computed from data of WHO/UNICEF Joint Monitoring Programme for Water Supply and Sanitation.

Water in Africa

Africa has the lowest safe drinking water coverage, with only 64 per cent of the total population having access to an improved water supply – that, is water that is affordable and of sufficient quantity, and available without the investment of excessive time or effort. The situation is much worse in rural areas, where coverage is only 50 per cent, than in urban areas, where coverage is about 85 per cent.[17] In global terms, 29 per cent of the world's urban population without access to improved water supply lives in Africa.

In 2003, 82 per cent of **sub-Saharan Africa's** urban population had access to an improved drinking water supply – a smaller proportion than urban dwellers in any developing region in the world. In 2002, only 45 per cent of rural sub-Saharan Africans had access to safe water. The sub-region as a whole did not experience any changes in the coverage levels during the 12 years between 1990 and 2002, remaining at 82 per cent throughout.[18] In some countries, however, urban populations are grossly underserved. In Cape Verde, Eritrea, Niger and Rwanda, only one-third of the urban population had access to an improved water source in 2001.

Lack of access to piped water affects some cities within the sub-region more than others. Some capital cities have had particularly low coverage, including Kampala, Uganda, at 15.1 per cent, and Kigali, Rwanda, at 35 per cent. Moreover, the proportion of households having access to piped water in Central and Western African capital cities is much lower than the sub-region's average, particularly in Luanda, Angola (13.1 per cent); Yaoundé, Cameroon (33.5 per cent); Ouagadougou, Burkina Faso (33.8 per cent); and Conakry, Guinea (39.2 per cent). Access to piped water in some secondary African cities is even worse: in Nazret, Ethiopia, the proportion is 16 per cent; and in Butembo, Democratic Republic of the Congo, the proportion is 14.4 per cent.[19] Lack of water source improvement is bound to continue affecting health, education, productivity, and income generation throughout the sub-region.

FIGURE 2.3.3 ACCESS TO PIPED WATER IN SELECTED CITIES IN SUB-SAHARAN AFRICA, 2003

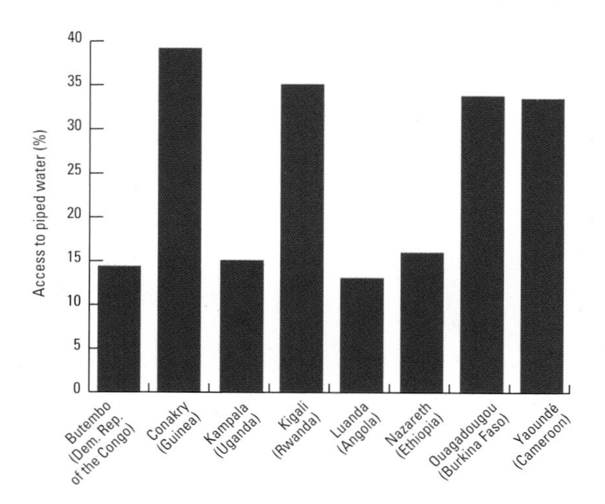

Source: UN-HABITAT, Global Urban Observatory, Urban Indicators Programme Phase III.
Note: Data based on Demographic and Health Surveys in various years.

In **Northern Africa**, access to safe water in urban areas was almost universal in 2002, at 96 per cent. In this sub-region, the urban population is expected to increase slowly and the number of people living in slums should decline. It is more likely that the reduction in the number of slum dwellers will aid the sub-region's performance on the other slum indicators, particularly overcrowding and sanitation, than on the water indicator, as other shelter deprivations are more prevalent than access to safe water. Additional government efforts are required to reach the remaining 4 per cent of the population that lacks access to an improved water source. Those who still lack safe drinking water may be more difficult to reach because their communities are physically isolated from others and their pervasive poverty limits their ability to improve their own facilities.

Data on piped water connections in urban areas reveal contrasting histories in Africa as a whole. In Northern Africa, piped water coverage increased from 83 per cent in 1990 to 96 per cent in 2003, due in large part to significant improvements in Egyptian and Moroccan cities. Despite an acute shortage of water in the sub-region, most residents of its cities and rural areas enjoy regular, affordable piped water access. The data from Northern Africa supports the theory that the world's "water crisis" is more of a political and governance crisis than a physical scarcity crisis.[20]

Water in Asia

Asia accounts for two-thirds of the world's population that lacks safe water: 670 million people in both rural and urban areas.[21] In urban areas, coverage is over 90 per cent in most sub-regions, but poor access to water facilities in urban areas is reported in various countries, particularly in Cambodia, where only 58 per cent of people living in cities have access to an improved water source and Lao People's Democratic Republic, where 66 per cent have access. The most impressive gains in access to improved water sources were made in Southern Asia's cities fuelled primarily by increased coverage in India from 88 per cent in 1990 to 96 per cent in 2002.

The proportion of the population with access to an improved water source in urban areas in Asia is very high – 93 per cent – as reported by the water utilities and ministries in charge of drinking water services. Piped water is more precious, since only 70 per cent of the Asian urban population had access to it in 2002, approximately half of whom had access to a working tap within the dwelling. The heterogeneity of the region is made clear by the contrasts observed in the level of piped water coverage in the different sub-regions. By far the lowest proportion of people having access to piped connections in urban areas is found in South-Eastern Asia, with just 45 per cent, whereas coverage in Western Asia reach-

es 79 per cent in urban areas. Impressive growth has taken place in several Indian cities, namely in Hyderabad, Amritsar, Akola and Hisar, where access to improved water sources increased from 48 per cent to 86 per cent between 1990 and 2003. The most impressive gains were recorded in the city of Hai Phong, Viet Nam, which provided improved water sources to twice as many people from 46.4 per cent of the urban population in 1990 to 99.6 percent in 2002.

■ Water in Latin America and the Caribbean

Water coverage estimates based on data from the WHO/UNICEF Joint Monitoring Programme suggest that 95 per cent of the urban population and 69 per cent of the rural population in Latin America and the Caribbean had access to improved drinking water sources in 2002. Today, around 20 million people in the region's urban areas are without access to improved water supply. While most countries in the region have over 90 per cent coverage in urban areas, some countries have relatively low coverage: Anguilla (60 per cent), Argentina (85 per cent), Belize (83 per cent), Dominican Republic (83 per cent), Ecuador (81 per cent), El Salvador (85 per cent), Jamaica (81 per cent), Haiti (49 per cent), Panama (88 per cent), Peru (88 per cent) and Venezuela (88 per cent). In Haiti less than half of the urban population has access to an improved water source.

Figures on piped water in the region are available only for major cities in a few countries: Brazil, Colombia and Guatemala. On average, urban areas in the three countries had some of the highest levels of piped water connections in the

TABLE 2.3.2 CITIES MAKING RAPID PROGRESS IN DRINKING WATER COVERAGE, 1999 -2003

Cities that increased coverage by at least 15% between 1990 and 2003

City/Country	Drinking Water Coverage (%) 1990	2003	% increase 1990-2003
Kigali, Rwanda	72.7	84.5	16
Ouagadougou, Burkina Faso	77.5	94.3	22
Fortaleza, Brazil	68.9	93.7	36
Goiânia, Brazil	85.7	98.8	15
Tijuana,Mexico	67.8	98.2	45
Agartala, India	79.2	99.6	26
Bitung, Indonesia	73.6	97.7	33
Cebu, Philippines	66.5	98.3	48
Da Nang, Viet Nam	84.7	96.6	15
Jaya Pura, Indonesia	47.1	94.1	100
Metro Manila, Philippines	83.8	96.9	16
Rajahmundry, India	83.6	99.6	19

Source: UN-HABITAT 2005, Urban Indicators Programme Phase III.

MILOON KOTHARI

WATER AND SANITATION: ONLY A HUMAN RIGHTS APPROACH WILL DO

Economic globalization policies – part of a global structural adjustment agenda that finds its most boisterous proponents among the wealthy nations of the world – have lent momentum to an ongoing movement toward privatization and commodification of basic services, such as water and sanitation. This phenomenon, now widely assumed to be irreversible, coupled with the inability of governments to provide their citizens with affordable access to such services, tends to have a disproportionately severe impact on those most vulnerable segments of the population, the poor and socially marginalized.

Water, essential to human life and all life on the planet, is part of the global commons and arguably the most quintessential of all collective resources. It is *not* a private commodity to be bought, sold or traded for profit – an exclusive luxury accessible to a few and elusive to the majority. This fundamental principle is clearly articulated in the General Comment No.15 (2002) of the UN Committee on Economic, Social and Cultural Rights, which says that:

"Water should be treated as a social and cultural good," and that "investments should not disproportionately favour expensive water supply services and facilities that are often accessible only to a small, privileged fraction of the population, rather than investing in services and facilities that benefit a far larger part of the population."

The human rights of people and communities to housing, water and sanitation – long recognized as indivisible, and guaranteed under international law – continue to be eroded as the processes of privatization become more entrenched and quicken in pace. While the promise of economic globalization to help alleviate want and reduce poverty may exist in the abstract, its basis on the Washington Consensus and reliance on a theory of presumed trickle-down benefits find little basis in history. The time has come to rethink current global economic and social policies, and the perverse and brutalized neo-liberal logic that underpins them, and reaffirm our commitment to the human rights principles and standards that offer the only real paradigm for improving the lives of millions of the poor.

The consequences of having inadequate or no access to water, while universally devastating, tend to be more acutely felt by women and children. When water is not readily available, it is principally women and children who are charged with the burdensome responsibility of its collection, often expending inordinate amounts of time and energy in the process. This has a detrimental impact on their health, security and education. While the lack of sanitation facilities affects both men and women alike, sanitation needs and demands tend to differ as a function of gender. Women have particular needs and concerns of privacy, dignity and personal safety, and the lack of sanitation facilities in the home can force women and girls to use secluded places, often at great distance from the home, thereby exposing them to heightened risk of sexual abuse. Furthermore, lack of accessible basic services can often lead to or further exacerbate tense and stressful relations within the home, increasing women's vulnerability to domestic violence.

Privatization of water and sanitation services warrants close scrutiny when assessing the impact of globalization on not only the right to adequate housing in particular, but in a broader sense on the extent of States' compliance with their legal obligations under various international and human rights treaties and guidelines. By transforming a basic social service and scarce resource into an economic commodity, the world's economic and policy planners are operating under the myopic macroeconomic assumption that existing water resources can be managed and consumed efficiently in accordance with competitive market principles. Let us not be naïve. A consideration of the three major criticisms of privatization will readily dispel any notion that the basis of such an assertion is to be found in reality: private businesses put too much emphasis on profits and cost recovery; services to vulnerable groups are inadequate and of poor quality; and private operators are not accountable to the public. What's more, the lack of capacity, or willingness, on the part of States to regulate the operations of private providers only magnifies the above outlined shortcomings of privatization.

There is an acute need to strengthen participatory monitoring mechanisms, as processes of privatization are extremely difficult to reverse once implemented, and corporations enjoy formidable legal recourse through multilateral trade agreements. Consequently, the expansion of any such agreements, such as the World Trade Organization General Agreement on Trade Services (GATS), which led to the privatization of social services and the entry of corporations into the arena of providing social goods such as water, will only serve to exacerbate an already adverse situation. The right to an effective remedy for anyone whose rights have been violated cannot be contracted away by the State nor denied by the operations of intergovernmental institutions. Investment or trade bodies should not adjudicate concerns that fall firmly within the ambit of human rights as if they were simply disputes between corporations and state actors. Any violation should, and must, be dealt with through the relevant human rights enforcement mechanisms that seek the integration of human rights obligations into national and international policy making, thereby establishing a clear and positive precedent for the future.

Of equal or greater importance is the need for sustained vigilance at each stage of this protracted campaign, to actively safeguard against the collateral erosion of other human rights during the ongoing effort to achieve the Goals. The effort at improving the living conditions of some, by way of slum upgrading projects, for example, must not lead to the breach of human rights of others, such as through forced eviction or the now rampant phenomenon of land-grabbing in all of its forms.

A human rights approach must both inform the normative discussion, as well as guide the processes surrounding efforts to achieve the Millennium Development Goals. Such a strategy, coupled with existing international human rights treaties, declarations and guidelines, provides a framework through which the formulation of responsible economic policies for the benefit of humankind can become a reality.

Miloon Kothari is the United Nations Special Rapporteur on Adequate Housing.

developing world in 2003, with the Colombian cities of Bogotá, Medellin, Neiva, and Valledupar reporting universal coverage. Slightly less than two-thirds – 71 per cent – of the households included in the data had piped connections in their homes, with the remaining third using a tap in the yard or plot.

Understanding access to safe drinking water in urban areas: The case of Addis Ababa

UN-HABITAT is working on refining the methodology used to measure access to "improved water supply". An urban inequities survey conducted in Addis Ababa, Ethiopia, showed that if the indicator includes variables for measuring the proportion of people with access to safe water that is affordable, provided in sufficient quantity and does not require excessive time or effort to acquire, the number of urban dwellers without sufficient access is much higher than officially acknowledged (see figure).

Statistics collected by the Government of Ethiopia in Addis Ababa, for instance, report that 88.5 per cent of the urban population has improved water provision;[24] however, UN-HABITAT's study, as part of its Monitoring Urban Inequities Programme (MUIP), demonstrated that the proportion of urban residents with an improved water supply can drop to 21.3 per cent if the operational definition that includes ease of access is combined with variables on sufficient quantity, affordability and time required to collect it.[25] The same study demonstrated that when data is disaggregated at intra-city levels, massive disparities are apparent. Official statistics belie the actual conditions under which people in poverty live. In Addis Ababa, the survey showed that the proportion of non-slum urban households with access to safe water was almost two times higher than the proportion of slum households with access to safe water.

FIGURE 2.3.4 ACCESS TO WATER DECREASES DRAMATICALLY WHEN COST AND QUALITY ARE CONSIDERED: THE CASE OF ADDIS ABABA

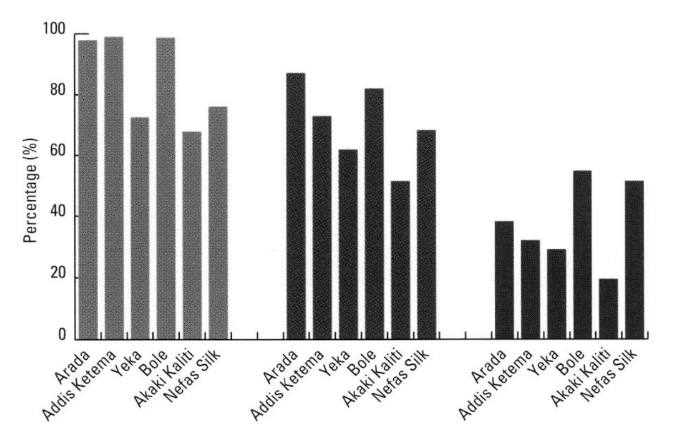

Access to improved water
Access to improved water when cost is factored in
Access to improved water when cost and quality are factored in

Source: UN-HABITAT, Addis Ababa Urban Inequity Survey 2003.

Endnotes

1 WHO/UNICEF Joint Monitoring Programme for Water Supply and Sanitation 2004.
2 Ibid.
3 UNEP 2003.
4 These countries are: Andorra, Barbados, Belize, Botswana, Chile, Costa Rica, Dominica, Egypt, Lebanon, Paraguay, Ukraine, and Zimbabwe. (WHO/UNICEF Joint Monitoring Programme for Water Supply and Sanitation 2004.)
5 These countries are Albania, Armenia, Azerbaijan, Bolivia, Colombia, Côte d'Ivoire, Dominican Republic, Gabon, Guatemala, Honduras, Kazakhstan, Malawi, Namibia, Pakistan, Sri Lanka, and Uzbekistan. (WHO/UNICEF Joint Monitoring Programme for Water Supply and Sanitation 2004).
6 UN-HABITAT 2005c.
7 The Virtual Water Forum Final Report, 3rd World Water Forum, March 2003: http://www.worldwaterforum.org.
8 Hansen & Bathia 2004.
9 UN-HABITAT 2005c.
10 Rakodi, et al. 2000.
11 Civil Society submission to the Government of Bangladesh for the development of the "*National Strategy for Economic Growth, Poverty Reduction and Social Development*", Poverty Reduction Strategy Paper. Bangladesh, 2003.
12 World Health Organization 1999.
13 Bartram, et al. 2005.
14 The remaining percentage, 18.8 per cent, could not be described with a single variable, owing to differences in definitions and methods of computation. UN-HABITAT Urban Indicators Programme Phase III, 2005.
15 UNFPA 2003.
16 UN-HABITAT 2005c.
17 WHO/UNICEF Joint Monitoring Programme for Water Supply and Sanitation 2000. The percentage reported by on water supply coverage in urban areas is 85 per cent, while according to UN-HABITAT's urban indicators, this percentage is 89 per cent.
18 UN-HABITAT 2005c.
19 Secretariat of the Third World Water Forum, 2003.
20 Hansen & Bathia 2004.
21 WHO/UNICEF Joint Monitoring Programme for Water Supply and Sanitation 2000.
22 UN-HABITAT 2005c.
23 WHO/UNICEF Joint Monitoring Programme for Water Supply and Sanitation 2004.
24 Addis Ababa City Council 2004.
25 UN-HABITAT 2004b.

2.4 The Silent Tsunami: The High Price of Inadequate Sanitation in Urban Areas

Ecuador T.NEBBIA/UNEP/STILL PICTURES

Since 350 B.C., when the Greek philosopher Aristotle posited a distinction between the public sphere of political activity and the private sphere associated with family and domestic life, debates about the right to privacy have dominated popular discourse. In today's world, privacy is increasingly determined by individuals' power and social status: the rich can withdraw from society whenever they wish, but those living in poverty cannot so easily escape their neighbours' gaze. This is particularly true with regard to sanitation. People living in poverty are subject to intrusion and observation around their most private affairs and habits; they are much less likely than their wealthier neighbours to have access to safe sanitation facilities behind closed doors. The absence of decent toilets in impoverished neighbourhoods violates residents' right to privacy and is an affront to their dignity. Being deprived of adequate sanitation facilities is the most direct and most dehumanizing – but least often acknowledged – consequence of poverty.

Improved sanitation: A basic principle

A household is considered to have access to "improved" sanitation if it has a human excreta disposal system, either in the form of a private toilet or a public toilet shared by a maximum of two households. In urban areas, access to improved sanitation is defined by direct connection to a public, piped sewer; direct connection to a septic system; or access to pour-flush latrines or ventilated improved pit latrines, allowing for acceptable local technologies.

Sanitation is not simply about disposal of human waste. It is a broad concept that encompasses the safe removal, disposal and management of solid household waste, wastewater, industrial waste and the like. "Improved" sanitation, however, refers to a *basic sanitation approach* that focuses on securing sustainable access to safe, hygienic and convenient facilities for human excreta disposal. The first, and least expensive, step toward a more broadly integrated sanitation system in any community, improved sanitation includes the need for privacy, safety, hygiene and convenience.

Linking inadequate sanitation and health

More than a quarter of the developing world's urban population lacks adequate sanitation. Globally, an estimated 2.6 billion people lack toilets and other forms of improved sanitation. UN-HABITAT analyses reveal that while the world's cities have made significant progress in improving people's access to water, access to improved sanitation lags far behind, particularly in sub-Saharan Africa, Southern Asia and Eastern Asia, where the proportion of the urban population having access in 2003 was only 55 per cent, 67 per cent and 69 per cent respectively.

Lack of access to an adequate toilet not only violates the dignity of the urban poor, but also affects their health. Urban poverty is often related to poor hygiene – the result of inadequate sanitation facilities combined with an inadequate or unsafe water supply. Every year, hundreds of thousands of people die as a result of living conditions made unhealthy by lack of clean water and sanitation options. The number of deaths attributable to poor sanitation and hygiene alone may be as high as 1.6 million per year – five times as many people who died in the 2004 Indian Ocean tsunami. Inadequate sanitation is therefore something of a "silent tsunami" causing waves of illness and death, especially among children. Although Millennium Development Goal 7, target 10, aims to halve the proportion of people without access to safe drinking water and basic sanitation by 2020, the relationship between water and sanitation is very often ignored when allocating resources; for every dollar invested in water supply, only 20 cents goes toward the provision of basic sanitation.[1] This explains the huge gap between water and sanitation coverage in the world: 83 per cent of the world's residents have access to safe drinking water, but only 58 per cent have access to improved sanitation.[2]

Research has made it clear that those without access to adequate sanitation are more exposed to diseases than other groups experiencing lack of safe water and other shelter deprivations. They are 1.6 times more likely to experience diarrhoea, and they have consistently higher rates of morbidity and mortality.[3] This has been amply demonstrated by several studies, including a 1996 study by the World Health Organization (WHO) in Asia, which concluded that sanitation and hygiene are among the most influential factors in reducing diarrhoeal diseases, particularly among the sick and children.[4] In 77 per cent of the 144 cases in the WHO study, positive health benefits in the form of reduced incidence of diarrhoea could be definitively attributed to improvements in sanitation systems, whereas only 48 per cent could be attributed to improvements in the water supply. Another multi-country study confirmed that it is possible to obtain a reduction of up to 37 per cent in cases of diarrhoea when access to improved sanitation facilities is provided to unserved populations.[5]

Lack of sanitation is not always a top priority among people living in poverty, whose needs for drinking water and sufficient food often take precedence. Poor sanitation is perceived as "tomorrow's priority" even if it has life-threatening consequences. Thus, slum dwellers in some African cities resort to disposal of excreta in plastic bags, which then get discarded carelessly in drainage channels, rubbish bins or in the streets of the neighbourhood itself. These so-called "flying" toilets in Nairobi or "mobile" toilets in Lusaka have negative health consequences that are not often apparent to the inhabitants. Low priority accorded to sanitation is also manifested by the fact that a limited number of households having latrines make appropriate use of them, and an even smaller number maintain them properly. For instance, a study conducted in Zambia, Zimbabwe and South Africa showed that only 17 per cent of the population maintain their latrines properly.[6] There is clearly a strong need to link sanitation and hygiene to education, awareness-raising and cultural attitudes toward waste disposal.

Attributing unhygienic practices only to lack of concern about inadequate sanitation, however, can misguide analysis and conclusions. Poverty and deprivation also play key roles. For instance, residents of the neighbourhood of Mbare in Harare, Zimbabwe, prefer to defecate wherever possible in the open because their pour-flush toilets are overused and poorly maintained. In that community, up to 1,300 people share one communal toilet with only six squatting holes, most of which are no longer flushable.[7] In Nepal, two-thirds (67 per cent) of the country's population defecates in the open despite the fact that half of them have access to latrines; they consider the existing latrines unsanitary and unsafe.[8] In many countries, children are afraid or reluctant to use latrines because they are perceived

Defining sanitation: To estimate or to underestimate, that is the question

Getting a clear picture of the global sanitation situation is complicated by the variety of terminologies used for data collection. An individual or a household can have access to "adequate" or "improved" sanitation depending on the definition used. Studies indicate that between 850 and 1.13 billion people worldwide lacked "adequate" sanitation, whereas about 400 million people lacked "improved" sanitation in 2004. These figures refer to the sanitation *hardware*, or technologies, used, including sewerage systems, toilets and hygienic latrines. More accurate estimations should also consider the *software*, or the hygiene conditions, as well, such as maintenance of latrines and provision of hand-washing facilities.

Monitoring adequate or improved sanitation is further complicated by the fact that "access" assumes the *use* of the facilities; yet, in many locations, facilities are not being used. A family may have *access* as defined for purposes of the indicator, but may fail to use the facility for practical, cultural, or social reasons. Defining "use" of the facilities, therefore, can be nuanced to "convenient use". In Nepal, for instance, the government considers a "functioning latrine" as one that is kept free of faeces and can be used by all family members, including children over the age of 5 years.

Differences in definitions and data reporting methods make direct comparisons difficult. Some countries have higher standards for defining adequate sanitation services, excluding ordinary pit latrines, or counting only ventilated improved pit latrines or flush toilets connected to a septic tank or sewerage system. In Uganda, for example, pit latrines are counted as sanitary, and the latest Demographic and Health Survey shows 80 per cent of households with access. But, if pit latrines are not counted, the population with access shrinks to a mere 3 per cent. These variations in standards make the quality of the data questionable in some countries. For instance, in Zambia, one of the least developed countries, more than 87 per cent of the population has access to adequate sanitation, while Brazil, a middle-income country that is far wealthier and much more developed than Zambia, reports access of only 83 per cent. Furthermore, data from surveys and routine reports can also have significant discrepancies, because some government reports rely on outdated data that fails to take into account new informal settlements and sanitation facilities that fall into disrepair. This explains why some of the data from individual countries shows rapid and implausible changes in coverage from one assessment to the next.

UN-HABITAT is working with the WHO/UNICEF Joint Monitoring Programme to standardize definitions to allow for more accurate global comparisons. Based on UN-HABITAT suggestions that most of the simple pit latrines and traditional latrines in use in cities are in fact unsanitary, the data has been revised for 2004. Figures for people without adequate sanitation have thus changed from 2.4 to 2.6 million worldwide. The WHO/UNICEF Joint Monitoring Programme agreed to reduce by half the share of the population with adequate access who use traditional, pit or simple latrines. For instance, the Government of Burkina Faso reported that 54 per cent of the urban population had access to improved sanitation. However, out of this percentage 84 per cent, had access to pit latrines, only half of which (42 per cent) were counted as improved sanitation. These changes were made in order to facilitate inter-regional comparisons across time.

Sources: Hansen & Bathia 2004; UK Parliamentary Office of Science and Technology; Bendohmme & Swindde 1999; UNICEF 1997; UN-HABITAT 2005c.

as dark, dirty, unsafe, or smelly. In many places, latrines are not available at all. Today, one out of every three children in the developing world do not have access to a toilet of any kind in the vicinity of their dwellings.[9]

Sanitation, hygiene and health are interconnected, linked by three main factors in low-income communities: lack of access to a safe, decent toilet; lack of awareness about the connections between defecating in the open and contamination of food and water sources; and cultural indifference to using the public environment for the disposal of human waste. The combined issues of extreme deprivation, lack of education and poor quality and maintenance of sanitation facilities are compounded by the fact that interventions, if any, are typically neither sustained nor systematic, owing to inadequate economic and financial sector policies, poor urban management, and lack of political will to respond to the needs of people living in poverty. Non-recognition of informal settlements by the authorities in charge of provision is also an issue in some cities.

Global trends in sanitation provision

In 2002, nearly half of the population of the developing world – about 2.5 billion people – did not have access to adequate sanitation. Of those lacking adequate sanitation, 76 per cent (1.98 billion) lived in Asia; 18 per cent (470 million) lived in Africa; and 5 per cent (130 million) lived in Latin America and the Caribbean (see box on definition).[10]

As with most other shelter provisions, sanitation coverage is significantly higher in urban areas than in rural areas. Data collected by the WHO/UNICEF Joint Monitoring Programme for Water Supply and Sanitation indicates that in 2002, 81 per cent of the world's urban residents had access to improved sanitation, compared to 37 per cent in rural areas. As expected, these values are lower for the developing regions, where the proportion of the population with adequate sanitation in urban areas is 73 per cent, and in rural areas is 31 per cent.[11] Estimations of the deficit depend heavily on the data sources and definitions used.

While 73 per cent of the urban population in the developing world has access to adequate sanitation facilities, more than **560 million city dwellers** are still deprived of a basic, decent toilet facility. The region with the lowest coverage of improved sanitation in urban areas is Africa (63 per cent), with a sub-regional variance of 55 per cent coverage in sub-Saharan Africa at one end, and 89 per cent in Northern Africa at the other end. At the continental level, Africa accounts for around one-fifth (21 per cent) of the world's population lacking improved sanitation. Asia has the second-lowest coverage in urban areas in the world (66 per cent). Because of the population sizes of China and India, along with other large nations in the region, Asia also

Manila HARTMUT SCHWARZBACH/UNEP/STILL PICTURES

accounts for the largest numbers of people without sanitation nearly 400 million people. Afghanistan has by far the lowest proportion of the urban population with access to improved sanitation (16 per cent) in the region.[12]

Latin America and the Caribbean has a relatively high sanitation coverage of 84 per cent; the region is performing better on sanitation than Asia and Africa, and it hosts only 12 per cent of the world's population without access to improved sanitation. More than 75 per cent of the urban populations in virtually all countries in the region have improved sanitation, with the exception of Haiti and Belize – two countries in which less than 50 per cent of the urban population has improved sanitation facilities.

In contrast, people in the **developed world** enjoy more or less universal provision of advanced sanitation facilities. Virtually all households have access to improved sanitation, as more than 98 per cent of homes are connected to piped, municipal sewage treatment systems. Those not connected to municipal sewage systems use septic tanks and similar solutions that are regulated, inspected during construction and regularly tested for performance. In Japan, for instance, 100 per cent of the population has improved sanitation: 81 per cent of the population is connected to municipal sewerage systems, and 19 per cent have household septic tanks (*johkasoh*) or use other improved facilities in agricultural communities.[13]

North America has the highest reported coverage for any world region as a whole: 100 per cent. Reporting mechanisms for Europe are extremely poor – contrary to what might be assumed – so available data for that region is not statistically representative.[14] Available data indicates that 99 per cent of the European population has access to improved sanitation, but in many of the region's new economies, the infrastructure for sanitation still needs to be developed or improved. Consequently, more than 80 million people in the region, or 10 per cent of the total population, do not have improved sanitation.[15]

Sanitation, hygiene and health are interconnected.

Urban sanitation data for regions of the developing world

Based on the definition of improved sanitation, and taking into consideration the limitations of measurement, it is possible to claim a modest improvement in the proportion of the world's urban residents with access to sanitation: from 68 per cent in 1990 to 73 per cent in 2003.[16] A significant proportion of the urban population in the developing world consequently does not use or have access to any type of sanitation facility; an estimated 10 per cent of the world's urban population, or 300 million people, defecate in the open or use unsanitary bucket latrines.[17]

The problem with these estimates is that different countries use different types and categories of latrines, making it difficult to code them for monitoring purposes. Apparently, not all of the reported facilities are improved; where they are, a large proportion of them are overcrowded, unsafe, lack provisions for cleaning hands and body, and are poorly maintained. In many urban settings, especially in densely populated areas, pit latrines do not significantly reduce the risk of faecal-oral diseases because of their unhealthy conditions. Studies have shown that latrines alone do not have a clear health impact unless the behavioural patterns associated with sound hygiene practices are also ensured. In Pakistan, for instance, the fact that only a negligible difference exists in the frequency of diarrhoeal episodes between households having latrines and those without latrines indicates that hygiene is inadequate.[18]

Sanitation in African cities

In **Africa**, the increase in the proportion of people with improved sanitation varies greatly, depending on the sub-region. A positive trend has been observed in Northern Africa, whereas signs of stagnation are clear in sub-Saharan Africa. In the former sub-region, the population enjoying improved sanitation services grew from 84 per cent in 1990 to 89 per cent in 2002, due primarily to an increase registered in the Egyptian cities of Cairo, Alexandria, Port Said and Beni Suef, and the Moroccan cities of Casablanca and Rabat. Yet, small cities did not experience the same growth in provision; some still have access comparable to cities in sub-Saharan Africa, where performance on sanitation indicators is as poor as on other development indicators. Centralized sewerage systems are the most appropriate solution, in terms of expense, infrastructure and health returns. In Northern Africa, increased sewerage provision is the most common solution in large, primary cities – national capitals and economic centres. In the sub-region's secondary cities, however, piped sewerage technology is not as common; in Assyut, Egypt, fewer than 20 per cent of the households are connected to piped, waterborne sewer systems.

In 2003, slightly more than half of the urban population in sub-Saharan Africa enjoyed adequate sanitation facilities (55 per cent) – roughly the same proportion that was recorded in 1990. The rate of urbanization in this sub-region has been the highest in the world: it experienced an 80 per cent increase in the number of urban residents between 1990 and 2003. Subsequently, the number of people lacking improved sanitation has drastically increased – from 77 million in 1990, to 132 million in 2001, to 160 million in 2005.[19] If no remedial action is taken, poor sanitation will continue to have a significant – and dismal – impact on the lives of people living in poverty in the cities of sub-Saharan Africa.

Evidence of success exists in a few cities that have managed to expand coverage between 1990 and 2003 significantly, including Kigali, Rwanda, and Ibadan, Nigeria which increased coverage from 47.8 per cent and 26.8 per cent in 1990 to 79.4 per cent and 67.3 per cent in 2003, respectively. Some countries, such as South Africa and Zimbabwe also have extensive sewerage networks. This example gives hope and direction to the achievement of sanitation target of the Millennium Development Goals. In contrast, sanitation facilities in some cities such as Addis Ababa in Ethiopia and Porto Novo in Benin have less than half of their population served with improved sanitation.

TABLE 2.4.1 IMPROVED SANITATION COVERAGE AMONG URBAN POPULATION BY REGION, 2003

	Access to improved sanitation, 2003 (%)	Urban Population 2003 (thousands)	Population lacking improved sanitation (thousands)	Distribution of urban population lacking improved sanitation in developing world (%)
Northern Africa	89.4	77,910	8,245	1.5
Sub-Saharan Africa	55.1	251,166	112,815	20.1
Latin America and the Caribbean	84.2	417,229	66,061	11.8
Eastern Asia	69.4	564,871	172,756	30.8
Southern Asia	67.0	448,738	148,084	26.4
South-Eastern Asia	80.0	228,636	45,727	8.2
Western Asia	94.9	124,370	6,322	1.1
Total			**560,011**	

FIGURE 2.4.1 DISTRIBUTION OF URBAN POPULATION LACKING IMPROVED SANITATION BY REGION, 2003

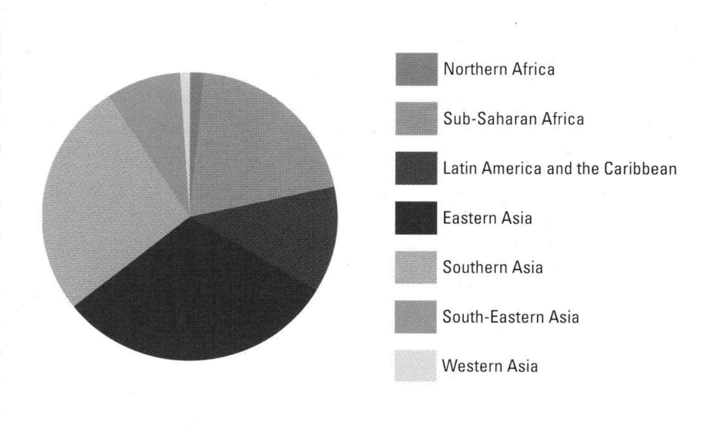

Source: UN-HABITAT (2006), Urban Indicators Programme Phase III and United Nations, World Urbanization Prospects; The 2003 revision.
Note: Access to safe water was computed from data of WHO/UNICEF Joint Monitoring Programme for Water Supply and Sanitation.

MAP 10 PROPORTION OF HOUSEHOLDS WITHOUT ACCESS TO IMPROVED SANITATION IN SELECTED CITIES, 2003

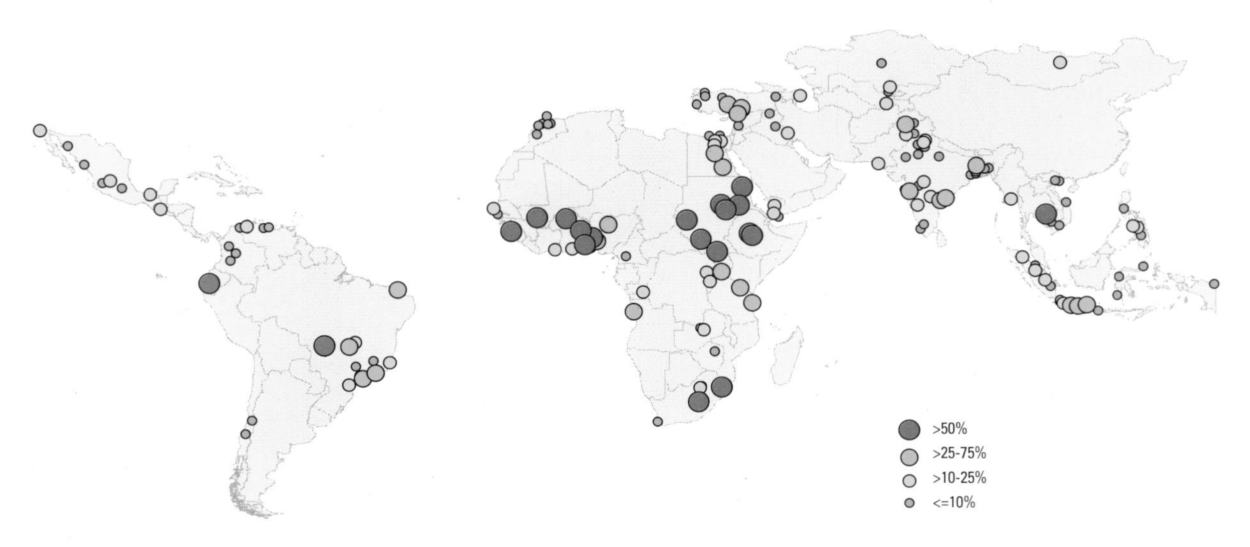

Source: UN-HABITAT 2005, Global Urban Observatory, Urban Indicators Programme, Phase III.

Sanitation in Asian cities

Time series analysis based on UN-HABITAT urban indicators shows that **Asia** has made major progress on the provision of improved sanitation in cities. The region is diverse and heterogeneous, however, with both advanced and poor economies.

South-Eastern Asia experienced the highest growth in improved sanitation coverage since 1990, increasing from 67 per cent that year to 79 per cent in 2002, particularly in mid-sized cities such as Bogor and Kediri in Indonesia, Cagayan de Oro in Philippines, and Hai Phong in Viet Nam. About 29 million people still lacked access to improved sanitation in 2005; by 2020, the deficit is expected to decrease to 25 million.

Guanabara Bay, Rio de Janeiro ©ELDER SALLES. IMAGE FROM BIGSTOCKPHOTO.COM

Southern Asia's coverage is also among the highest, with an increase of 12 percentage points, although it started with the lowest baseline in Asia – 54 per cent in 1990. Several Indian cities – namely Akola, Kanpur and Kharagpur – made rapid progress, increasing coverage by at least 25 per cent between 1990 and 2002. However, Southern Asia also has countries with low proportions of urban dwellers with access to improved sanitation: Nepal (68 per cent); and Afghanistan (16 per cent).[20]

Growth in access to improved sanitation in Eastern Asia was rather moderate, increasing from 64 per cent in 1990 to 69 per cent in 2003, owing in large part to the increase in China and Mongolia's largest cities.[21] Deficits in sanitation facilities in the two countries remain high: in China, 33 per cent of the urban population still lacks improved sanitation, as does 54 per cent of the urban population of Mongolia.

In Western Asia, coverage was quasi-universal in 1990, at 96 per cent, but since then, it has been difficult for countries to reach the poorest of the poor. In some of the middle-income countries of the sub-region, a great deal of the infrastructure is in place, but much of it is in poor condition and does not function reliably.[22]

The sanitary situation of the countries of the **Commonwealth of Independent States** in Asia is not well documented. According to data from the WHO/UNICEF Joint Monitoring Programme, there are significant disparities in access to improved sanitation between urban and rural areas: 81 per cent and 45 per cent, respectively. Multiple Indicator Cluster Surveys conducted by UNICEF in the region reveal that the only available government statistics are limited to the proportion of the population served by centralized sewerage systems.[23] Since such systems are generally limited to central districts of the region's large cities, smaller cities are not normally covered, and only 10 per cent of households in rural areas are covered.[24] Almost all the population not served by centralized sewerage systems uses individual sanitation facilities, such as pit latrines and pumped latrines. Observations made in various parts of the region suggest that most pit latrines, for individual household use and in public institutions such as schools and hospitals, are in poor condition, especially in economically disadvantaged areas.

Sanitation in Latin America and the Caribbean

Latin America and the Caribbean has relatively high sanitation coverage but with vast internal variations between rural and urban environments, and among individual countries and cities. According to data compiled by the WHO/UNICEF Joint Monitoring Programme, the proportion of the rural population lacking improved sanitation was almost half of the urban proportion: 44 per cent versus 84 per cent, respectively. In various countries of the Caribbean, improved sanitation coverage is universal, particularly in Antigua, Bahamas, Barbados, Cuba, Surinam, and Trinidad and Tobago.[25] UN-HABITAT urban indicators data reveals high coverage in various cities, namely in Bogotá and Medellin in Colombia; Caracas in Venezuela; Chillan in Chile; and Guadalajara in Mexico.

FIGURE 2.4.3 THE UNEQUAL DISTRIBUTION OF IMPROVED SANITATION IN ROSARIO, ARGENTINA

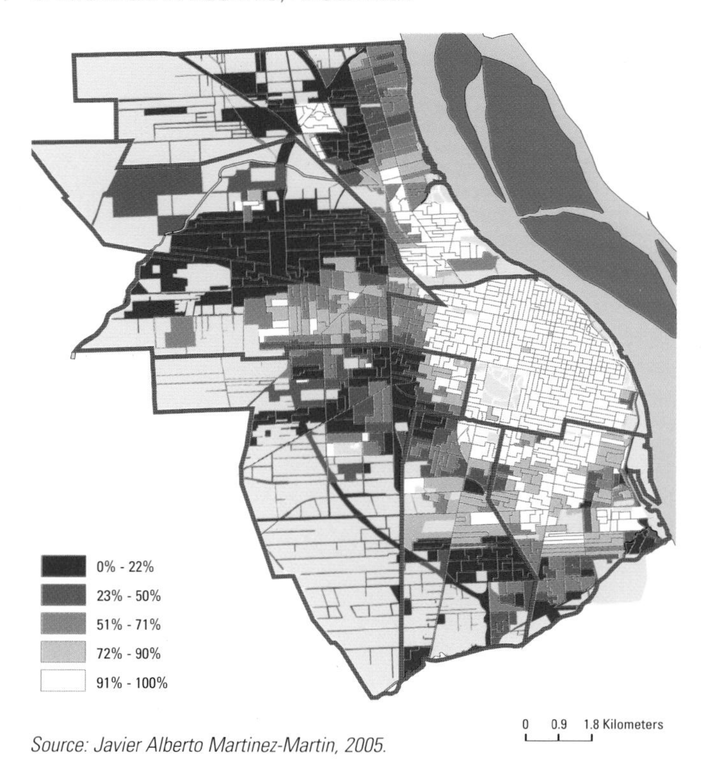

Source: Javier Alberto Martinez-Martin, 2005.

FIGURE 2.4.2 PROPORTION OF URBAN HOUSEHOLDS WITH ACCESS TO IMPROVED SANITATION

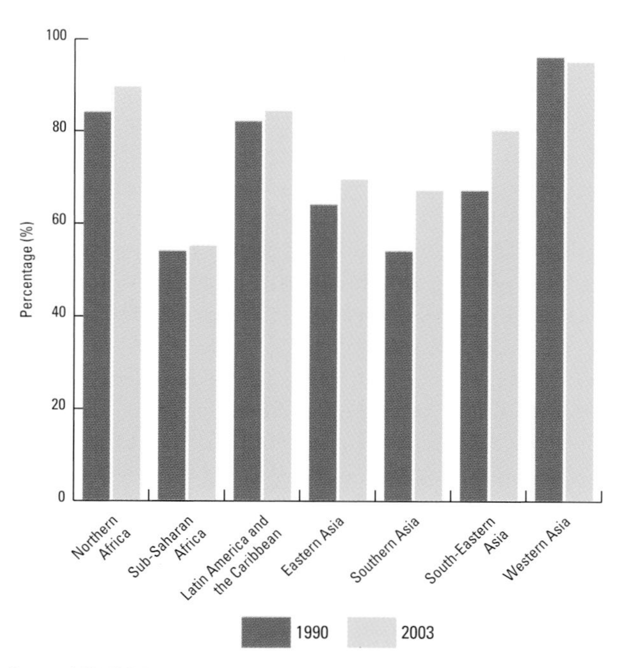

Source: UN-HABITAT Urban Indicators Programme Phase III.
Note: Data for 1990 not available for some regions.

FIGURE 2.4.3 CITIES WITH LOW SANITATION COVERAGE: PROPORTION OF URBAN POPULATION WITH ACCESS TO IMPROVED SANITATION, 2003

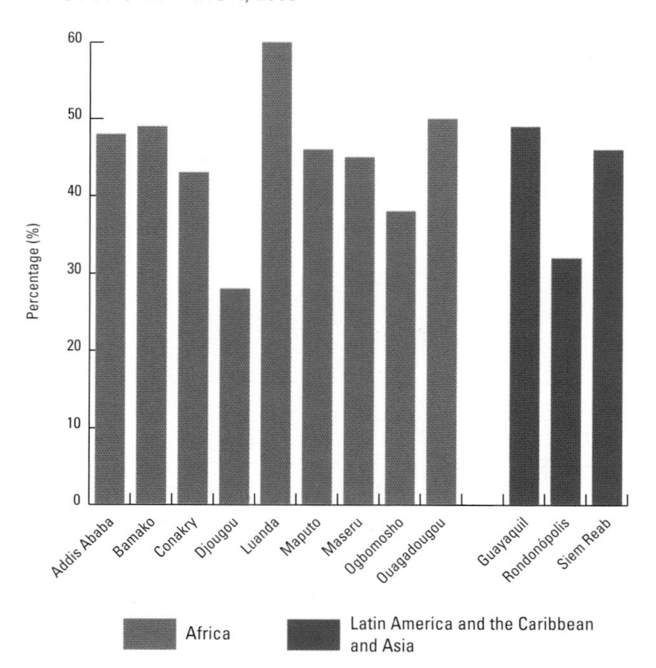

Source: UN-HABITAT 2005, Urban Indicators Programme Phase III.

Despite improvements in the overall quality of life in some developing countries, many cities, such as Rondonopolis in Brazil and Guayaquil in Ecuador, have been largely unsuccessful in creating reliable and adequate sanitation services. Studies suggest that in most smaller urban centres, the proportion without adequate sanitation provision is even higher; most urban centers in low- and middle-income nations have no sewers at all and have little or no other public support for good-quality sanitation.[26] However, significant improvements have been recorded in cities such as Guatemala City, from 33 to 85 per cent between 1990 and 2003. The city of Fortaleza in Brazil offers a valuable lesson in how development of sanitation infrastructure can have a positive health outcome. The city experienced significant reduction in infant mortality rates from 74 per 1,000 births to 28 per 1,000 births in 2001 - the same period in which sanitation coverage increased from one-third to more than half the urban population.

Endnotes

1. From 1990 to 2000, sanitation received only 20 per cent of the $16 billion invested in water by national governments and external support agencies. (United Kingdom Parliamentary Office of Science and Technology 2002.)
2. WHO/UNICEF 2004.
3. Evans 2005.
4. South-East Asia Regional Office of WHO, quoted in Evans 2005.
5. Esray 1996.
6. Manase, et al. 2001.
7. Ibid.
8. Conference Proceedings for the South Asian Conference on Sanitation, Dhaka, Bangladesh, 2004.
9. UNICEF 2004.
10. WHO/UNICEF 2000 assessment.
11. WHO/UNICEF 2004.
12. WHO/UNICEF 2000 assessment.
13. Report on Urban Indicators submitted by the Japanese Government to UN-HABITAT, May 2005. Refer as well to the Report on Urban Indicators prepared by the Canada Mortgage and Housing Corporation 2005.
14. In the WHO/UNICEF 2000 assessment, the coverage data represented 15 per cent for 1990 and 44 per cent of the region's total population. For the UN-HABITAT urban indicators, no reporting mechanism was developed, with the exception of that by the Dutch Government.
15. WHO 2005.
16. Data is based on Urban Indicators Programme, Phase III. The percentage in 1990 is the average of 29 cities in Africa, 54 cities in Asia and 25 cities in Latin America and the Caribbean. The percentage in 2003 is the average of 55 cities in Africa, 78 cities in Asia and 31cities in Latin America and the Caribbean.
17. As mentioned earlier, a large proportion of people with latrines prefer to defecate in the open. A study conducted by the World Bank in Mumbai concludes that toilets for slum dwellers, even when available are in such poor condition that people refuse to use them. (Nitti & Sarkar 2003).
18. Conference Proceedings, South Asian Conference in Sanitation, Pakistan, 2004.
19. Estimations based on the UN-HABITAT Urban Indicators Data 2005 and the United Nations Urbanization Prospects 2004.
20. UN-HABITAT 2005c.
21. WHO/UNICEF 2004.
22. Ibid.
23. According to UNICEF, 54 per cent of the population in five of the Commonwealth of Independent States countries has access to centralized sewerage systems.
24. Cherp 1999.
25. WHO/UNICEF 2004.
26. Satterthwaite, et al. 2005.

Sanitation: A women's issue

No issue touches the lives of women – particularly poor urban women – as intimately as that of access to sanitation. In low-income settlements where there are no individual toilets, women have to queue for long periods to gain access to public toilets; some have to bear the indignity of having to defecate in the open, which exposes them to the possibility of sexual harassment or assault.

Communal water tap in Soweto, Johannesburg RASNA WARAH

Although men also suffer from the burden of poor sanitation, they are more likely to resort to other means to relieve themselves. In many slums, men urinate and defecate along railway tracks and in open spaces. But women – whose anatomy, modesty and susceptibility to attack does not allow them to discreetly relieve themselves in public – have no choice but to wait until dark, usually early in the morning when there is less risk of being accosted. "Going to the toilet" for these women often means squatting in a private spot or waking up before dawn to queue at public toilets.

One woman interviewed in a Mumbai slum explained what it means to have no toilet: "We use the toilet outside our settlement, five minutes away. We have to stand in a queue for half an hour. That is why the men all go under the bridge and only the women use the toilets. Children also go out in the open."

A disproportionate share of the labour and health burden of inadequate sanitation falls on women. For women living in slums, a long wait at the public toilet can mean that children are left unattended, or that a household chore is delayed. Unhygienic public toilets and latrines threaten the health of women, who are prone to reproductive tract infections caused by poor sanitation. For women who are menstruating, the need for adequate sanitation becomes even more acute. Moreover, because it is generally women who are responsible for the disposal of human waste when provision of sanitation is inadequate, they are more susceptible to diseases associated with contact with human excreta.

Despite all this, the sanitation crisis affecting women has not been given a high priority on the agendas of human rights and women's organizations. United Nations and other international bodies tend to confine women's issues to reproductive health and education. Few, if any, governments focus on the impact of inadequate sanitation on women. This could also be partially explained by the fact that improving access to sanitation was only recently recognized as a pressing internationally agreed target – in 2002 at the World Summit on Sustainable Development – so the issue has not been on the public agenda for long. Although women's lack of access to water in both rural and urban areas and its health implications – including severe back pain caused by carrying heavy vessels of water over long distances – has been the subject of several studies, women's lack of access to sanitation has not received the same attention. Preliminary UN-HABITAT analyses indicate the need for further study of the issue, as they show that lack of sanitation in slums increases health risks among all slum residents, women and children in particular.

Because rural women – no matter how poor – do not have to face the same dilemma as their urban counterparts when it comes to sanitation, poverty reduction efforts, which are currently focused on rural areas, particularly in sub-Saharan Africa and Asia, do not factor in women's access to sanitation in urban areas. Most rural households have access to at least one toilet – even if it is a crude pit latrine – which means that women in rural areas rarely queue to go to the toilet and are less likely to share toilets with dozens of other people. They are also more likely to keep the toilets clean, as their family's health often depends on it.

Sources: Warah 2005a; UN-HABITAT 2003a; Hardoy, et al. 2001; Mukherjee 2001.

LENA SOMMESTAD

STRONGER ACTION IS NEEDED TO ACHIEVE THE SANITATION TARGET

Kibera, Nairobi THIERRY GEENEN, FOR NAIROBI RIVER BASIN PROJECT, UNEP

Today millions of poor people suffer from diseases and parasites because sanitation has not received enough political attention. Young children die of diarrhoea and women are denied security, privacy and dignity. Approximately 2.6 billion people in the world are lacking sanitation services.

A sanitation target was set at the World Summit on Sustainable Development, held in Johannesburg in 2002. Sanitation was also discussed during the 2005 UN-HABITAT Governing Council and during the 30th session of the Commission on Sustainable Development in New York in April 2005. I am pleased that sanitation has finally attracted the attention of policymakers worldwide. However, to achieve the sanitation target, we need stronger action and more innovative approaches. In particular, women's role and interests must be fully taken into account.

The present growth of the world population is almost entirely in the urban centres of the developing world. Solutions for improved urban sanitation therefore require urgent innovation in urban areas. However, sanitation infrastructure planning rarely takes into account a gender perspective. Women and children are hardly ever considered priority target groups for sanitation provision. This needs to be improved. The social context of sanitation and the needs of families must be built into solutions if they are to be sustainable.

Women, along with children, are the ones who suffer most from lack of water and sanitation. Girls face lack of sanitation in schools. This is a significant barrier to education that must be identified and removed. Women normally have the main responsibility for work tasks associated with health and sanitation, such as water collection, washing and cleaning of the house and the latrines. Women, children and elderly people are disproportionately affected by illness caused by polluted water.

The challenge surrounding equity and access to sanitation is closely linked to the fact that sanitation systems are seldom planned using the elements of sustainability and ecological principles. This is why sustainable sanitation approaches have been developed.

The essential features of sustainable sanitation are proper containment allowing for sanitation and recycling, closing of the nutrient and water loops, protection of downstream health and environment, local management and financing, affordability, and equitable services for rich and poor, women and men, old and young. Sustainable sanitation is pursued in very close cooperation with the people who will use these systems. Advanced solutions are coupled with local knowledge.

The Government of Sweden contributes to the development of an ecological sanitation approach. We are working through SIDA – the Swedish International Development Cooperation Agency – to support the development of ecological sanitation in several developing countries. The Swedish support aims to create a global confidence in ecological sanitation as a reliable, cost-effective and sustainable alternative to conventional systems.

For decades, sanitation has been neglected in many parts of the world. This has had severe consequences, in particular for the poor, and in particular for women. Now, we have a great challenge ahead. Sanitation should be a human right. Sustainable solutions are key. By empowering women and children, we can make a difference.

Lena Sommestad is the Swedish Minister for Sustainable Development.

2.5 Owners without Titles: Security of Tenure in Cities of the Developing World

Evictions: The most severe consequence of insecure tenure

Mass evictions of slum dwellers in various parts of the developing world in recent years have raised fears that security of tenure and housing rights are becoming increasingly precarious in the world's cities. A global survey[1] in 60 countries found that 6.7 million people had been forcibly evicted from their homes between 2000 and 2002, compared with 4.2 million people between 1998 and 2000.[2] Some experts have described the unprecedented rise in the number of evictions in the last five years as a global "epidemic".[3]

Although forced evictions are an extreme consequence of insecure tenure, their increasing prevalence in recent years point to trends that suggest that attitudes of local and national governments towards the urban poor are becoming increasingly intolerant. This can be attributed to to a variety of factors, including globalization, which is putting pressure on national and local governments to "beautify" or "clean up" their cities in order to become more competitive in a global economy that has seen the gap between the rich and poor widen and dramatically increased the price of urban land, pushing lower-income groups to the edge of cities to unplanned and poorly serviced areas.

Evictions are particularly prevalent in sub-Saharan Africa and Asia; most are carried out to make room for large-scale development projects and infrastructure, such as dams or roads, or to accommodate city "beautification" programmes. Cities that have experienced mass forced evictions in recent years include Beijing, Lagos, Abuja and Nairobi. Even when evictions are "justified" – as when they are carried out in the public interest, to build roads or other infrastructure necessary for urban development or when they are carried out in order to "protect" slum dwellers from hazards – they not do not take place in conformity with the rules of international law.[4] Most evictions are carried out without legal notice and without following due process. Evicted people not only lose their homes (in which they have invested a considerable portion of their savings), they are often forced to relinquish their personal belongings as well.

It is not uncommon for evicted families to sleep out in the open around the demolished site without food or basic amenities. Children and women are particularly vulnerable in such situations. Incidents of rape and killing of victims during and after eviction exercises have been reported in many places.

When evictions take place, they not only destroy homes, but also entire communities, which can lead to urban unrest and insecurity. Evictions result in loss of income and disrupt highly integrated and complex networks of the informal economy.[5]

Media and other reports suggest that the magnitude of urban evictions is currently highest in sub-Saharan African cities, although rural evictions are also quite common. This could be partly because urban populations in African cities have so far not been able to organize themselves politically in large enough numbers to be able to resist evictions or demand rights from their governments. In addition, many African governments inherited outdated, elitist laws from colonial powers that discriminate against the urban poor; these policies have led to the creation of "apartheid-type" cities, with the neighbourhoods of the rich and the poor clearly demarcated.[6]

Strong civil society action in Asia and better legislation has had a significant impact on improving the tenure status of slum dwellers, but with pressures to "globalize" mounting, particularly in the region's more economically successful industrializing cities, this trend could be reversed in the near future. Moreover, escalating land and house prices in Asian cities could lead to economic evictions as lower-income groups are pushed out of the city simply because they can no longer afford to live there.

In Latin America, progressive slum upgrading and regularization programmes have increased tenure security among the urban poor, but evidence suggests that tenure security is not reaching the most vulnerable groups. For instance, a study in Brazil shows that poor blacks and mulattos are least likely to live in adequate housing with secure tenure, and are most likely to live in slums (see figure).

However, improving the tenure of existing urban populations is not enough; measures must also be undertaken to prevent the growth of new slums and informal settlements where tenure security is at risk. This requires a parallel approach to increase the supply of planned, legal and affordable land on a scale equal to present and future demand.

Tenure security: The thin line between legality and illegality

Non-empirical evidence suggests that between 30 per cent and 50 per cent of urban residents in the developing world lack any kind of legal document to show they have tenure security.[7] Development agencies, academics and practitioners in urban issues concur that informal growth has become the most significant mode of housing production in cities of the developing world. In fact, gaining access to housing through legal channels

Family being evicted during slum clearance, Minas Gerias, Belo Horizonte, Brazil DYLAN GARCIA/STILL PICTURES

is the exception rather than the rule for most urban poor households.[8] In many cases the majority of inhabitants live with tenure systems that are "informal", which means that their occupation of land and/or housing is either illegal, quasi-legal, tolerated or legitimized by customary or traditional laws, which can either be recognized or simply ignored by the authorities.

Slums – the generic term used to classify informal, illegal or unplanned settlements – are the invisible "zones of silence" on tenure security. Little is known about the formal or informal tenure systems slum dwellers enjoy – or don't enjoy – as official censuses and households surveys do not at present measure tenure security as a development indicator, even though informality – or "illegality" is perhaps the most significant factor in the physical and demographic growth of cities in the developing world and is the main mechanism through which poor people in cities gain access to land and housing.[9]

The status of slum dwellers in developing countries is made more ambiguous by the fact that they are often not included in national censuses and household surveys, which means that their tenure is neither recorded nor guaranteed. Often, new informal settlements are not enumerated, and even when their inhabitants are included in censuses, they normally appear as "owners" of the dwelling they occupy, even though surveys and studies have shown that large proportions of slum dwellers are actually tenants or are "owners without titles". In 2003, the Inter-American Development Bank estimated that around 60 per cent of the urban poor in Latin America were home owners even though very few had land or housing titles.[10] In 2005, the Central Statistical Bureau in Indonesia reported that "out of those that own their home, only 32 per cent can show legitimate proof in the form of a certificate from the national land agency".[11] In Nicaragua, one-third of the urban population was recorded as being "owners with no deed" in 2001.[12] The high prevalence of "owners without titles" conceals the real number of people living in informal settlements and significantly distorts figures and estimates reflecting the magnitude of urban dwellers who live without secure tenure in cities.

Security of tenure is critical to the livelihood of slum dwellers and should encompass a minimum package of rights, which could progressively evolve towards a higher order of rights. This formalization process can be accomplished through an incremen-

FIGURE 2.5.1 PERCENTAGE OF URBAN HOUSEHOLDS WITH ADEQUATE HOUSING IN BRAZIL, 1992-2003

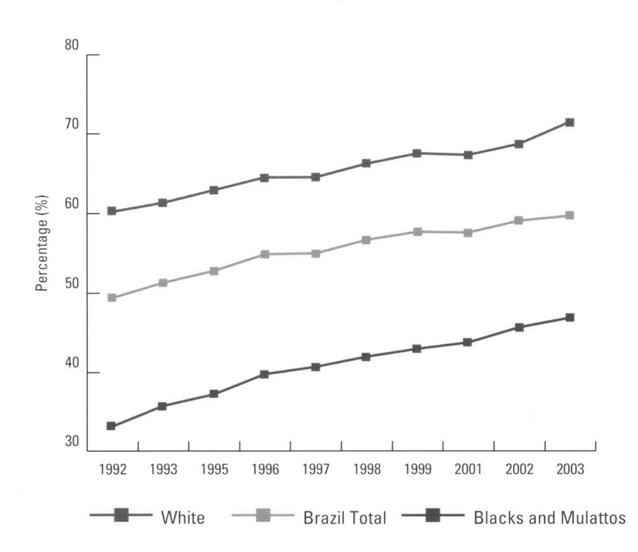

Source: Morais, Maria da Piedade, rapporteur Target 11, MDGs, IPEA based on PNAD microdata, IBGE, 1992 to 2003.

Monitoring secure tenure

UN-HABITAT defines secure tenure as the right of all individuals and groups to effective protection from the State against forced evictions. Under international law, "forced eviction" is defined as "the permanent or temporary removal against their will of individuals, families and/or communities from the homes and/or land which they occupy, without the provision of, and access to appropriate forms of legal or other protection".

For global monitoring purposes, UN-HABITAT proposes to adopt two more component indicators to measure secure tenure at the individual/household level:

Component Indicator 1: Proof of Documentation.
This component indicator assumes that documentation may be considered as proof of occupancy and therefore could provide certain levels of security. In most developing countries, tenure security in informal settlements is achieved incrementally over time through the accretion of various documents, such as utility bills, voter registration forms, ration cards and municipal tax receipts.

Component Indicator 2: Perception of security/insecurity of tenure.
This indicator measures the individual's or household's perception of their own tenure situation. It is based on the experience and perceptions of those who are most directly confronted with the reality of evictions in a country or city.

The two component indicators are complemented by a diagnostic of the policy environment that helps to determine the factual status of tenure security in a specific context. This is done through a qualitative measurement of the legal, institutional, administrative and policy environment governing security of tenure.

tal process of tenure upgrading that evolves from *de facto* tenure (taking into account a variety of socially accepted norms in land and housing tenure) to *de jure* tenure. This approach also allows governments to build technical and administrative procedures over time and within their own resource capacity.

The precarious status of land and housing tenure among slum dwellers can also be progressively strengthened through an institutional and social construct in which the accretion of various documents plays a key role in the process. The process can start initially with the occupant providing simple proof of occupancy, such as utility bills, voter registration forms, ration cards and municipal tax receipts. Gradually, documentation could evolve towards more consolidated forms of occupancy rights, and eventually to formal tenure regularization and the provision of legal rights, such as freehold or long-term leases, if these are possible and desirable.

In formal, advanced systems, tenure rights are reflected in laws and regulations governing housing and property rights. In developed countries, security of tenure is in most cases guaranteed, and people enjoy a higher order of rights that enables them to sell, rent, improve, develop, sub-let or inherit land or property. In virtually all developed countries, these rights are further embedded in infrastructure, land administration and land recording mechanisms. Rights derived from land and property are defined in a way that makes them easy to identify in terms of boundaries, demarcations, registration and transactions. Individuals (owners and tenants) have a clear understanding of the potential that land and housing offers in terms of use, appropriation and trade. UN-HABITAT uses the concept of "tenure advantage" to describe these more advanced rights of individuals and households.[13] However, it is understood that slum dwellers can also sell, rent or improve the land or property they occupy, but in a less secure environment. Tenure advantage rights are also known as "transferability rights" that have a direct bearing on the livelihoods of people, as they are extensively used as security from which capital can be derived. As property markets become more active, transferability operations increase because public and private infrastructure supports the activities. Land is extensively used as security, multiplying the opportunities to obtain capital. As the land systems and markets become more stable, more complex products appear, such as land and credit being placed in secondary markets and stocks.

Ownership is not always the solution

Although ownership is typically regarded as the most secure form of tenure, evidence from around the world suggests that ownership is not the norm in both the developed and developing world, and is not the only means through which tenure security can be achieved. Home owners are a minority in most countries of the developing and developed world. In Central Europe, for instance, more than half the inhabitants rent the houses in which they live. Yet, their tenure is extremely secure. And despite a significant increase in ownership in North America that saw home ownership rise - from 64 per cent to 69 per cent from 1993 to 2003 - a third of the region's inhabitants still do not own their own home.[14] Even in the developing world, studies have shown that ownership is neither necessary nor sufficient to generate tenure security. A Demographic and Heath Survey conducted in Senegal in 2005 shows that 45 per cent of the inhabitants owned their homes, while 42 per cent were tenants. Only 40 per cent of the so-called owners claimed to have title deeds, while 25 per cent had a certificate of occupation or a receipt of purchase. A significant proportion – 13 per cent – had no formal or informal authorization to occupy the dwelling. Of the tenants, only 14 per cent had a formal contract and a staggering 68 per cent did not have any kind of document to prove tenancy. Yet 76 per cent of both owners and tenants said they enjoyed security of tenure.[15]

The view that ownership is the only path to security has gained credence among development agencies and practitioners who argue that urban poverty can be drastically reduced if slum dwellers acquire ownership rights that can enable them to secure loans to improve their housing and to invest in enterprise.[16] Land and housing are perceived as market assets that have the potential of generating bottom-up economic and social rewards, thereby reducing poverty. Diametrically opposed to this way of thinking is the perception that access to land is a fundamental human right necessary for a secure livelihood.

Most governments and development agencies consider tenants living in informal settlements as having "insecure tenure", whereas "owners" are automatically classified as "secure". Countries from various regions, such as Bosnia and Herzegovina, Cambodia, Cameroon, Côte d'Ivoire and Gabon have used indicators such as "the percentage of land parcels having titles", the "percentage of the population with access to property", or the "number of households owning their lodging" to measure tenure security.[17] The Economic Commission for Latin America and the Caribbean (ECLAC) also associates insecure tenure among the urban poor with lack of ownership.[18] This mode of coding seems to respond to ideologies and belief systems that view rights to land and property as being dependant on market forces and affordability, rather than as fundamental rights guaranteed by governments.

Yet many examples from around the world show that whilst titling has benefited many slum communities, and deserves a place in tenure policies, it has not necessarily increased access to credit or prevented growth of new informal settlements. Empirical evidence does not support the view that full titling lifts the poor out of poverty; in many cases, an incremental approach – based on the right to a secure livelihood – has proved to be more effective in the long term. In some cases, large scale titling programmes can actually contribute to legitimizing and exacerbating unequal systems of land and property distribution.[19] For example, in Buenos Aires, Argentina, a study found that the allocation of property rights across slum households is usually not random, but based on wealth, family characteristics, political patronage and other mechanisms that mark differences between those who have property rights and those who do not.[20] Slum upgrading projects in other countries have also been known to play into the hands of illegal structure owners and negligent landlords, who lay claim on upgraded dwellings in order to extract more rent from tenants or to sell them off to higher-income groups.[21] Upgrading policies based on ownership and large scale granting of individual titles are also extremely expensive and cumbersome, especially in countries where titling systems are slow, laborious, inflexible and generally unaffordable – not to mention prone to corrupt practices that harm rather than benefit the urban poor. In the Philippines, for instance, establishing legal ownership takes 168 procedures and between 13 and 15 years.[22] All these problems are compounded by the fact that little is known about the severity or range of insecure tenure within cities of the developing world, which makes it difficult to make appropriate interventions at the policy level or in the implementation of upgrading and regularization programmes.

In developing countries, customary ownership, religious systems governing land issues and informal agreements between owners (both public and private) and tenants play a more important role in securing tenure than titling. In most cases, land in both rural and urban areas is neither registered, not is there an official title for it. UN-HABITAT estimates indicate that less than one-third of land in developing countries is accounted for in official land records and registries[23] and questions regarding ownership are tackled through customary, communal or religious laws governing land. According to World Bank estimates, in Africa formal tenure extends to only between 2 per cent and 10 per cent of all residential land.[24] Unfortunately, in some parts of the world, particularly in sub-Saharan Africa, customary law actually works against the interests of women, who are prevented from inheriting land or property. While an increasing number of sub-Saharan African countries have recognized women's equal rights to land and property, thus complying with international human rights standards and obligations, there are still some countries, such as Kenya, Lesotho, Zambia and Zimbabwe, where discrimination in customary and personal law matters (such as inheritance) is still permitted in these countries' constitutions.[25]

The Challenge of Measuring Tenure in Informal Urban Settlements

Since many local authorities are reluctant to recognize the existence of informal growth in their cities, or are not predisposed to address it in a systematic manner, they do not develop appropriate means to measure and monitor the level of informality in urban areas. In fact, in 2005, UN-HABITAT's Urban Indicators Programme found that around 100 cities in Africa, Asia and Latin America, representing more than 70 per cent of a global sample of cities, acknowledged that they did not know to which extent urban growth could be attributed to informal settlements; 40 cities, or 20 per cent of the sample, provided some general data as a percentage of the total urban growth; less than 20 cities, or 10 per cent of the sample, made available accurate information in square kilometers, as requested by the Programme.[26]

Consequently, the number of people lacking secure tenure is not known in most cities and countries of the developing world. The lack of official data on informal growth is symptomatic of the poor capacity of local authorities to plan the urbanization process. Instead of learning to accept inevitable urban growth in informal settlements and slums, governments, like the proverbial ostrich, have chosen to bury their heads in the sand, hoping the problem will go away. That is why methods to capture and measure informal settlement growth are not perceived as necessary. It is also the reason why the extent and scope of tenure insecurity is not known in most cities and countries of the developing world.

There have, however, been exceptions to this rule. Some governments have addressed urban tenure security in their plans, providing land to urban dwellers before occupation. Others have integrated tenure security in their housing programmes

El Alto, Bolivia.

Informal growth has become the most significant mode of housing production in cities of the developing world.

and projects, but in a very sporadic way. Many governments have responded to informal occupations through remedial actions of regularization; however, their interventions have neither been systematic nor politically disinterested. The majority of the governments have opted to ignore informal settlements altogether, either because of their unwillingness to accept in-migration and urban growth, or their incapacity to cope with the accelerated process of urbanization.

The problem is compounded by the fact that while United Nations and other agencies have been testing and developing systems of monitoring global poverty, disease, illiteracy, unemployment, and other indicators over the past five decades, the operationalization of the secure tenure concept, as part of a global monitoring system, remains challenging. Indeed, at the present time, it is neither possible to obtain household-level data on secure tenure, nor to produce global comparative data on various institutional aspects of secure tenure.[27] Although a growing global network of organizations, such as the Centre on Housing Rights and Evictions (COHRE), Amnesty International and Human Rights Watch, among others, are trying to establish monitoring systems on issues such as evictions and other housing rights abuses, even they admit that a comprehensive eviction-monitoring system remains elusive as many gaps in coverage remain because for every reported eviction, there is an unknown number of unreported cases.[28] In the absence of a monitoring framework that provides a reference point and guidance, it is statistically difficult to prove whether tenure has improved or deteriorated, as evidence remains mostly anecdotal or based on media reports.

As long as mainstream systems of data collection and analysis (censuses and surveys) do not recognize secure tenure as a unit of analysis, data will not be periodically produced. ECLAC, for instance, recognizes the difficulties in determining secure tenure in Latin America and the Caribbean and can only estimate the level of informality in the region's the cities, which it places at between 10 and 15 per cent of the urban population in Argentina and Uruguay, between 20 and 40 per cent in Mexico and Peru and between 50 and 70 per cent in Ecuador and Honduras.[29] The World Bank estimates that more than 50 per cent of the peri-urban population in Africa and more than 50 per cent in Asia has some form of informal tenure. But these are at best estimates as few countries produce data on secure tenure, which can be used as basis for global and regional monitoring or to assess progress or setbacks.

Toward a Global Monitoring Strategy on Secure Tenure

In the last thirty years, security of tenure has been part of the conceptual, institutional and technical discussions about land and housing policies. In some moments of this saga, tenure security has received a great deal of attention, particularly during international conferences, political declarations and the preparation of technical reports. These discussions, however, have not resulted in greater efforts to integrate tenure security in policy reforms and urban interventions. In fact, no mechanism currently exists to monitor secure tenure as part of Millennium Development Goal 7, target 11 on improving the lives of slum dwellers.[30]

UN-HABITAT and its partners are working on the preparation of a global monitoring system[31] that could in the future provide a framework to assist governments at local and national levels to produce estimates at the household level on how many people have secure tenure, using a consistent methodology in terms of definitions, indicators and variables. The monitoring system would serve to track changes in land and residential secure tenure to measure how the right to adequate housing is progressively realized and how slum dwellers are improving their living conditions. It would also be an advocacy and policy instrument to bring together policy formulation, action and monitoring activities; otherwise, policy actions will continue to be formulated independently of results, without clearly indicating if there is efficient, equitable and sustainable progress in attaining target 11.

Endnotes

1 Much of the data on evictions in this chapter is drawn from various reports by the Centre on Housing Rights and Evictions (COHRE).
2 COHRE 2003.
3 COHRE 2004.
4 The most comprehensive interpretation of the scope of protection against forced evictions was made by the UN Committee on Economic, Social and Cultural Rights in its general comment No 7, adopted in 1997, while the right to adequate housing is enshrined in many international human rights instruments, most notably the Universal Declaration of Human Rights and the International Covenant on Economic, Social and Cultural Rights.
5 United Nations 2005c.
6 These "apartheid cities" appear to be more prevalent in former British colonies, such Kenya and Zimbabwe, where the migration of the indigenous population to cities was highly regulated and often prohibited during the colonial period.
7 World Bank 2003.
8 De Soto 2000.
9 Fernandes, et al. 1998.
10 Inter-American Development Bank 2004.
11 Government of Indonesia 2005.
12 Demographic and Health Survey, Nicaragua, 2001.
13 Bazoglu & Moreno 2005.
14 US Census Bureau, American Housing Survey (1993 and 2003).
15 UN-HABITAT 2005c.
16 De Soto makes a strong argument for ownership in *The Mystery of Capital* (2000).
17 Lee & Ghanime 2004.
18 Economic Commission for Latin America and the Caribbean 2005.
19 Quan, et al. 2005.
20 Goytia 2005.
21 Huchzermeyer 2006.
22 UN-HABITAT 2004c.
23 Augustinas 2003.
24 World Bank 2003.
25 UN-HABITAT 2004d.
26 UN-HABITAT training workshops in five regions of the developing world. Urban Indicators Programme, Cluster B Urban Data, Nairobi, 2005.
27 Bazoglu & Moreno 2005.
28 COHRE 2004.
29 Clichevsky 2003.
30 A list of 18 targets and more than 40 indicators corresponding to these goals ensure a common assessment and appreciation of the status of the Millennium Development Goals at the global, national and local levels. Among the indicators for monitoring progress on the Goals, *secure tenure* was given a prominent place (indicator 32, "the proportion of households with access to secure tenure"), as part of Target 11 "*by 2020, to have achieved a significant improvement in the lives of at least 100 million slum dwellers*" of Goal 7 "Ensure Environmental Sustainability".
31 UN-HABITAT currently undertakes this monitoring strategy in collaboration with several partners, namely DFID, the World Bank, and the governments of Canada, USA and Belgium.

One of the affected areas in Harare prior to the May 2005 evictions.

The same area after the May 2005 evictions.

Source: IKONOS: Copyright INTA Space Turk 2005; QUICK BIRD: Copyright Digital Globe 2005; Image processing and analysis: UNOSAT.

Demolition of a backyard extension.

Evicted family.

Evictions fail to address the root cause of urban poverty in Zimbabwe

In May 2005, with little or no warning, the Government of Zimbabwe embarked on an operation to "clean up" its cities. "Operation Murambatsvina", or Operation Restore Order, started in the capital Harare, and rapidly evolved into a nationwide demolition and eviction campaign carried out by the police and the army. Popularly known as "Operation Tsunami" because of its speed and ferocity, it resulted in the destruction of homes, business premises and vending sites in several parts of the country. A July 2005 report by the UN Special Envoy on Human Settlements Issues in Zimbabwe, Mrs. Anna Tibaijuka, estimated that some 700,000 people in cities across the country had either lost their homes, their source of livelihood or both as a result of the Operation and a further 2.4 million were indirectly affected in varying degrees.

Operation Restore Order took place at a time of persistent budget deficits, triple-digit inflation, critical food and fuel shortages and chronic shortages of foreign currency. It was implemented in a highly polarized climate characterized by mistrust, fear and a lack of dialogue between the government and local authorities, and between the government and civil society. Although the economic crisis was precipitated by a variety of factors, including increasing isolation by Western powers, many of Zimbabwe's problems precede the country's independence.

Zimbabwe achieved independence in 1980 amid promises of peace and prosperity. While the government successfully provided social services, such as education and health care, and increased wages for the black majority during the early years of independence, underlying socio-political and economic problems were left unresolved and eventually produced a national crisis. Of these, the land question was the most problematic. While the liberation war was fought over land, historical inequity was embedded in the constitutional settlement agreed upon at independence that preserved colonial patterns of land ownership. To make matters worse, a failed attempt at structural adjustment in the 1990s led to massive retrenchment of civil servants, closure of manufacturing industries, inflation and deterioration of basic services. In February 1998, peasants took matters into their own hands by staging illegal – and politically motivated – invasions of commercial farms, forcing the government to initiate a "fast-track land reform programme" in 2000.

It is against this background that Operation Restore Order took place. Ironically, while the government tried to appease the country's rural population, it took a rather elitist approach with its urban citizenry by imposing stringent by-laws and standards that deemed many dwellings in the city "illegal". As the report of the UN Special Envoy states, "The nationalist elite seemed to have perpetuated the colonial mentality of high standards for a few at the expense of the majority. In the end, while the liberation struggle was against the 'white settlers' and the economic and political power they monopolized, the government was not able to reverse the unequal and exploitative nature of colonial capitalism itself."

Like many former British colonies in sub-Saharan Africa, urban planning in the country formerly known as Rhodesia typically reserved the city core for whites, while leaving an undeveloped buffer space around the central business district. Towns were often pre-planned and imposed on localities, without much attention being given to existing constraints. The indigenous population was either relocated to black townships on the outskirts of the city or to rural "reserves" to make room for European settlers.

The indigenous African population moved to towns and cities in large numbers only after attaining independence when policies prohibiting their movement to cities were abolished or discarded. This resulted in a major shift of populations from rural to urban areas. Within a decade of independence, Zimbabwe's urban population rose from 23 per cent in the 1980s to 30 per cent by the early 1990s. However, stringent by-laws and standards adopted from the colonial administration ensured that Zimbabwe's cities remained largely immune to the explosive growth of slums and squatter settlements that are characteristic of other African cities. Official statistics compiled by UN-HABITAT show that in 2001 only 3.4 per cent of the urban population in the country lived in slums, a figure much lower than that of even industrialized countries that had about 6.2 per cent of their population living in slum-like conditions, and dramatically lower than that of other African cities, where between 30 per cent to 70 per cent of the urban population lives in slums.

The acquisition of peri-urban farms during the fast-track reform programme in 2000 provided one of the first opportunities for the urban poor to occupy land in the vicinity of the city, many of which were in the form of "backyard extensions" of legal dwellings. These extensions provided affordable rental housing to the city's poor and were a source of much-needed income for the owners. Most of these extensions within the cities have now been demolished, affecting hundreds of thousands of women, men and children who are sinking deeper into poverty and rendered more vulnerable. A follow-up report by the UN Office for the Coordination of Humanitarian Affairs (OCHA), for instance, found that some of the worst affected were women and children living with HIV/AIDS. (Zimbabwe has one of the world's highest HIV prevalence at about a quarter of the total population.) A survey by the Bulawayo-based Matabeland AIDS Council, for instance, found that many of those displaced by the "clean-up" campaign could not continue with their treatment and were in dire need of drugs.

It is for these reasons that the UN Special Envoy's report recommends, among other things, that outdated laws be suspended or reviewed in order to align them to the social, economic and cultural realities facing the majority of the country's population, namely the poor. It also recommends that the international community draw lessons from the Zimbabwe crisis for the entire continent of Africa by ensuring that policies aimed at reducing poverty do not have the opposite effect.

Sources: United Nations 2005c; UN-HABITAT 2003a; IRIN News 2005.

3

Part Three

Where We Live Matters

This Part provides concrete evidence of how inadequate housing and lack of basic services threaten the health, education and employment opportunities of slum dwellers. Using data that goes beyond the conventional urban–rural dichotomy, the Part presents, for the first time, disaggregated information at slum and non-slum levels that help us to understand the connection between living conditions and human development. This connection is fundamental to appreciate both the vulnerability of slum dwellers and the levels of poverty and deprivation that they experience, particularly in relation to social and health outcomes.

Mexico City, Mexico TERRAZAS GLAVAN MONICA/UNEP/STILL PICTURES

3.1 The Social and Health Costs of Living in a Slum

The Urban Rural Divide ©FEDOR SIDOROV, IMAGE FROM BIGSTOCKPHOTO.COM

How do inadequate water supply and overcrowding in slums impact child mortality rates? Is urban insecurity related to inequality within cities? How do conflicts in rural areas exacerbate slum formation in urban areas? Why are women who live in slums more likely to be infected with HIV than their rural counterparts? Are the Millennium Development Goals being met in the slums of the world? How does one's physical address influence one's health, education and employment opportunities? Does it matter where we live?

Anyone who has dealt with real estate agents knows the mantra "location, location, location": place and progress are inextricably intertwined. This is especially so for the world's urban poor. Indeed, as the following chapters illustrate, where we live can have a significant influence on whether or not we are likely to be healthy, educated, employed, safe, or impoverished. UN-HABITAT analyses of recent survey data show that people who live in slums face serious threats to their well-being. In some cases, living in a crowded, unsanitary slum is even more life-threatening than living in an impoverished village. Some studies have also shown that job applicants from slum communities are less likely to be interviewed than those living on "the right side of town". In other words, living in a slum often means being more vulnerable to a host of social and economic threats that make the achievement of the Millennium Development Goals in cities both a major challenge and an urgent need.

The Millennium Development Goals have been accepted internationally as a common development framework. At their core, the Goals aim to bring the vast majority of the world's population out of a poverty trap that robs them of their health, dignity and aspirations for fulfilling their human potential. UN-HABITAT has been assigned the responsibility of assisting Member States of the United Nations to monitor and attain Millennium Development Goal 7, target 11: *by 2020, to have achieved a significant improvement in the lives of at least 100 million slum dwellers.*

The inclusion of the slum target in the Millennium Development Goals indicates a recognition by the international community that urban poverty is a growing challenge. However, national and international data and poverty reduction strategies still do not acknowledge the deprivation levels in slums and consistently underreport health, literacy and other development indicators. For instance, while aggregate health statistics suggest that urban dwellers have better health status compared to those living in rural areas, UN-HABITAT indicators show that there is a large and growing gap between the health status of high-income urban residents and those living in poverty at the margins of society.

The internationally agreed-upon slum target has been largely ignored in country and agency reports on progress on the Millennium Development Goals, due in part to the lack of intra-city data disaggregated across slum and non-slum areas. A review of the existing strategies to improve in the lives of slum dwellers reveals a gap in addressing the situation of the urban poor in national and international programmes. Most national reports underestimate the level of urban poverty; moreover, the measurement of poverty in both rural and urban areas is based on income, which often does not provide an accurate picture of the scale and multidimensional nature of poverty experienced by the urban poor. The crisis that slum dwellers are facing has been masked by the common practice in social science to analyse the human settlements dimension by categorizing information according to "urban" and "rural". In country reports, all urban households – rich and poor – are averaged together to provide single estimates of poverty, education, health, employment, and human settlements, leading to an underestimation of the urban poor and the conditions in which they live.[1] Another aspect that gets lost in urban averages is intra-city inequality. Studies show that the decreased mortality recorded in urban areas in the 1990s was primarily a result of high-income residents living longer, indicating widening health disparities between the rich and the poor. This trend is particularly prevalent in Latin America.

Data produced by the United Nations, World Bank and other agencies presents urban poverty on a regional scale and generally links it to theoretical projections not based on actual surveys. For instance, World Bank projections indicate that the locus of poverty will move to cities only after 2035.[2] This projection serves as an "early warning system", much like the warnings issued by the World Health Organization (WHO) in the 1980s about the impending AIDS crisis. However, other agencies have a more realistic view of the scale of urban poverty and believe that the crisis has already begun.

As discussed elsewhere in this Report, although the number of slum dwellers is not an accurate measurement of the number of urban poor – poverty can manifest itself in non-slum areas; conversely, not all people who live in slums are poor – slums are a physical dimension of urban poverty. It is, therefore, crucial to know *how many slum dwellers there are*, *where they are located*, and *what their basic needs* are in terms of shelter, water, sanitation, health, education, employment, and the like. Part One of this Report highlights the numbers and locations of slum dwellers around the world; this section addresses their needs in terms of specific Millennium Development Goal targets and indicators.

UN-HABITAT's analyses of disaggregated urban data point to some key findings. Child mortality rates in poor urban and rural communities, for instance, are much higher than those of high-income urban communities. Furthermore, families living under conditions of severe shelter deprivation experience, in some countries, a child mortality rate three times higher than that of families that enjoy full use of safe water, improved sanitation, durable housing, and decent living conditions. The fact that inequalities based on socio-economic disparities are so persistent in urban areas of developing countries implies that reliance on global average statistics to allocate resources between rural and urban areas could be dangerously misleading. Lack of basic shelter services, as a correlate of poverty, is the expression of various social and health issues such as low education, wide gender inequalities, poor maternal and child health, and hunger. Poor living conditions also contribute to a host of diseases and infections, such as diarrhoea, acute respiratory infections, malaria and HIV/AIDS. In terms of education, studies indicate that a majority of parents settling in slums postpone sending their children, especially girls, to school, until they are able to manage other expenses, such as food, rent and transport.

In this Report, UN-HABITAT aims to show that improvement in the lives of slum dwellers leads to progress on the achievement of all of the Millennium Development Goals. By improving slums – or preventing their formation – governments are also eradicating poverty and hunger, increasing literacy, combating HIV/AIDS, reducing child mortality, improving the environment, and promoting gender equality. This calls for the localization of the Goals: local policy needs to be informed about the consequences of persistent inequalities in cities and the myriad problems associated with the living conditions of poor urban communities.

Endnotes

1 Fry, et al. 2002.
2 Ravallion 2001.

3.2 Hunger: The Invisible Crisis in Cities

Kibera, Nairobi HIROSHI SATO

Eradicating extreme poverty and hunger is the first Millennium Development Goal. Reducing the proportion of people suffering from hunger in the world is therefore acknowledged as essential to achieving all of the other Goals. The United Nations *Millennium Development Goals Report 2005* states that there were 815 million hungry people in the developing world in 2002 and that "most of the world's hungry live in rural areas and depend on the consumption and sale of natural products for both their income and their food". The report adds that sub-Saharan Africa and Southern Asia are the worst-affected regions and that "hunger tends to be concentrated among the landless or among farmers whose plots are too small to provide for their needs".

Although hunger is most often associated with low agricultural output, drought and famine in rural areas, various studies have shown that hunger is not always related to food production or availability; rather, in urban areas, other factors, such as low incomes, inadequate access to basic services and poor living conditions, play more significant roles.

What makes hunger in cities unique?

In rural communities, exogenous factors such as geography and climate are major determinants of food availability and dietary intake. Rice is generally consumed in the humid tropics, while millet is more frequently eaten in arid regions. People who live in mountainous areas are limited to barley and potatoes at the highest altitudes, but a variety of cereals can be produced in lowland valleys. Pastoralists are more likely to rely on their animals for food. Unless sophisticated market systems have been developed, the inhabitants of a particular ecological zone consume only what they can produce locally. There may also be dramatic differences in the types and amounts of food available in different seasons.[1]

Whereas crop patterns, size of land and the time and quality of the harvest often determine food availability for the family of the subsistence farmer, disposable income and food prices largely determine the amount and types of foods consumed by low-income families in urban areas.[2] In cities, hunger is usually the consequence of people's inability to purchase food that is both sufficient and nutritious. An assessment of the "food basket" of slum households shows that it is mainly composed of items low in calories and vitamins,[3] making these households more prone to malnutrition.

Even in situations where a country produces enough food to feed everyone, hunger may persist in urban areas. In fact, the situation of the urban poor can be worse during famines and droughts than the situation of villagers; international food aid distributed during difficult times is concentrated in rural areas, while in cities, prices for essential food products produced within the country soar during such times, adversely impacting the ability of low-income people to purchase food. When inflation hits food supplies, poor urban families may be forced to use up to 70 or 80 per cent of their disposable income to purchase food, which often means that they have little money left over to pay for non-food items, such as rent, school fees and transport. Thus, variations in income or food prices directly translate into rising rates of malnutrition in urban areas. In poor urban communities, even seasonal variability in income or food availability can lead to seasonal swings in malnutrition.

When inflation hits food supplies, poor urban families may be forced to use up to 70 or 80 per cent of their disposable income to purchase food.

The link between inadequate shelter and hunger

UN-HABITAT analyses indicate that hunger and malnutrition is particularly high in slums and in rural areas, where access to adequate housing and basic services, such as safe water and sanitation, is poor or non-existent. The poor living conditions prevalent in slums and in rural areas impact people's ability to avert hunger and malnutrition in various ways. Households' sources of drinking water and methods of waste disposal impact children's nutritional status, as diarrhoea and other diseases resulting from inadequate water and sanitation can prevent young children from absorbing nutrients and growing properly. Without an adequate and safe supply of water, a household's personal, domestic and food hygiene are compromised and the risk of contamination and diseases – including diarrhoea and acute respiratory infections – increases. Overcrowded slum households are also more likely to use inadequate sanitation and to share toilet facilities with many other households, which increases the risk of diarrhoea and respiratory infections. At the community level, lack of waste management and wastewater treatment increases the prevalence of diseases such as diarrhoea, acute respiratory infections and malaria, all of which impact nutritional status and overall health.

When levels of child malnutrition are used to measure hunger, evidence suggests a strong link between malnutrition and slums. That is, places that report a high prevalence of child malnutrition typically have correspondingly high levels of slum incidence.[4] For instance, countries such as Bangladesh, Ethiopia, Guatemala, Haiti, India, Nepal and Niger – all of which have a high incidence of slums – are also those with among the highest prevalence of malnourished children. In these countries, 4 out of 10 children in slums are malnourished – a proportion 20 times higher than that of developed countries. In some countries, incidence of child malnutrition in slums is almost the same as that of rural areas. In Ethiopia, for instance, child malnutrition in slums and in rural areas is 47 per cent and 49 per cent, respectively, compared with 27 per cent in non-slum urban areas. Similar findings are reflected in Niger, where child malnutrition is 50 per cent in slums, 52 per cent in rural areas and 35 per cent in non-slum urban areas.

In general, malnutrition is much higher in rural and slum areas than in non-slum urban areas, even in countries with low levels of slum incidence. In Morocco, slum and rural children are twice as likely to be malnourished as their non-slum counterparts; while 7 per cent of children in non-slum areas are malnourished, 14 per cent of children in both slum and rural areas are malnourished. The greatest inequalities exist in Brazil and Côte d'Ivoire, where child malnutrition is three to four times higher in slums than in non-slum areas (19 per cent versus 5 per cent, and 37 per cent versus 10 per cent, respectively).

The relationship between malnutrition and poor living conditions is illustrated by various studies that show that malnutrition levels decrease when investments are made to improve services and infrastructure in low-income areas. The greatest decline in malnutrition in Eastern Asia, for instance, happened when China significantly improved its food distribution networks and health facilities, and provided increased access to improved drinking water. A study conducted in India in the 1950s attributed much of that country's rise in life expectancy (from about 25 years to 50 years) in the first half of the 20th century to the prevention of recurrent famines that had characterized the subcontinent's history; this was achieved by stabilizing food supplies with railroads, road networks, irrigation, food distribution markets and political security.[5]

The urban penalty

There is increasing evidence of what UN-HABITAT refers to as the "urban penalty": a number of key health indicators for vulnerable urban populations are as bad as or worse than those of rural populations. Despite the improved coverage of health services and basic service delivery in some countries, certain population groups have been left behind and opportunities remain unevenly distributed. This is particularly true in slum settlements around the world, which are as disadvantaged as rural populations, especially in least-developed countries with high urban growth rates.

Hunger eradication strategies must embrace multiple interventions, not only those related to food availability, but also those related to shelter. Access to adequate housing, safe water and adequate sanitation *do* improve the nutritional status of slum dwellers and rural populations, with or without an increase in food availability. This justifies a comprehensive approach that includes strong linkages between slum upgrading and the sustainability of programmes delivering health and nutrition services.

Fruit vendor in Kabul RASNA WARAH

Endnotes

1 Bidinger, et al. 1986.
2 Ibid.
3 See Demographic and Health Survey comparative report, ORC Macro 2004.
4 For a better understanding of hunger and food deprivation in cities, UN-HABITAT has analysed Demographic and Health Surveys and Multiple Indicator Cluster Surveys data on child nutrition in Africa, Asia and Latin America and the Caribbean. Child malnutrition is assessed by the proportion of children underweight, and associated variables. Underweight, defined as low weight for age, takes into account both acute malnutrition (wasting) and chronic malnutrition (stunting). A child can be underweight for his or her age because he or she is has suffered from "wasting", "stunting" or both (UNICEF 2003). Wasting may be the result of inadequate food intake or recent episodes of illness causing loss of weight and the onset of malnutrition. Among adults, this is defined as food consumption insufficient to meet minimum levels of dietary energy requirements. Stunting reflects failure to receive adequate nutrition over a long period and may also be caused by recurrent and chronic illness; it represents a measure of the long-term effects of malnutrition in a population (Food and Agriculture Organization of the United Nations 2002). For global trends, malnutrition is assessed for both adults and children. However, due to data limitations, UN-HABITAT intra-city differential analyses focus on children, and to some extent mothers, while recognizing that hunger also affects adults.
5 Davis 1951.

FIGURE 3.2.1 SLUM INCIDENCE AND PROPORTION OF UNDERWEIGHT CHILDREN IN SELECTED COUNTRIES

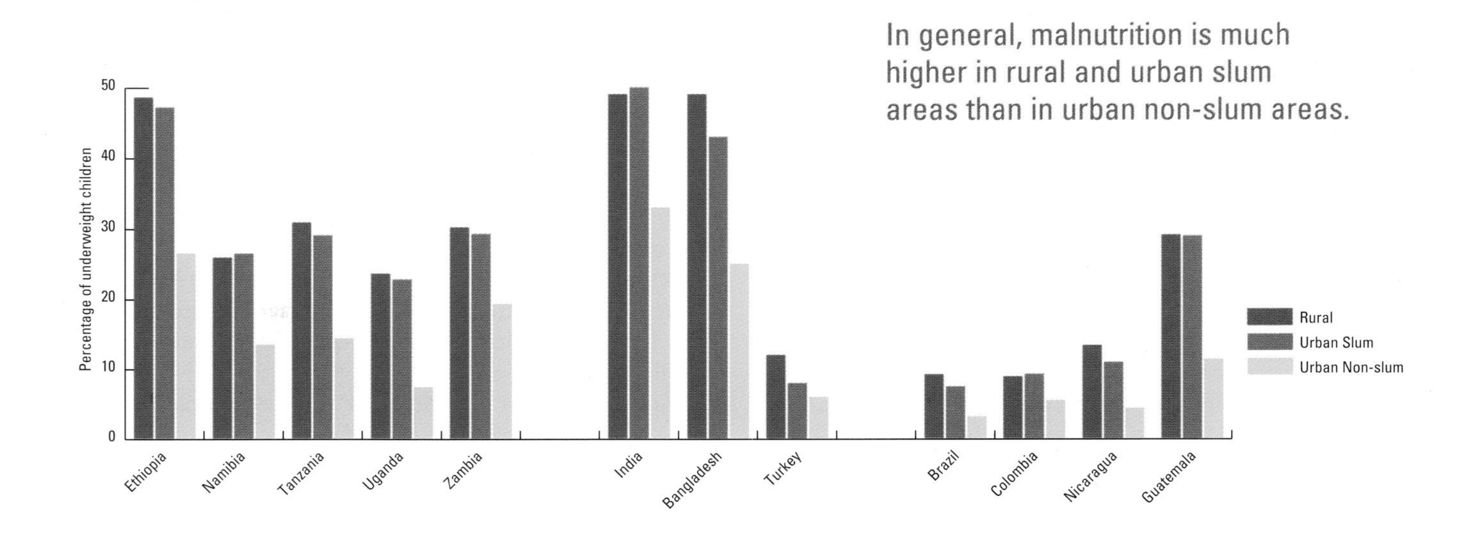

UN-HABITAT, 2005 Urban Indicators Programme, Phase III.
Source: Demographic and Health Surveys 1995-2003.

Agrocities: Combating hunger in urban areas

In many cities, particularly in the developing world, urban agriculture has helped increase food security and reduce hunger among vulnerable populations. Evidence suggests that urban agriculture – the practice of growing, raising, processing and distributing food in and around an urban area – contributes significantly to urban food supply and household food security, particularly among low-income groups.

Surveys conducted in the late 1990s by a range of institutions in 24 cities, mainly in Africa and Asia, and urban areas of Bulgaria, Romania and the Russian Federation, showed that households involved in growing some of their food made up anything from an important minority to a large majority of all households in any given city. The surveys also showed that poor households that practised urban agriculture ate more meals and had more balanced diets than those households that did not rely on urban agriculture for their food supply. Self-provisioning also helped urban poor households to save money that they might previously have used to purchase food.

In many cities, urban agriculture is a main or supplementary source of income or employment among low-income households. Generally, the higher the market value of the produce, the greater its contribution to household income. A survey conducted in Lome, Togo, for instance, showed that market gardeners earned 10 times the monthly minimum wage.

Moreover, the quantity of food supplied through urban agriculture comprises a significant amount of the total food consumed in cities and is worth tens of millions of dollars. In the late 1990s, milk produced each year in Dar es Salaam, Tanzania, was estimated to be worth more than $10 million. In the mid-1990s, rain-fed maize in and around Zimbabwe's capital city, Harare, was valued at $25 million and covered more than 9,000 hectares.

Less is known, however, about the quality and safety of food produced in urban or peri-urban areas where soil and water contamination levels are higher than in rural areas. In cases where quality has been tested, the results have been mixed. Findings for fish in Calcutta and for vegetables in Accra, for instance, showed no difference in contamination levels between urban-based and rural-based supplies but produce from industrial sites in Poland showed higher levels of heavy metals. A recent newspaper report in Kenya also warned that an increasing number of "urban farmers" in and around the capital of Nairobi were vandalising sewage networks to fertilize their vegetables, a practice that is affecting the quality of vegetables produced. Experts say that since sewage networks often receive non-domestic waste, they could contain high levels of heavy metals and organic matter that does not decompose easily, which could seriously affect the health of consumers of sewage-fed vegetables.

Rosario's kitchen gardens

Urban agriculture is not only an important source of food and income among urban poor households – it can also support a wide range of economic activities related to the production, sale, marketing and consumption of produce. In Rosario in the province of Santa Fe in Argentina, for instance, the Urban Agriculture Programme has been implementing an urban kitchen gardens project that has helped improve food security in the city, generated income for urban poor families and transformed uncultivated land into productive spaces.

The Urban Agriculture Programme of Rosario not only provides training to urban poor families and identifies vacant public and private spaces that could be used to grow chemical-free vegetables, but has also established a food production system that practises low-input agriculture using appropriate technologies that are easy for the urban poor to adopt. It has also ensured that the vegetables produced are sold in strategic markets within the city and that urban agriculture is institutionalized as a local government policy that is incorporated in the city's strategic plan.

Since the programme started in 2001, 791 urban kitchen gardens have been set up, providing employment to over 5,000 families. An additional 10,000 families are directly linked to the production of chemical-free vegetables that feed over 40,000 people in this city with a population of 1.3 million. The programme has also led to the creation of a network of 350 groups that participate in local fairs where the produce is marketed and sold.

The formal recognition of urban agriculture as a legitimate urban land use policy has enabled the municipality to set up a register and a Geographical Information Systems (GIS) databank on potentially productive land areas in the city. It has also improved the tenure status of the urban poor through user-rights agreements and tax incentives for land owners who make their land available for urban agriculture. The demand for the vegetables has also increased as the local fairs are the only places where residents can access organic produce.

For the urban poor, the kitchen gardens are not only an important source of employment but also a source of nutritious, chemical-free food. As a result of the success of the programme, proposals are being developed to incorporate urban agriculture in future settlement and housing plans.

Sources: Mougeot 2005a & 2005b; www.bestpractices.org; Musa 2005.

3.3 The Urban Poor Die Young

Santos Praia, Brazil TOPHAM PICTUREPOINT

Child mortality is closely linked to poverty, and child mortality rates are reliable indicators of human and economic development in countries. Millennium Development Goal 4 aims to reduce child mortality by cutting the worldwide under-five mortality rate by two thirds between 1990 and 2015. Doing so will require a special focus on the most vulnerable young children and families – those living in rural areas and in urban slums. Inadequate shelter and poor living conditions in slums are related to a host of health risks, including exposure to infectious diseases and indoor air pollution that shorten the life span of slum dwellers. This chapter describes the major health risks for slum dwellers and argues that even simple improvements in their living conditions can save lives.

Intra-city disparities in child mortality

Child mortality rates in developing countries are 10 times higher than those in the developed world. In 2003, sub-Saharan Africa and Southern Asia had the largest share of children who died before reaching their fifth birthday.[1] Child mortality rates appear to be closely related to urban poverty levels, and particularly to the incidence of slum households, as defined by the five shelter deprivations described in Part Two of this Report. Where child mortality rates are high, the proportion of slum households is typically also high. In such countries, child mortality is highest in slums and rural areas and is lowest in non-slum urban areas.

Mortality rates often reflect inequalities in access to shelter, health care and education.

Five diseases – pneumonia, diarrhoea, malaria, measles, and HIV/AIDS – account for more than 50 per cent of all child deaths. The chances of contracting any one or a combination of these diseases are compounded by poor living conditions and poor access to health services. Mortality rates often reflect inequalities in access to shelter, health care, employment, and education among different socio-economic groups. High mortality rates in slums are also compounded by the fact that millions of slum dwellers live on hazardous sites that are prone to natural disasters, such as floods, or that are located in or near toxic areas, such as garbage dumps, quarries or factories. Children are particularly at risk of illness and death as a result of environmental exposure to hazards and toxins, as they tend to have greater contact with the soil and contaminated water than adults, and, by virtue of their low body weight, they are more quickly and adversely harmed by any toxins that they ingest.

FIGURE 3.3.1 UNDER-FIVE MORTALITY (DEATHS PER 1000 BIRTHS) BY TYPE OF RESIDENCE IN SELECTED COUNTRIES

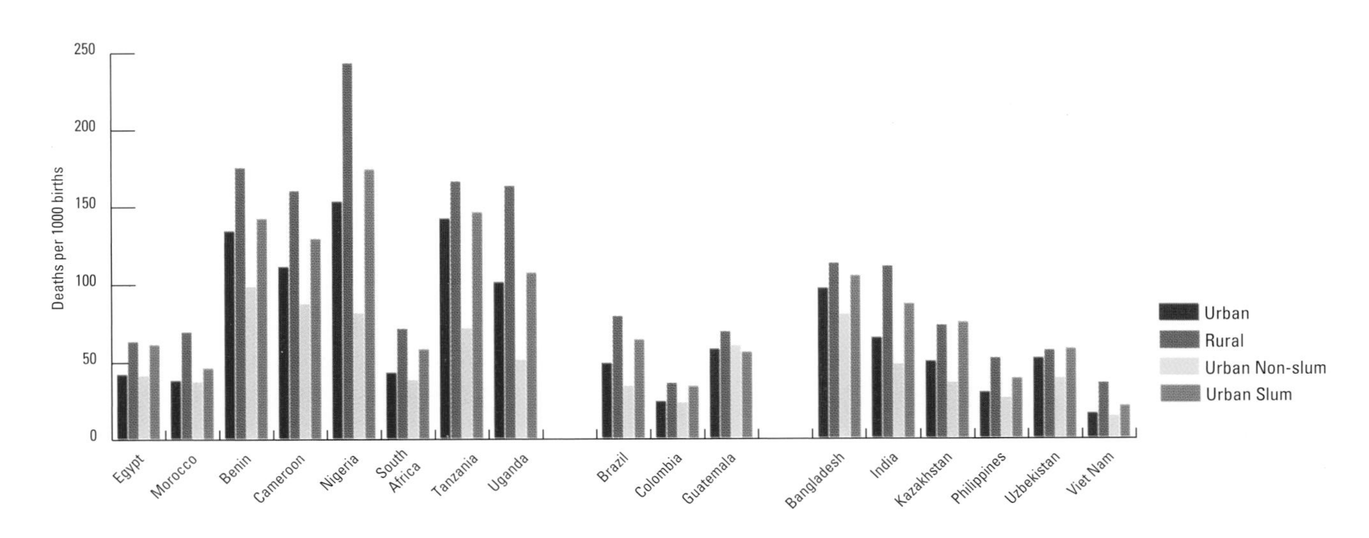

Source: UN-HABITAT 2005c.
Note: Computed from Demographic and Health Surveys (DHS) data 1995-2003.

The ratio of child deaths in slum areas to child deaths in non-slum areas is consistently high in all developing countries, even in countries that have made progress toward reducing child mortality overall. Several studies show that mortality differentials across groups tend to narrow only if policies focus explicitly on increasing equity in access to healthcare and safe housing. Without such a focus, improvements in the average rate may not reflect real improvements for both disadvantaged and advantaged socio-economic groups.[2] In other words, only when governments develop health policies that address the needs of the most vulnerable populations do child mortality rates decline.

Much evidence supports the theory that inequality breeds ill health. The World Health Organization (WHO) concedes that "being excluded from the life of society and treated as less than equal leads to worse health and greater risks of premature death. The stresses of living in poverty are particularly harmful to [pregnant women], babies, children and old people."[3] Medical research confirms the fact that income distribution is a more powerful determinant of health and mortality than the overall wealth of nations.[4] The lessons learned from the public health experience of developed countries point to some patterns. In the United Kingdom, differences in the health experiences of various groups have been shown to result more from the social disparities that shape health than from the quality of the national health system. Income, unemployment, education level, quality of housing, eating habits, and the work environment have emerged as major health indicators related to social inequality in the United

Where child mortality rates are high, the proportion of slum households is typically also high.

FIGURE 3.3.2 UNDER-FIVE MORTALITY (DEATHS PER 1000 BIRTHS) BY TYPE OF RESIDENCE IN SELECTED CITIES

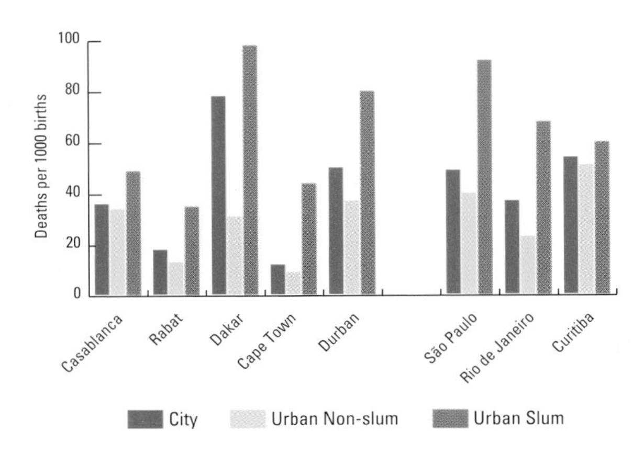

Source: UN-HABITAT 2005, Urban Indicators Programme Phase III.
Note: Computed from Demographic and Health Surveys (DHS) data 1995-2003.

Rubbish dump, Bangladesh TOPHAM PICTUREPOINT

Income distribution is a more powerful determinant of health and mortality than the overall wealth of nations.

Kingdom.[5] Urban social ecology studies in different developed countries also show that health is strongly related to access to medical care. One study found a strong correlation between mortality and income disparities in the United States, where access to medical care is often dependent on the ability to pay. However, in Australia, Canada and Sweden, where medical care is more affordable and is often provided for free to vulnerable groups, mortality was not related to income.[6]

Degrees of socio-economic inequality and corresponding child mortality rates vary throughout the world. While Northern African countries, such as Morocco and Egypt, report low average child mortality rates in a context of few social inequalities, in Latin America, Brazil displays low overall child mortality rates in a context of high degrees of socio-economic inequality. In Morocco, child mortality is only 24 per cent higher in slums than in non-slum areas of cities, while in Brazil, child mortality rates in slums are twice the non-slum rates and are comparable to slum/non-slum ratios in countries such as Ethiopia. Brazil has managed to decrease child mortality with advances in its public health system, but children living in slums are still at much greater risk than their non-slum counterparts.

Large cities tend to display wider inequalities than smaller cities or towns, even in countries with low levels of socioeconomic inequality, such as Morocco. In Morocco's capital city of Rabat, the under-five mortality rate is 2.7 times higher in slums than in non-slum areas. Cities that display high levels of inequality, such as Rio de Janeiro, Brazil, and Cape Town, South Africa, also show huge disparities between slum and non-slum areas. The under-five mortality rate in Rio de Janeiro's slums is three times higher than the rate in non-slum areas of the city, while in Cape Town, children under the age of five living in slums are five times more likely to die than those living in high-income areas.

Immunization is no substitute for healthy living conditions

Persistently higher rates of child mortality in low-income settlements and slums than non-slum urban areas point to deficiencies in current approaches to curbing mortality, which have focused primarily on immunization against deadly childhood diseases. Three-quarters of all children in developing regions are now immunized against measles, and immunization levels are high in both rural and urban areas, including slums. Immunization has continued to be a prevalent approach to decreasing child mortality around the world and is supported by the international community and individual governments, but immunization alone appears insufficient for children who live in slum conditions.

In countries in which most children receive the measles vaccine, measles-related deaths have dropped dramatically or been eradicated. The remaining child mortality rates reflect instead deaths related to illnesses such as pneumonia, diarrhoea, malaria, and HIV infections, with malnutrition as an important contributing factor. In countries that have been successful in immunizing children, policies to reduce child mortality must now address the significant environmental and social factors that contribute to the death of children under five.

Some countries, including Niger, Nigeria, India, Pakistan, and Haiti, report that measles is still among the five main causes of child deaths, particularly in slums and rural areas. In these countries, the reduction of high mortality rates will require substantial resources to immunize children against measles as well as improve living conditions that contribute to the incidence of diarrhoea, pneumonia and malaria. The immunization coverage in Niger is as low as 33 per cent, and the country's child mortality rate is among the highest in sub-Saharan Africa (270 per 1000 live births), with wide inequalities in coverage between non-slum (86 per cent), slum (35 per cent) and rural areas (28 per cent). In Haiti, coverage is comparable to some of the poorest countries in sub-Saharan Africa, at around 50 per cent for both rural and slum areas. Overcrowding, inadequate water and sanitation and poor hygiene all contribute to the prevalence of infection and disease among children. Each of these environmental factors is more prevalent in slums than in non-slum urban areas.

Diarrhoea: The silent killer in slums

Infectious agents enter the body through four main pathways: air; food, water and fingers; skin, soil, and inanimate objects; insect vectors; and mother-to-child transmission.[7] Children living in slums are likely to come into contact with contaminated air, food, water and soil, and to be exposed to conditions in which parasite-carrying insects breed. Two conditions – pneumonia and diarrhoea – are prevalent among children in slums and are responsible for a large proportion of child deaths, each killing more than 2 million children in developing countries each year.[8] Despite their impact on children's health, pneumonia and diarrhoea – and the conditions within children's living environments that cause them – are not typically given high priority in interventions aimed at reducing child mortality. In some Asian countries, slum dwellers are more likely to suffer from diarrhoea than both the non-slum and the rural population. For example, in Bangladesh, the prevalence of diarrhoea among slum dwellers is 25 per cent – double the rural and the non-slum level.

One factor that may explain the high level of diarrhoea in slum areas is the existence of pit latrines sometimes shared by hundreds of families. The use of ventilated pit latrines alone is not a health hazard – in rural households, they have an insignificant relationship to the prevalence of diarrhoea – but in urban areas, the number of latrines may not be sufficient for the number of households, leading to unsanitary conditions that increase the risk of coming into contact with contaminated faecal matter and spreading the bacteria that cause diarrhoea. In Nigeria and Cameroon, the use of pit latrines in urban areas is strongly related to the prevalence of diarrhoea. This opens the debate on whether the current practice of defining ventilated pit latrines as an acceptable form of "improved sanitation" in urban areas is still valid.[9]

Another contributor to the high rates of diarrhoea in slums may be the fact that in many cities, slum households are not connected to municipal drinking water supplies. Families may have to rely on water sold by vendors or from other sources that may be contaminated. In slums, the risk of contamination from unhygienic latrines, lack of solid waste disposal, poor drainage, and inadequate wastewater treatment is also high and can contribute to the spread of a variety of water-borne and water-related diseases.

The higher the incidence of slums in cities, the greater the prevalence of diarrhoeal infections among the urban population. In the Sudanese capital of Khartoum – where the slum population comprises 80 per cent of the urban population, the highest slum incidence for a capital city – the prevalence of diarrhoea is 33 per cent, compared with 29 per cent in rural areas. In Khartoum's slums, the prevalence is even higher, at 40 per cent. The importance of disaggregating urban data is illustrated by statistics collected in Nairobi, Kenya. In 1998, the prevalence of diarrhoea among slum children in Nairobi was 27 per cent, compared with 19 per cent in rural areas. Official figures, however, show that the average prevalence of diarrhoea among children in Nairobi is 12 per cent, a figure that masks the high proportion of children suffering from diarrhoea in the city's slums.

The incidence of diarrhoea among children living in slums is higher than that of rural children, regardless of household income. Demographic and Health Surveys show that children from the highest-income groups within slums have higher rates of diarrhoea than children of the poorest rural families. This suggests that the living environment of slum children, in which they are exposed to contaminated water, soil and air, is a more important determinant of whether or not a child will have diarrhoea than the ability of his or her parents to afford health care.

FIGURE 3.3.3 DIARRHOEA PREVALENCE AMONG CHILDREN UNDER FIVE YEARS IN SELECTED COUNTRIES

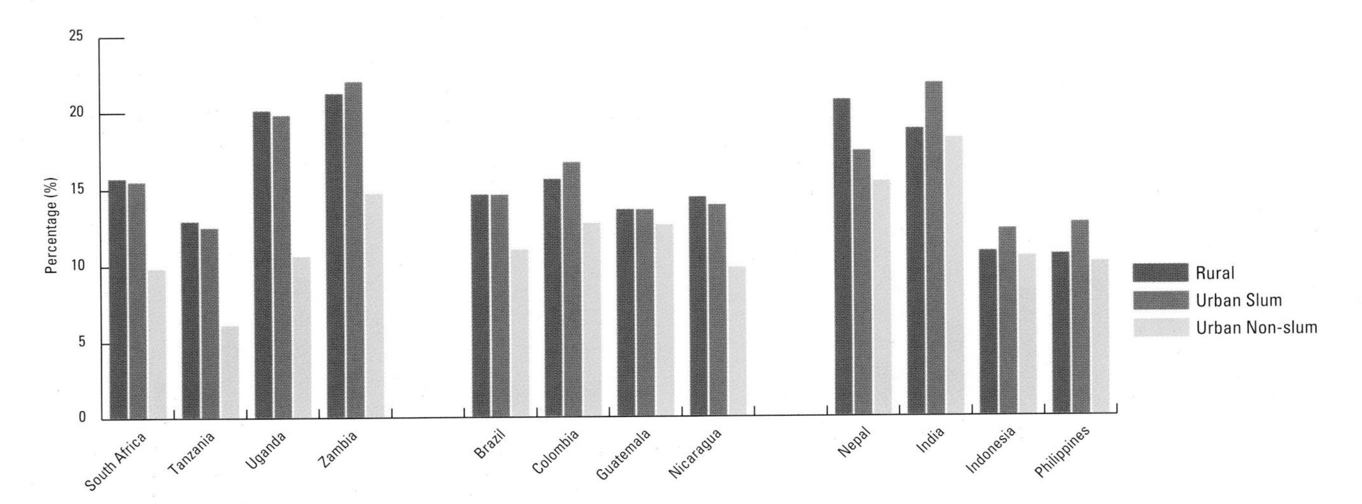

Source: UN-HABITAT 2005, Urban Indicators Programme, Phase III. Based on Demographic and Health Surveys 1995-2003.

Indoor air pollution and acute respiratory infections contribute to child mortality in slums

The concept that the living environment has a direct relationship to child mortality is further supported by data on air pollution and acute respiratory infections. Acute respiratory infections, primarily pneumonia, account for about 18 per cent of deaths among children under five.[10] Effective treatment with oral antibiotics can help in preventing bacterial infections that cause pneumonia, but many poor families in developing countries cannot get to health facilities or purchase medication fast enough. Data from 29 countries shows less than half of the children with respiratory infections are taken to health care providers. In West Africa, less than one-third of infected children have access to health care providers and the medication they need to survive.

In cities of both the developed and the developing world, high levels of indoor and outdoor air pollution caused by motor vehicles, industrial emissions and use of solid fuels for cooking have led to an increase in respiratory illnesses. According to some reports, more than 400,000 people die each year in the Chinese capital of Beijing from pollution-related diseases.[11]

In slums, high exposure to indoor air pollution caused by use of solid or biomass fuels, poor ventilation and overcrowding have also led to higher rates of respiratory illnesses. Families living in overcrowded, poorly ventilated housing without adequate sanitation and safe water are constantly exposed to infectious air-borne diseases.[12] The prevalence of acute respiratory illnesses is, therefore, much higher in slums and rural areas than in non-slum urban areas.

Battery factory in Dhaka slum, Bangladesh JORGEN SCHYTTE / STILL PICTURES

Incomplete and inefficient combustion of solid fuels results in the emission of hundreds of compounds, many of which are health-damaging pollutants or greenhouse gases that contribute to global climate change. Linkages among household solid fuel use, indoor air pollution, deforestation, soil erosion and greenhouse gas emissions have become increasingly important in understanding the impacts of domestic energy use on the local and global environment, and on health. In addition to their local and global environmental impacts, biomass and coal smoke contain a large number of known health hazards.

FIGURE 3.3.4 PROPORTION OF UNDERWEIGHT CHILDREN UNDER AGE FIVE IN SELECTED CITIES IN LATIN AMERICA AND ASIA

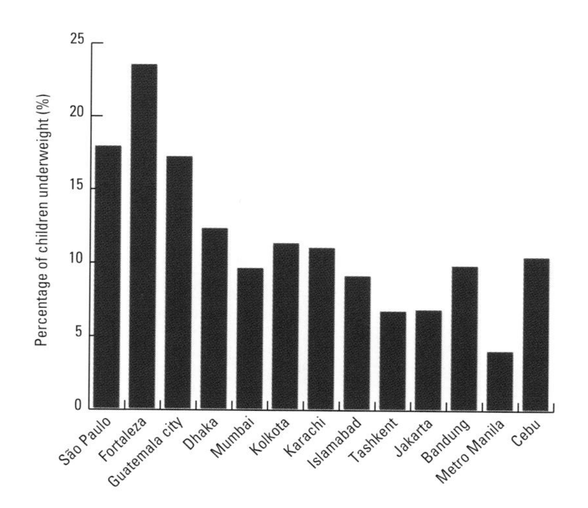

Source: Demographic and Health Surveys 1995-2003.

FIGURE 3.3.5 PROPORTION OF CHILDREN UNDER AGE FIVE WITH ACUTE RESPIRATORY INFECTIONS IN SELECTED AFRICAN CITIES

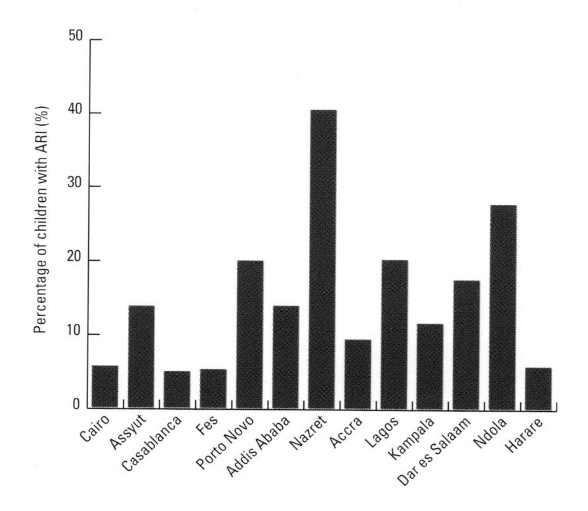

Source: Demographic and Health Surveys 1995-2003.
Note: Acute respiratory infections recorded in the two weeks before the survey.

Exposure to indoor air pollution from the combustion of solid fuels, which especially affects women and small children who are more likely to spend more hours indoors, has been implicated, with varying degrees of evidence, as a causal agent of several diseases in developing countries. Every year, 1.6 million people die from exposure to indoor air pollution, 1 million of whom are children.[13]

Approximately one half of the world's population relies on biomass – wood, charcoal, crop residues, and dung – and coal as the primary source of domestic energy for cooking and heating. Solid fuel use is especially common among low-income households in Africa and South-Eastern Asia. Slum dwellers are up to 10 times more likely to use solid fuels for cooking than those living in non-slum areas. Indoor air pollution can also lead to illness in non-slum households that have enclosed, poorly ventilated cooking areas and are situated among other households using solid fuels.

In overcrowded areas, the potential for viral, bacterial, fungal, and parasitic epidemics is also high. Diseases such as meningitis childhood tuberculosis and adult respiratory infections appear to be closely associated with overcrowding in deprived areas. Conclusive evidence from a study conducted in São Paulo, Brazil, demonstrated a strong relationship between tuberculosis and household overcrowding, particularly in smaller housing units, which suggests that the disease requires prolonged contact. Poor ventilation and crowded living conditions predispose household members to respiratory and skin infections. The lack of a separate kitchen and the use of solid fuel for cooking in an overcrowded house contributes to the high prevalence of acute respiratory infections in slums and urban areas in general.

Endnotes

1 United Nations 2005b.
2 Cornia & Mechinii 2005.
3 Wilkinson & Marmot 2003.
4 World Bank 2005.
5 Crombie, et al. 2005.
6 Ross, et al. 2001.
7 Henry & Chen 1983.
8 World Health Organization 2002.
9 UN-HABITAT is currently working with agencies such as WHO and UNICEF on redefining "improved sanitation" in urban areas on the basis that while a pit latrine may be a safe and adequate form of sanitation in a rural setting, in overcrowded slums where sharing of latrines by dozens of people is common, they are neither safe nor adequate.
10 World Health Organization 2002.
11 Watts 2005.
12 See also Henry & Chen 1983.
13 United Nations Department of Economic and Social Affairs, Statistics Division 2005.

FIGURE 3.3.6 PREVALENCE OF ACUTE RESPIRATORY ILLNESSES AMONG CHILDREN IN SLUMS, RURAL AREAS AND NON-SLUM URBAN AREAS, IN SELECTED COUNTRIES

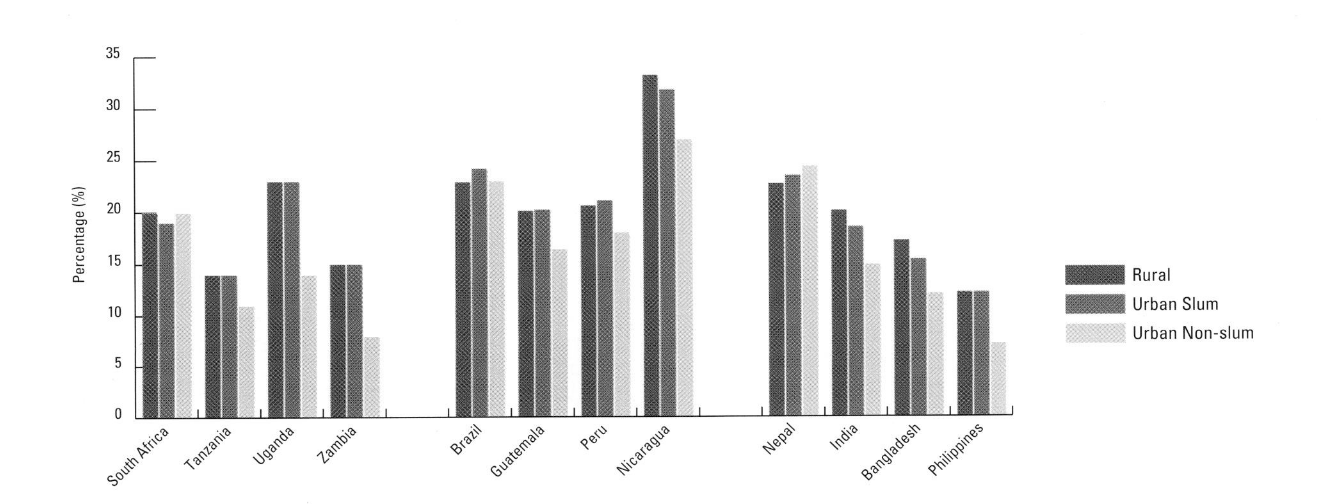

Source: Demographic and Health Surveys 1995-2003.

3.4 HIV/AIDS and Urban Poverty

Where the HIV/AIDS pandemic is rampant, it most deeply affects three linked populations: the mobile, the urban and the poor. Halting and reversing the spread of the disease will require special attention to the needs and struggles of those living in poverty in cities – the most vulnerable, and the most at risk.

Historically, migration has served as a major contributing factor to the spread of HIV/AIDS. Communicable diseases usually spread faster and farther as road and transport networks expand; in fact, disease patterns often follow major highways, seaports and airports.

In the case of HIV/AIDS, trends indicate that the disease first appears in cities and then diffuses to rural areas along major road networks. In Côte d'Ivoire, for instance, HIV first appeared in the capital city, Abidjan, then spread outwards to villages throughout the country. In Southern Africa, HIV/AIDS has been known to spread through transport routes that begin in Zambia and end in South Africa via Zimbabwe, Malawi and Mozambique. The highway that stretches from the Kenyan coastal city of Mombasa through Uganda and Rwanda has also been cited as an HIV danger zone with blame attributed mainly to the risky sexual behaviour of truck drivers along the route. According to a 1994 study, about 50 per cent of truck drivers arriving in the Rwandese capital of Kigali from Mombasa and Nairobi were HIV positive.[1] Tourist centres that attract travellers and sex workers are also conduits for the transmission of the disease.

Kibera, Nairobi SEAN SPRAGUE/STILL PICTURES

Urbanization has emerged as an increasingly important factor in the spread of HIV/AIDS, particularly in sub-Saharan Africa

FIGURE 3.4.1 HIV PREVALENCE AMONG MEN AND WOMEN AGED 15-49 IN URBAN AND RURAL AREAS IN SELECTED SUB-SAHARAN AFRICAN COUNTRIES 2000-2004

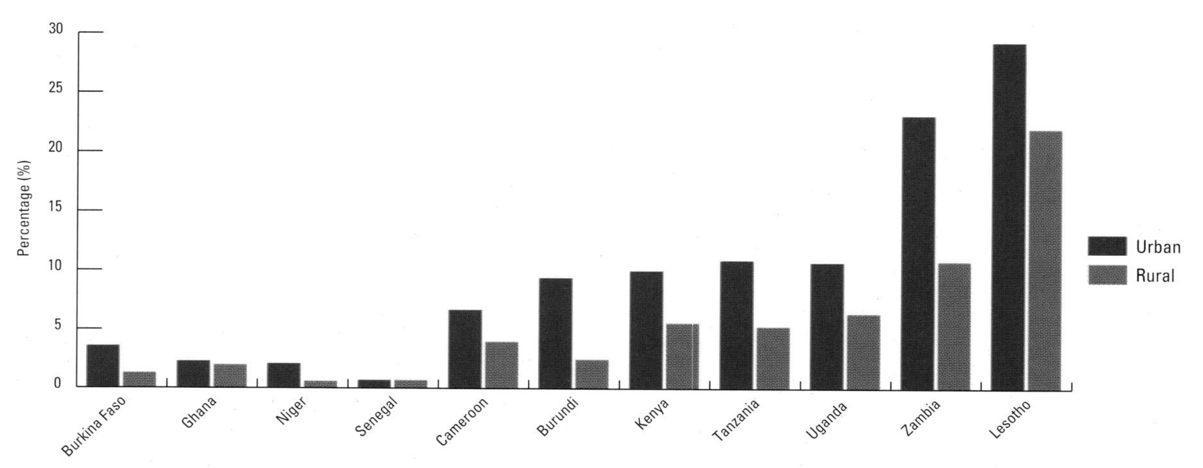

Source: MEASURE DHS, various surveys 2000-2004.

Safe Sex Campaign, Accra, Ghana DUCLOS-UNEP/STILL PICTURES

HIV/AIDS: A largely urban phenomenon

In recent years, urbanization has emerged as an increasingly important factor in the spread of the disease, particularly in sub-Saharan Africa, which is home to more than 60 per cent of all people living with HIV, or more than 25 million people.[2] Recent Demographic and Health Surveys in seven African countries – Burkina Faso, Burundi, Ghana, Kenya, Mali, Niger and Zambia – show that in all countries, HIV prevalence was higher in urban areas than in rural areas, and was also higher among urban women than among rural women. In all countries, women were disproportionately affected, reflecting a general trend in the region. Stark differences were found in Burundi – the least urbanized country in the region – where the percentage of the urban population infected was almost four times the rural percentage, for both men and women. In Kenya, Tanzania and Zambia, HIV prevalence among urban populations was almost twice that of rural populations, whereas in Ghana there was not much difference between rural and urban prevalence; Ghana also had relatively low HIV prevalence nationally, with approximately 2 per cent of the population infected.[3]

Data from other regions shows that HIV tends to be concentrated in the larger cities, where opportunities for commercial sex work are higher and where intravenous drug use is more prevalent. In Argentina, for instance, HIV is concentrated in the largest cities, and an estimated 65 per cent of HIV infections occur in the capital, Buenos Aires, alone. Similarly, in Bolivia, the epidemic is concentrated largely among commercial sex workers in cities such as Santa Cruz and La Paz.[4]

Urban poor disproportionately affected

The urban poor are disproportionately affected by HIV/AIDS in both developed and developing regions. HIV prevalence appears to be closely linked to levels of poverty in both the United States and Canada; in the former, the epidemic is disproportionately lodged among African Americans and is affecting increasing numbers of women, while in the latter,

In Latin America, HIV appears to be concentrated in large cities.

ALEKSEY KLEMENTIEV, IMAGE FROM BIGSTOCKPHOTO.COM

research shows that indigenous people are twice as likely to be infected as non-indigenous people.[5]

Heterosexual intercourse accounts for most of the HIV diagnoses among African American women in the United States, but the risk factor involves the often-undisclosed risk behaviour of their male partners. Recent research in a low-income area of New York City, for example, has shown that women are more than twice as likely to be infected by a husband or a steady boyfriend as by casual sex partners.[6] This trend was also found in the Chinese city of Guangzhou, where some 72 per cent of women with sexually transmitted diseases said they had only had sex with their husband or regular partner. Women now account for more than one quarter of new HIV infections in India; most of those tested said they were in long-term relationships.[7]

In some countries of Eastern Europe and Central Asia, economic transition, liberalization and rising inequality have contributed to the spread of the epidemic. An estimated 1.4 million people are living with HIV in these regions – a greater than nine-fold increase in less than 10 years. The Russian Federation has the largest number of people living with HIV in the region and accounts for some 70 per cent of all HIV diagnoses officially registered in Eastern Europe and Central Asia.[8] Intravenous drug use appears to be a more significant factor in the spread of the disease than commercial sex, although the prevalence of drug use among commercial sex workers is high, increasing their chances of contracting the virus.

High levels of unemployment and low wages are related to transactional sexual activity among the urban poor, and many women are resigned to using sex to meet their basic needs for food, shelter and clothing. Low socio-economic status increases the risk of transactional sex among women and raises their risk of experiencing coerced sex.[9] Indeed, economic hardship not only compounds women's sexual vulnerability, but is also associated with early sexual debut and pregnancy, extramarital sexual activity, and multiple sexual partnerships, all of which have serious implications for the spread of the disease.[10]

Clearly, the inability of many people living in poverty in cities to satisfy their basic needs has implications for the spread and management of HIV/AIDS transmission, which in sub-Saharan Africa is primarily via heterosexual relations. High-risk sexual behaviour, fractured family networks and poor access to health services appear to account for the high prevalence of HIV/AIDS in Africa's urban areas. The African Population and Health Research Center's work with collaborators has shown that the poorest women in Kenya's capital city, Nairobi, initiate sex one year earlier than their rural counterparts, and three to four years earlier than their wealthier city counterparts.[11]

Similarly, the proportion of the urban poor with multiple sexual partners is significantly higher than for the rural poor. This may be attributable to the extreme poverty in slum communities interacting with the centrality of money to urban survival. Because men in urban slums generally have low incomes, they pay very little for their sexual transactions, in a sense forcing women to retain multiple partners in order to make ends meet.

More troubling is the fact that, in the face of these realities, the urban poor often do not use condoms for protection against sexually transmitted diseases. In India, for instance, research reveals that one quarter of street-based sex workers do not use a condom if their clients decline to use one. Despite their riskier sexual practices, and greater knowledge of and access to condoms, urban dwellers in Kenya are only slightly more likely to use condoms than those who live in rural areas – 10.8 per cent versus 8.3 per cent, respectively.[12] These findings are robust even for married women. Nairobi's poor are actually less likely to use condoms than the rural poor in Kenya, further increasing their risk of contracting HIV.

The risky sexual outcomes observed among the urban poor are not simply a result of low HIV/AIDS awareness. Rather, the manifestation of deprivation in urban settings appears to disadvantage residents more than in rural contexts. High unemployment, low and unstable wages, small and congested living spaces, and fractured family and social relationships all contribute to the urban poor's vulnerability, which forces them to resort to sexual behaviour they might otherwise avoid.[13] Conversely, the rural poor often do not face the same magnitude of challenge and survival difficulty as their urban counterparts, in the sense that rural residents may have fewer housing expenses and may be able to grow part or all of their own food.

The deprived conditions in urban slums also serve to encourage children and adolescents to experiment with sexual activity at an early age.[14] Parents in poor urban settings worry about their children being socialized into sex at a young age. Children are exposed to prostitution in their communities and

Slum conditions increase risk of HIV infection in Nairobi

Given their poor access to proper medical facilities, people living in rural areas in developing countries are often assumed to have worse health outcomes than people living in urban areas. While this assumption generally holds true in most countries, in Kenya, evidence suggests that those in urban slums are worse off than their rural counterparts and are more vulnerable to infectious diseases such as HIV/AIDS.

A survey conducted in 2000 by the African Population and Health Research Center (APHRC) found that compared to other areas in the country, slum residents in the capital city Nairobi suffer worse health and reproductive health conditions than their non-slum counterparts. Not only are morbidity risks for all major childhood diseases (fever, cough, diarrhoea) higher for slum children compared with children elsewhere, but slum children also have less access to immunization, and subsequently suffer higher mortality rates than children in rural areas. Vulnerability to HIV/AIDS was also significantly high in slums, particularly among girls and women.

The survey also revealed a marked difference in perceptions of how to avoid contracting HIV. Despite being the most widely known sexually transmitted infection in slum communities, HIV/AIDS awareness campaigns appeared not to have reached some of the most vulnerable women. The APHRC survey showed that a substantial percentage of uneducated and never-married women who reside in the city's slums (the group most likely to supplement household income with commercial sex) were ill-informed about the disease. The lack of awareness about HIV/AIDS prevention among the urban poor is reflected in national data on HIV prevalence. According to UNAIDS/WHO, in 2002, HIV prevalence in Kenya was more than two times higher in urban areas (14.3 per cent) than in rural areas (6.3 per cent). As in many African countries, HIV prevalence in Kenya is also higher among women (8.7 per cent) than among men (4.5 per cent). However, more recent data shows a marked decrease in infections nationally, from 10 per cent in 2003 to 6.1 per cent in 2005, which is mainly attributed to increased awareness and the establishment of voluntary testing centres (VCTs) countrywide.

Poverty-driven commercial sex, crime, domestic violence, child abuse, unwanted pregnancies and unsafe abortions are some of the most socially damaging consequences of urban poverty. In Nairobi, urban poverty is spatially manifested in dehumanizing, overcrowded slums that lack the most basic services, including toilets and health care facilities. Social exclusion and the breakdown of social support structures, such as the family and the community, have contributed to high crime levels and to increase in sexually risky behaviour. A recent report indicates that rape and incest are increasingly becoming urban phenomena in Kenya and are particularly prevalent in slums.

With an annual growth rate of between 7 per cent and 4 per cent over the last two decades, Nairobi remains one of the fastest-growing cities in Africa. The population has grown more than tenfold since 1960 to over 2.8 million people today. The growth of Nairobi reflects a pattern countrywide: with nearly 40 per cent of its population already living in urban areas, the country will become half urban in less than a decade. Rapid urbanization and a lack of increase in service provision, poor urban management, inefficient revenue collection, and policies that favour the rich at the expense of the poor have all contributed to increasing poverty and slum growth in the city. Official data shows that while absolute poverty increased from 48 per cent to 53 per cent in rural areas between 1992 and 1997, poverty in Kenya's urban areas increased by a much larger margin: from approximately 29 per cent in 1994 to 50 per cent in 1997. An official slum survey shows that 35 per cent of Nairobi's households live in slums, although other sources estimate a much higher slum population of between 40 and 60 per cent of the city's total population.

Nairobi's slums have been described as among the most dense, unsanitary and insecure slums in the world. Recent studies indicate that only 24 per cent of slum households in the city have access to piped water; slum residents pay significantly higher charges for water than other Nairobi residents, adding to their financial burden. In some slums, more than 200 people share a single toilet.

Social isolation, poor or non-existent basic services, overcrowding, low incomes, illegal status, and a generally dehumanized existence all combine to make slums in Nairobi vulnerable to a host of health and environmental hazards, which are manifested in higher prevalence of HIV/AIDS and other infectious diseases, including HIV-related tuberculosis. Combating HIV/AIDS and other infectious diseases in slums – by providing better access to water, sanitation, adequate housing, health facilities, information, and education – will not only improve the health and dignity of the urban poor, but will also help them become more economically productive and improve their livelihoods, leading to a general reduction in urban poverty in the city.

Sources: African Population and Health Research Center (APHRC) 2002; Government of Kenya/UNCHS 2001; Central Bureau of Statistics 2000; Nalo 2002; UNAIDS/WHO 2004a; UN-HABITAT/DPU 2003, Chamber of Justice and others 2005, Daily Nation 2005.

Extreme deprivation in poor urban settings in sub-Saharan Africa

- Because of extreme levels of poverty and the unique social characteristics of urban poor settings, the urban poor are, to a large extent, more likely than their rural counterparts to initiate sex very early and to have multiple sexual partners.
- Urban poor women initiate sex one year earlier than the poorest women in rural areas and three to four years earlier than their wealthiest counterparts in urban areas.
- The proportion of Nairobi's poorest who engage in multiple sexual partnerships is more than three times greater that of the city's wealthiest residents.
- Married women living in Nairobi's informal settlements are at least three times as likely as their rural counterparts to have multiple sexual partners.
- Even though the urban poor exhibit riskier sexual behaviour, their condom use rates are low (10.5%), and do not vary significantly relative to those of the rural poor (8.3%).

Disadvantages Associated with Slum Settings Compared to Rural Settings

- No privacy exists for parents to have sexual intercourse. This does not only foster an interest in sexual activity among children at young ages, but also denies parents moral authority over their children as it relates to sex.
- The financial ability to meet immediate basic needs of food and shelter overshadows the risk of contracting HIV/AIDS and other sexually transmitted diseases (i.e., women opt to engage in commercial sex to buy food for their children despite knowing the dangers to which they are exposing themselves).
- Because of widespread prostitution in urban poor communities, many young girls living in these communities consider prostitution a viable livelihood regardless of its risks.
- There are higher proportions of single men and women in slum settlements than in any other community, and this contributes significantly to the high levels of risky sexual behaviours among residents of these communities.

Source: African Population and Health Research Center (APHRC).

Big City Life ©GALINA BARSKAYA. IMAGE FROM BIGSTOCKPHOTO.COM

their parents' own sexual activities at home, owing to cramped living quarters and lack of privacy. Parents argue that being seen or heard having sex impacts their dignity and robs them of the moral authority over their children. Perhaps most disturbing, the economic deprivation in poor urban communities appears to have commercialized sex even for adolescent girls, who have little else to trade but their bodies. When the economic situation gets especially desperate, parents sometimes draft their young daughters into contributing their share of household expenses.

There are multiple links between poverty and risky sexual behaviour among young people. Research in Southern Africa has shown that poorer young people have less knowledge of HIV/AIDS and begin having sex at younger ages than their wealthier peers. Poverty and lack of parental resources are cited as primary reasons for young women to trade sex for goods or favours or to engage in relationships that involve financial support. Condom use is reported to be consistently lower in these types of sexual encounters.[15] In a national survey in South Africa, young people aged 15 to 24 living in poor informal settlements had more than double the HIV prevalence of those residing in wealthier urban areas: 20 per cent versus 9 per cent, respectively.[16] In this age group, 79 per cent living in informal urban settlements reported being sexually active as compared to 53 per cent of those living in formal urban areas. In another large survey in South Africa, researchers showed that young people in poor informal urban areas had a much higher HIV prevalence rate than those living in urban formal areas: 17 per cent versus 10 per cent, respectively. HIV prevalence was three times higher among young women than among young men.[17]

The worst orphan crisis is in Africa, where 12 million children have lost one or both parents to AIDS; by 2010, this number is expected to climb to more than 18 million. Many of these children end up on city streets.

The socio-economic impact of HIV/AIDS

The loss of income-earning family members to AIDS has significant socio-economic implications for the urban poor. In slums, where there are large numbers of female-headed households, the loss of a parent can be devastating to children, who may be forced to drop out of school, become street children or engage in prostitution to meet the needs of younger siblings.

The worst orphan crisis is in Africa, where 12 million children have lost one or both parents to AIDS; by 2010, this number is expected to climb to more than 18 million. Many of these children end up on city streets, where their chances of escaping poverty are even lower. A recent study in Cambodia found that one in five children in AIDS-affected families had to start working to support their families. Many had to leave school or forego necessities such as food, medicine and clothing.[18]

At the national level, the epidemic's economic impact on societies has been devastating. In sub-Saharan Africa, many of the worst-affected countries are also among the poorest. Zambia's gross domestic product shrank more than 20 per cent from 1980 to 1999,[19] around the same period when almost a quarter of its urban population and one-tenth of its rural population became infected with HIV.

The epidemic's demographic impact is profound: if current infection rates continue, up to 60 per cent of Africa's 15 year-olds will not reach their 60th birthday.[20] AIDS threatens economic security and development because the disease primarily affects people in the prime of life, between the ages of 15 and 49. The International Labour Organization (ILO) projects that the labour force in 34 African countries will shrink by 5 per cent to 35 per cent by 2020 because of AIDS. This has serious repercussions for the continent's ability to achieve sustainable economic growth and poverty reduction. All of the factors involved in urban poverty must be confronted in order for cities, countries and the international community to make progress toward meeting the Millennium Development Goal 6 target of halting and reversing the spread of HIV/AIDS.

Endnotes

1 Bwayo et al. 1994.
2 UNAIDS 2004.
3 Demographic and Health Surveys in these countries conducted between 2001and 2003. Data derived from UNAIDS/WHO *AIDS Epidemic Update*, December 2004.
4 UNAIDS/WHO 2004b.
5 Ibid.
6 McMahon et al. 2004.
7 UNAIDS/WHO 2004b.
8 Ibid.
9 Hallman 2004.
10 Carael & Allen 1995.
11 Zulu et al. 2002.
12 Ibid.
13 A detailed report of this research is forthcoming in *Social Sciences and Medicine* – "Urban-rural differences in the socio-economic deprivation-sexual behaviour link in Kenya" by Dodoo, Zulu & Ezeh.
14 Dodoo et al. 2003.
15 Eaton et al. 2003.
16 Shisana 2002.
17 Pettifor et al. 2004.
18 UNAIDS 2004.
19 Ibid.
20 Ibid.

3.5 Education and Youth Employment: Debunking Some Myths about the 'Urban Advantage'

"If we are serious about reaching the Millennium Development Goals by 2015, we must involve young people today. We must invest in them; we must learn from them; we must be their partners." - UN Secretary-General, Kofi Annan[1]

We live in a youthful world. Almost half of the global population is under the age of 24; 1.2 billion people on the planet are younger than 15.[2] While the overall share of children and youth in the global population is shrinking as fertility rates decline, in absolute numbers, there are more young people today than ever before. Fully 85 per cent of the world's working-age youth,[3] those between the ages of 15 and 24, live in the developing world – primarily in Southern Asia and Africa. Within developing regions, it is the least developed countries that remain younger than the rest of the world: in 2005, the global median age was 28 years, but in 10 least developed African countries, the median age was 16 or younger.[4]

Youth embody a significant proportion of the world's human capital, but more than 500 million of them live on less than $2 per day. And while more young people are attending school today than ever before, 113 million children are still not enrolled and 130 million youth remain illiterate.[5]

Issues affecting children and youth are often framed as problems germane to underdeveloped rural areas rather than cities. Indeed, in general, cities appear to foster the healthy development of children and youth, providing easier access to education, health care and employment for young men and women than is available in rural villages. However, not all who grow up in cities benefit from the so-called "urban advantage", as data collected by UN-HABITAT and its partner agencies reveals. This chapter presents data on the stark differences for young people within cities: those living in slums, and those living in non-slum urban areas.[6]

Intra-city inequalities in access to education

Available data indicates that school enrolment rates are in general much higher in cities than in villages. In countries such as Burkina Faso, rural communities lag far behind their urban counterparts, with 21 per cent enrolment in rural areas and 73 per cent in cities. In Burkina Faso, living in an urban area has a clear advantage, regardless of whether one is rich or poor. Inequalities in access to school facilities can partly explain this urban-rural differential, but surveys in other countries show that while school enrolment rates in rural areas are dependent on the availability and accessibility of school facilities, the availability of schools in urban areas is not sufficient cause for children to be enrolled in school. Families in slum communities, in particular, often cannot afford to send their children to school because the combined costs of school fees, textbooks and uniforms are prohibitive. In Kenya, for example, the government mandated free primary education in 2003, but students must still purchase uniforms and supplies, and pay fees to take exams, making it difficult for low-income families to send their children to school and ensure their progress. Even in slum areas served by several schools, the number may not be sufficient, further prohibiting children's access to quality education. A study in the Nairobi slum of Kibera in 2003 found that while 14 public primary schools were situated within walking distance of the slum, the schools could only accommodate 20,000 of the more than 100,000 primary school-age children living in the area.[7]

Lack of access to school for poor children in cities is exacerbated by the fact that most national and international literacy and education programmes have focused in recent years on reducing the urban-rural gap in education. Although much remains to be done in rural areas, it is important to recognize that in the past decade there has been a significant increase in enrolment in rural areas and a decrease in enrolment in impoverished urban communities.

The problem is evident in poverty-stricken areas of many African cities, where primary school enrolment is decreasing. In Eastern and Southern Africa, the most significant progress in school enrolment in the late 1990s was concentrated in rural areas, leaving many poor urban families behind. In Tanzania net enrolment ratios increased in both rural and non-slum urban areas, but actually decreased in slum areas, as indicated

First Day in School, Patuakhali City, Bangladesh JORGEN SCHYTTE/STILL PICTURES

by UN-HABITAT's analyses of urban survey data. Similar situations have evolved in Zambia and Zimbabwe, as well, but the disparity is not confined to sub-Saharan Africa. In Guatemala in 1999, only 54 per cent of children living in slums were enrolled in primary education, versus 73 per cent in non-slum urban areas and 61 per cent in rural areas. The same situation was observed in Brazil in the late 1990s. Studies indicate that a majority of parents settling in slums postpone sending their children, especially girls, to school, until they can manage other expenses, such as food, rent and transport.

Causes of social inequality in basic education vary from country to country, but there is a common set of constraints to be considered, including poverty; the embedded costs of education; shortage of school facilities; unsafe school environments, especially in poor urban neighbourhoods; and cultural and social practices that discriminate against girls, including requirements that they provide domestic labour, marry and have families at a young age, and limit their independent movement to proscribed areas. More barriers to education exist for girls than for boys around the world. Where resources are limited and school systems are less responsive to the needs of girls, they risk losing important opportunities to fulfill their potential and improve their lives.

■ The gender gap in urban education

Eliminating gender disparities in access to education is essential to the achievement of the Millennium Development Goals – particularly Goal 3 on promoting gender equality and empowering women. Girls have historically had less access to educational opportunities than boys in many countries; in 2005, the United Nations Children's Fund (UNICEF) found that girls in 54 countries still did not have equal access to basic education.[8] Countries in several regions have made progress toward the goal of gender parity, however. In the 1990s, the

FIGURE 3.5.1 NET ENROLMENT RATE (PRIMARY) BY TYPE OF RESIDENCE IN SELECTED COUNTRIES

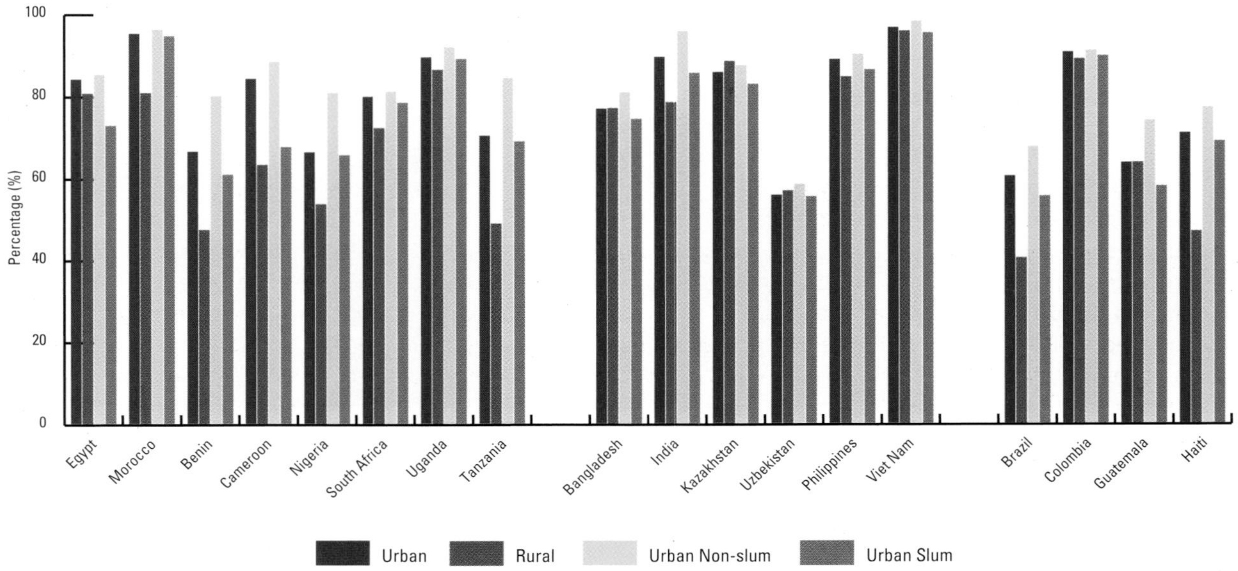

Source: UN-HABITAT 2006, Urban Indicators Programme Phase III.
Note: Computed from Demographic and Health Surveys DHS data 1995-2003.

gender gap in primary school enrolment narrowed, most evidently in regions where the gap was wide, such as Northern Africa. In the developed regions and in Eastern Asia, the gender disparity has reversed, with more girls than boys now enrolled at the primary level.

Progress indicated by regional estimates has been uneven within regions. Where girls are still at a disadvantage, resources and school facilities are limited and enrolment is altogether low. In many countries with low overall enrolment, fewer than 50 per cent of primary school-aged girls are enrolled. Female illiteracy rates are still high in these parts of the world, particularly in urban poor and rural areas, where many girls drop out of school too early to be able to acquire the necessary skills to function as literate individuals. Demographic and Health Survey data points to four main reasons why girls discontinue their education: lack of finances, early marriage and pregnancy, domestic work responsibilities, and poor performance. Only a small proportion of girls and young women who had left school – fewer than 10 per cent – indicated that they stopped attending because they had graduated.

Lack of finances

The direct financial costs of sending all children to school are often too high for families living in poverty in cities. Faced with household expenses, urban families may cut back by not paying school fees, and daughters are typically the first casualties of this choice. Girls are more likely than boys to suffer from limited access to education, especially in urban poor and rural areas. Secondary analysis of survey data shows that on average, the single most common reason young women reported for leaving school was inability to pay the associated fees. In the urban areas of Uganda and Zambia, for instance, 74 per cent and 51 per cent, respectively, of young women between the ages of 15 and 24 gave inability to pay as the main reason they stopped going to school.

While primary school tuition fees have now been abolished in many countries, public secondary education remains competitive and tuition-based in many parts of the developing world, limiting the number of students who can continue their education. Even for primary school, nearly all developing countries still require families to pay fees of various kinds – in many cases, these fees amount to more than the former tuition costs. Fees for uniforms, materials and other educational expenses have been shown to affect girls' chances of going to school more than boys', as they add to the already high costs of sending girls to school. Among some impoverished urban communities, it is common for families to choose to educate their boy children in their village of origin where schools are less expensive; girls, on the other hand, remain in the city to help parents with housework. This is reflected in the age pyramid of slum areas, which shows that slum communities have more girls than boys between the ages of 5 and 14 years. (See age pyramids in chapter 1.2, for example.)

Immigrants in Paris: Dreams go up in flames

Between April and September 2005, three fires ravaged residential buildings in Paris, killing 48 African immigrants, primarily from Senegal, Côte d'Ivoire and Mali. Most of the victims were children; many were undocumented.

The immigrants lived in cheap hotels and apartment houses ill-equipped for emergencies, lacking smoke detectors, fire extinguishers, emergency exits, and, in one case, even running water with which to put out the blaze. Some of the families had been placed in the substandard accommodations by social service agencies while waiting for their residency papers to be processed. Others entered the tenements on their own, squatting in the only shelter they could find.

For refugees from African slums seeking a better life in Europe, Paris offers little relief from the insecurity and destitution they experienced at home. Officials estimate that more than 200,000 people are homeless or living in temporary shelter in the city. Subsidized social housing units are scarce – in 2004, more than 100,000 families were on waiting lists for 12,000 available units. Some families languish in overcrowded and filthy provisional dwellings for 14 years or longer while they wait to be accommodated in social housing. Such long waits are not uncommon for immigrants. A government study found that nearly 30 per cent of immigrant applications had been pending for more than three years, two times the national average.

Although *droit au logement*, or the right to housing, is ensconced in French law, access to a decent, affordable place to live remains elusive for the lowest-income and minority residents. Legislation passed in 1991 requires that major cities dedicate 20 per cent of their housing stock to the social sector, but many contend that the law is not adequately enforced.

Finding appropriate housing remains challenging even for families who can afford market rental rates. In 2002, the housing vacancy rate in Paris was 6.2 per cent, the lowest since the late 1960s. Those few units that are vacant tend to be substantially older than occupied ones. In the ageing and dilapidated buildings in which the fires occurred, only one exit was available – via the central wooden staircases, which burned quickly and left families stranded on the upper floors.

The Paris city government plans to renovate 1,000 identified substandard apartment blocks, in addition to building 60,000 units of housing each year to help quell the crisis. Tenants' advocates, however, maintain that more than 120,000 new units are needed each year. For immigrants awaiting both housing and legal resident status in the tenements of Paris, every day in a building with faulty wiring, inadequate plumbing and only one way out brings the risk of another tragedy.

The problems for immigrants in Paris are deeper than substandard housing, as demonstrated by the riots that swept the city in October and November 2005. Young residents of minority communities throughout Paris and its suburbs responded with violence to the accidental deaths of two teenage boys of African origin, setting cars and buildings ablaze for more than two weeks. The frustration and anger expressed in the riots grew out of the marginalization of ethnic and religious minorities, the majority of whose members live in run-down high rise housing estates in poor neighbourhoods. Growing resentment over unemployment in their communities and the overriding sense that they are targeted by police and excluded from opportunity in France, has forced many immigrants to ask themselves whether they are really better off in their adopted lands.

Lack of opportunity and social exclusion remain major political and social issues for immigrant communities in France. The youth who spoke out during and after the riots protested vehemently over two questions employers consistently asked during job interviews: the applicants' ethnic origins and their *address*. Employers were known to discriminate against those who lived in stigmatized suburbs. (A similar study in Rio de Janeiro found that living in a favela appeared to be a bigger barrier to gaining employment than being dark skinned or female.) Unemployment among immigrant communities in France is estimated to be around 40 per cent, 30 per cent higher than the national average. A recent study found that white male applicants were 5 times more likely to get job offers than those with Arab-sounding names or those whose physical home address was among area postal codes that were deemed "undesirable".

The disparities in housing and employment opportunities between immigrants (most of whom are French nationals) and the local population has prompted the French government to create more health, education and employment programmes aimed at young people living marginalized, low-income neighbourhoods. Stigmatization and exclusion of neighbourhoods from the rest of society appears to have exacerbated the crisis in Paris. The French city of Marseilles for instance, was immune from the riots largely because the poor are not physically isolated within the city; there low-income and higher-income communities are more integrated.

Sources: Ford 2005; BBC News 2005b; Bennhold 2005; Norris and Shiels 2004; Langley 2002; BBC News 2005c, TIME 2005; Perlman 2005.

FIGURE 3.5.2 PROPORTION OF WOMEN AGED 15-24 WHO STOPPED GOING TO SCHOOL BECAUSE OF INABILITY TO PAY SCHOOL FEES

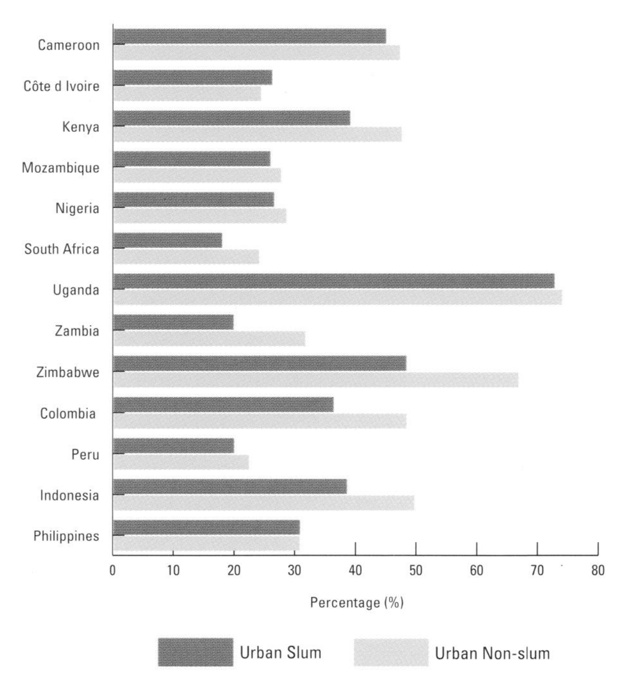

Source: UN-HABITAT Global Urban Observatory 2005.

Early marriage and pregnancy

A significant proportion of girls who discontinue their education in the higher grades of secondary school in urban areas leave school because of pregnancy. In many countries, especially in Eastern and Southern Africa, this proportion is particularly high. In Zambia, 17 per cent of the girls who dropped out of school in urban areas did so because they were pregnant; in the Central African Republic, 16 per cent of female dropouts cited pregnancy as the reason for leaving school, as did 12 per cent of female dropouts in Uganda's urban areas. Sexual harassment and abuse in schools further impacts the dropout rate; in some places, parents are inclined to withdraw girls from school to prevent them from getting pregnant or contracting HIV. In some countries, such as Chad and Nigeria, girls in urban areas often stop going to school to get married: 28 per cent of girls who left school in Chad cited marriage as the reason, as did 18 per cent of female dropouts in Nigeria.

Within urban areas, stark differences exist. Young women and men residing in slum areas are more likely to have a child, be married or head a household than their counterparts living in non-slum areas in cities. In Uganda, for instance, a 1999 Demographic and Health Survey revealed that 34 per cent of young people in slum communities headed households, versus five per cent in non-slum communities. The same pattern was observed in Kenya (27 per cent versus 14 per cent) and Côte d'Ivoire (16 per cent versus two per cent). The prevalence of HIV/AIDS in poor urban communities in these countries may be contributing to their higher rates of youth-headed households.[9]

Domestic work

Some young women stop going to school to help their families with domestic chores, including taking care of children. This phenomenon is particularly prevalent in slums where, in the absence of extended family, girls are taken out of school to do domestic work, such as fetching water, while their parents struggle to earn an income for food, housing and other necessities. In Mali and Chad, more than 10 per cent of young women in slums cited "helping the family" as the main reason why they stopped going to school. Family demands on girls' time place many obstacles in the way of gender equality in access to education. Other studies show that going to school is seen as a hindrance to the performance of household chores; parents perceive the cost of lost labour to be greater than the cost of keeping girls out of school. These perceived opportunity costs are usually much higher for girls than for boys, since girls are expected to do more domestic work than boys. By the age of 10, girls in Bangladesh and Nepal may be working up to 10 hours a day in productive activity inside and outside the home, while Ethiopian girls of primary school age often work 14 to 16 hours a day.[10]

Poor performance

The combined social and cultural factors that make it difficult for girls to enroll in and complete school also contribute to their dissatisfaction with and poor performance in school. Domestic responsibilities, marriage and motherhood, and financial constraints present strong challenges to girls' ability to maintain regular attendance and succeed when they do attend. Surveys indicate that a significant proportion of young women in urban areas stop going to school because of poor performance; the obstacles they face induce many to drop out before they complete their education or pass key national examinations.

A significant proportion of young women drop out because they "do not like school". This is the case for more than 30 per cent of the young women in slum communities in Mali and Guatemala who had left school. In Egypt, Nicaragua, Central African Republic, and Burkina Faso, more than 20 per cent of the young women in slum communities who dropped out of school reported that they did so because they did not like school. Schools in many countries are not girl-friendly and in some cases, they are even hazardous for girls. Failure to provide adequate sanitary facilities, such as toilets and running water, causes inconvenience for boys, but can make the situation disastrous for girls. Menstruating girls will not attend school if basic toilet facilities are not available. Even when toilets are available, they are often poorly serviced and maintained, as is the case in 30 per cent of schools in India. In many places, schools fail to provide separate toilet facilities for boys and girls, putting girls at risk of sexual harassment.

Young women and men residing in slums are more likely to have a child, be married or head a household than their counterparts living in non-slum areas.

Youth are employed in the growing informal sector

In cities of the developed world, more jobs are being created in the financial sector and in information management as a result of globalization, while in the developing world, trends point toward an increasing "informalization" of the urban economy, as the formal sector fails to provide adequate employment opportunities for the number of young people and adults seeking work. According to the International Labour Organization[11], approximately 85 per cent of all new employment opportunities around the world are created in the informal economy. In some countries, employment in the urban informal sector has risen sharply over the past decade. Lithuania, for example, experienced a 70 per cent increase in urban informal employment as a percentage of total employment between 1997 and 2000. The Economic Commission for Latin America and the Caribbean estimates that urban informal employment in that region increased from 43 per cent in 1990 to 48.4 per cent in 1999.[12]

The informal economy can afford youth a necessary pathway to legitimate work by conferring experience and self-employment opportunities. Tracking how many youth participate in the informal sector is difficult for a number of reasons, however, and limited data currently exists.[14] But some trends are beginning to emerge. UN-HABITAT analyses indicate that the majority of young people working in the urban informal sector live in slum areas. For example, in Benin, slum dwellers comprise 75 per cent of informal sector workers, while in Burkina Faso, the Central African Republic, Chad and Ethiopia, they make up 90 per cent of the informal labour force.

Gender differences in employment

In slum communities, early involvement in family responsibilities may explain the high employment rates of young men and the low employment of young women. Youth residing in slum areas are more likely to have a child, be married or head a household than their counterparts living in non-slum areas. In Uganda, 34 per cent of young men living in slum areas head a household compared with 5 per cent of young men living in non-slum areas. Family responsibilities at a young age often compel young men to seek and obtain jobs.

Education in Africa UNEP/STILL PICTURES

A study in the Nairobi slum of Kibera found that while 14 public primary schools were situated within walking distance of the slum, the schools could only accommodate 20,000 of the more than 100,000 primary school-age children living in the area.

FIGURE 3.5.3 PERCENTAGE OF YOUNG WOMEN AND MEN WORKING IN THE INFORMAL SECTOR IN SELECTED AFRICAN COUNTRIES

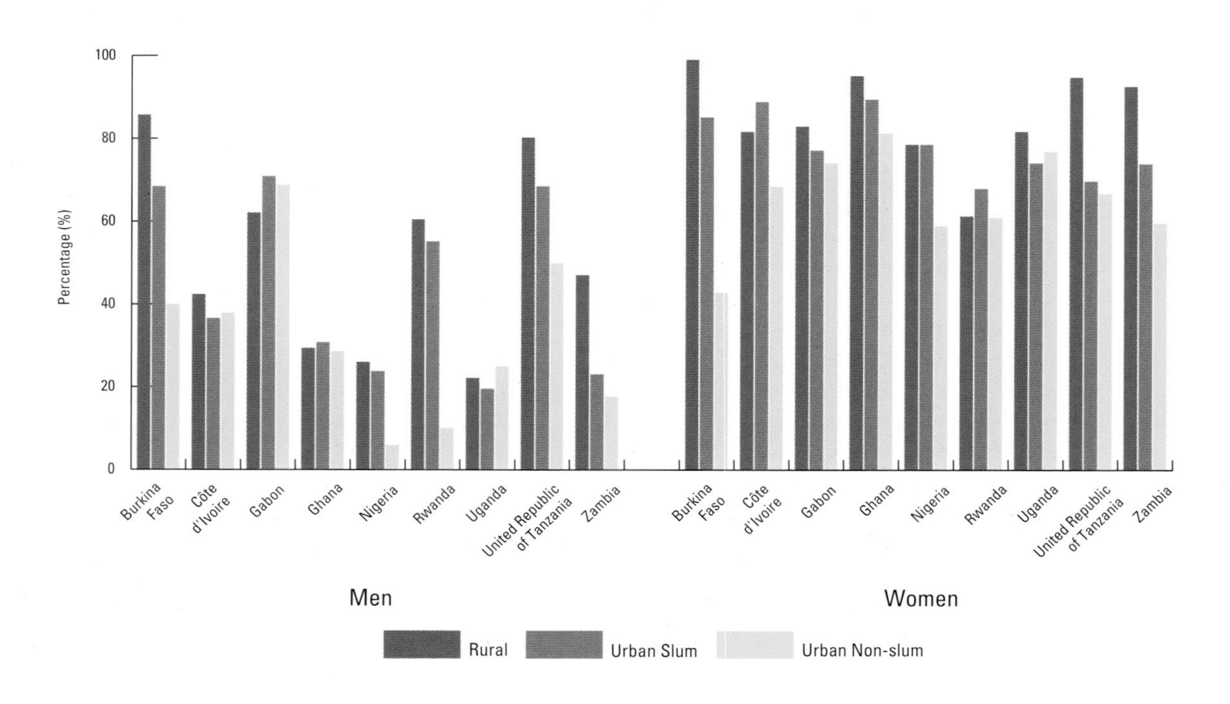

Source: UN-HABITAT Global Urban Observatory 2005.

FIGURE 3.5.4 PERCENTAGE OF YOUNG WOMEN AND MEN WHO HAVE FAMILY RESPONSIBILITIES IN SELECTED COUNTRIES

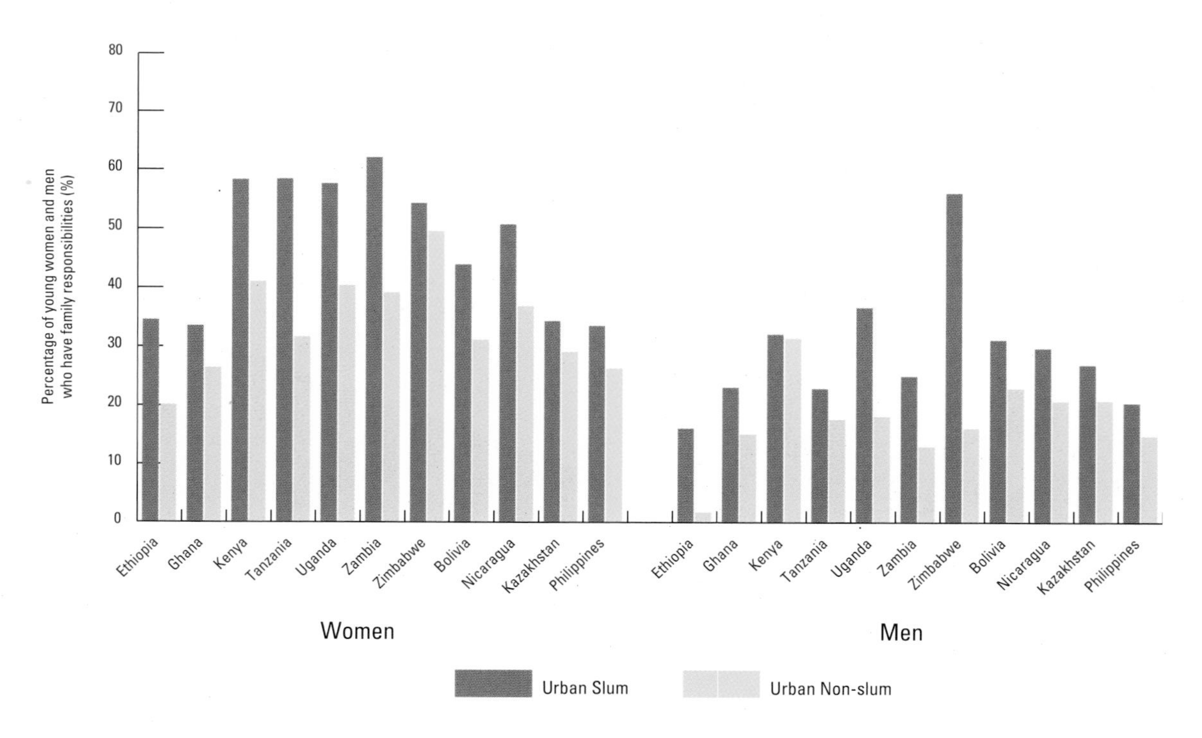

Source: UN-HABITAT 2005, Global Urban Observatory.

FIGURE 3.5.5 PERCENTAGE OF YOUNG WOMEN AND MEN WORKING IN THE INFORMAL SECTOR IN SELECTED ASIAN COUNTRIES

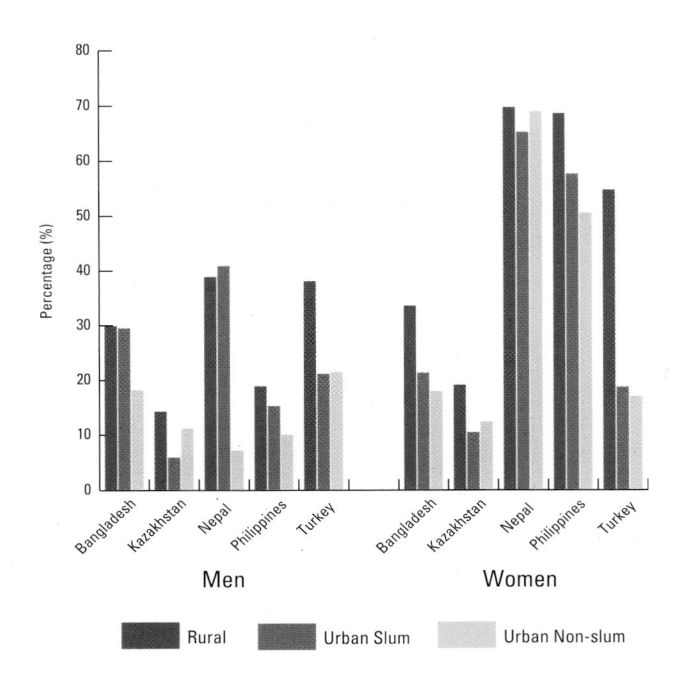

Source: UN-HABITAT 2005, Global Urban Observatory.

On the other hand, young women living in slums are less likely to seek paid employment, as early marriage and childbearing forces them to stay at home. Six out 10 young women living in Uganda's slum communities have a child or are married – double the number in non-slum communities. The majority of young women in slums tend to have children at an earlier age than their non-slum counterparts. In the absence of an extended family to help with taking care of children, the sick and the elderly, young women living in slums are more likely to stay at home to look after children and do household chores. This limits their opportunity to look for jobs away from home, particularly in the formal sector.

Consequences of youth unemployment

When youth seeking work fail to find productive, decent livelihoods, they can become socially excluded and enter a cycle of poverty, experiencing high rates of unemployment across their life spans.[13] The importance of helping youth find productive and decent employment has therefore become a primary motivation of international youth policymaking and development efforts.

Many countries in the developing world are experiencing distinctive "youth bulges", which occur when young people comprise at least 40 per cent of the population. There has been increasing concern among policymakers that the frustrations accompanying long-term unemployment among large populations of young men in urban areas may feed political and ideological unrest and provoke violence. As demonstrated by the riots in Paris in late 2005, high youth unemployment, particularly within marginalized ethnic minorities, can create urban unrest, which can challenge government authority and endanger national stability. More importantly, high levels of unemployment among youth, particularly in urban areas, indicate that cities are unable to absorb labour, which in the long term has a direct impact on economic growth and poverty reduction.

Endnotes

1 United Nations Secretary-General Kofi Annan, in his introduction to the exhibition, "Chasing the Dream: Youth faces of the Millennium Development Goals". New York, 12 August 2005.
2 United Nations General Assembly Economic and Social Council 2004.
3 "Youth" are defined by the United Nations as all persons between the ages of 15 and 24. In this Report, "young people" refers to all persons younger than 24, "children" refers to persons between birth and 15, and "youth" follows the conventional definition.
4 United Nations Department of Economic and Social Affairs, Population Division, 2005.
5 Ibid.
6 Inequalities in education and employment are assessed in this Report with data from Multiple Indicator Cluster Surveys administered by the United Nations Children's Fund (UNICEF) and Demographic and Health Surveys, funded by the U.S. Agency for International Development (USAID), collected between 1995 and 2003. These surveys include items on school attendance and literacy, youth employment and child labour that can be disaggregated by type of residence: urban and rural; slum and non-slum. UN-HABITAT has also used some data from Labour Force Surveys, funded by the International Labour Organization, and Living Standard Measurement Surveys, funded by the World Bank.
7 Lugano & Sayer 2003.
8 UNICEF 2005.
9 See chapter 3.4 for an analysis of the impact of HIV/AIDS on slums.
10 Watkins 2000.
11 International Labour Organization 2004.
12 Ibid.
13 International Labour Organization 2004.

3.6 Cities: The Front Lines in the Battle for Sustainability

Protest against traffic pollution, Turin, Italy ANGELO DOTO/UNEP/STILL PICTURES

As urbanization continues unabated, the global community[1] is confronting the need to think creatively about cities and their potential for leadership in harmonizing human settlements with ecological preservation and sustainability. Sustainable cities – those that enable all of their residents to meet their own needs and prosper without degrading the natural world or the lives of other people, now or in the future[2] – are products of careful planning in the context of their residents' daily lives.

Issues of sustainability are being addressed differently in different parts of the world, according to the policymaking and environmental priorities of cities and countries. In cities of the developed world, energy consumption remains a major concern, and many urban areas are being redeveloped with an emphasis on compact neighbourhoods, clean transportation options and the use of green technologies. Cities such as

Although cities have been much maligned as generators of waste and pollution, consumers of vast amounts of the world's natural resources and contributors to overall environmental degradation, examples from around the world demonstrate that cities have the potential to combine safe and healthy living conditions with remarkably low levels of energy consumption, resource use and waste.

Coco Taxis, Havana, Cuba ©ALEX BRAMWELL. IMAGE FROM BIGSTOCKPHOTO.COM

Vancouver are also analysing their "ecological footprints"[3] and adopting strategies to reduce their dependence on outside suppliers, such as increasing urban agriculture and localizing more of their food supply.[4] In cities of the Commonwealth of Independent States, with economies in transition, the priority is tackling the legacy of underused urban areas, decaying infrastructure and deteriorating housing stock. Some countries in Central Europe are also addressing air pollution and health by imposing heavy taxes on leaded fuel and phasing out its use altogether.[5]

In cities of the developing world, the need to accommodate rapid urban growth, provide essential infrastructure and services, control air pollution (especially in the rapidly industrializing cities of Asia) and improve the living conditions of the urban poor are emerging as new challenges.[6] Some cities, such as Singapore and Curitiba, have adopted careful urban planning and "greening" policies to significantly reduce air pollution and the use of private motorized transport. Singapore has been so successful at preserving its old-growth tropical rainforest, protecting and planting green spaces, and promoting clean rapid transit that it has become the only large city in the world that acts as a carbon sink, soaking up more carbon dioxide than it produces.[7] Elsewhere, in places such as Thailand, governments are embarking on major slum upgrading programmes that will also have a positive impact on the urban environment.

The most innovative cities in the world fulfill the ideals of Millennium Development Goal 7 – ensuring environmental sustainability – by integrating environmental stewardship and urban planning to achieve long-term stability and success. Although cities have been much maligned as generators of waste and pollution, consumers of vast amounts of the world's natural resources and contributors to overall environmental degradation, examples from around the world demonstrate that cities have the potential to combine safe and healthy living conditions with remarkably low levels of energy consumption, resource use and waste. Cities also offer enormous environmental opportunities and advantages.[8]

Cities provide economies of scale. High densities mean low per capita costs for the provision of piped water, water treatment and collection and disposal of garbage and human waste.

The high concentration of people in cities can lead to a reduced demand for land relative to population. Urban areas take up less than 1 per cent of the total land in most countries (and approximately 3 per cent of the earth's total surface area). Although urban sprawl is encroaching onto agricultural land in some nations, this can be avoided with coordinated urban and regional planning and effective land use management.

Cities offer great potential for limiting the use of motor vehicles if combined with adequate development of public transport systems. High concentrations of commuters make energy-efficient mass transit viable and affordable, and proximity ensures that more trips can be made on foot or by cycling, given the appropriate infrastructure.

The concentration of production and consumption in cities offers a range of possibilities for the efficient use of resources – through reclamation and wastewater recycling, for instance.

Cities that are unable to integrate economic growth with good planning and environmental care, on the other hand, can and do pollute the environment, contribute to the reduction of biodiversity, undermine the natural resource base, and increase the scale and depth of poverty. Many cities face challenges to implementing long-term plans for sustainability. Some of the most serious challenges centre on economic sustainability and poverty reduction, environmental degradation, social injustice and exclusion, and failures of governance.[9] These challenges are linked to specific problems within cities that preclude their ability to improve the built and natural environments for their residents. Urban data on sustainability indicators reveals the scope of the issues worldwide. A common thread through all of the research is the importance of engaging the urban poor: cities that do not recognize the impact of environmental problems on their poorest citizens, or the environmental costs of unplanned development, remain unsustainable.[10]

Osaka, Japan BINSYO YOSHIDA/UNEP/STILL PICTURES

Air pollution

The highly urbanized developed regions of the world are responsible for the greatest per capita emissions of greenhouse gases caused by burning fossil fuels. In 2002, people in the developed regions emitted 12.58 metric tons of carbon dioxide (CO_2) per capita, compared with 2.07 metric tons per capita in the developing world.[11] Heavy use of motor vehicles is largely to blame; in Canada, for instance, cars and trucks were the largest source of greenhouse gas emissions in the country in 2002, contributing 25 per cent of such emissions.[12] Even as developed countries work to limit air pollution, global emissions of CO_2 – the principal greenhouse gas – are predicted to rise by more than 60 per cent in the period between 1997 and 2010. The greatest increase – 65 per cent – will come from developing countries, and primarily from China.[13]

In the rapidly industrializing cities of Asia, ambient air pollution is on the rise as industrial and motorized transport emissions increase, and as dependence upon fossil fuels persists. China is home to 16 of the world's 20 most air-polluted cities and is the world's second-largest producer of greenhouse gases after the United States. Over the past 10 years, the concentration of pollutants in China's air has increased by 50 per cent. Urban outdoor air pollution, mainly from vehicle exhaust and industrial emissions, is responsible for the deaths of 3 million people around the world each year – most of them in developing countries.[14] In Beijing alone, more than 400,000 people die each year of pollution-related illnesses.[15] China is the largest producer and consumer of bituminous coal – the main contributor to its air pollution – and more than 64 per cent of its citizens use coal in their homes.[16]

Indoor air pollution from the burning of biomass fuels – firewood, charcoal, crop residues and animal dung – is another major challenge to environmental health and sustainability in developing countries, and is a growing problem in cities. Almost half of the world's population, 2.4 billion people, depend upon biomass fuels for their daily energy needs, nearly all of them in developing countries. That number is expected to rise by 200 million by 2030.[17] The burning of biomass releases toxic gases and compounds into the air, including carbon monoxide and methane, leading to a host of chronic respiratory diseases, lung cancer and pneumonia in those exposed to the smoke and particulate matter.[18] Women and children are disproportionately harmed by the burning of biomass, as they are more exposed to the dangers of indoor air pollution caused by cooking using fuels that emit toxic gases and particulates. Every year, 1.6 million people die from exposure to indoor air pollution, 1 million of whom are children.

UN-HABITAT analyses have shown that the prevalent use of biomass or solid fuels in poorly ventilated slum households has increased acute respiratory illnesses among children in Asia and Africa, where the use of solid fuels among low-income households is common. The per capita proportion of biomass use is highest in sub-Saharan Africa, but the greatest numbers of people who depend on the highly polluting fuels live in China and India.[19] In the slums of sub-Saharan Africa and Asia, use of solid fuels among urban households with extreme shelter deprivations averages 74 per cent and 60 per cent, respectively. Adopting cleaner-burning charcoal briquettes made of recycled ash and agricultural waste, along with more modern cooking equipment, could prevent up to 2.8 million premature deaths each year.[20]

China is home to 16 of the world's 20 most air-polluted cities and is the world's second-largest producer of greenhouse gases after the Unites States.

Traffic deaths

Motor vehicles are not only a major cause of air pollution – they are also responsible for most fatal accidents in urban areas. Traffic deaths are symptomatic in many places of failed urban planning and inadequate roads and transport systems. In the developing world, more and more motor vehicles are crowding onto roads not designed to handle them. Where people have limited transport choices, private cars and shared taxis may be the only viable options for getting where they need to go, putting them at risk of accidents. Since 1990, there has been a four-fold increase in the number of motor vehicles in China and Thailand – a common trend across the developing world. Increased auto, bus and motorcycle traffic has led to higher rates of accidents and fatalities in developing countries, where pedestrians, bicyclists and traditional vehicles share the roads. Global traffic deaths rose from approximately 990,000 per year in 1990 to nearly 1.2 million per year in 2002, with 85 to 90 per cent of the fatalities occurring in low- and middle-income countries.[21, 22]

In Europe and North America, traffic deaths have been declining since the 1970s, but they have risen sharply in Latin America, Asia and Africa. Between 1975 and 1998, traffic fatalities increased by 237.1 per cent in Colombia, 243 per cent in China and 383.8 per cent in Botswana. By 2020, if the current trend continues, traffic deaths will increase by 83 per cent in the world's low-income countries, even as they decrease by 30 per cent in high-income countries. Many who die in traffic-related accidents are pedestrians – between 1977 and 1994, 64 per cent of the traffic fatalities in the city of Nairobi were pedestrians. Better urban infrastructure, pedestrian-friendly streets and well-planned transport systems that provide safe options for getting around the city are needed to curb the rise in traffic deaths.[23, 24]

Inadequate access to water and sanitation

Ninety-five per cent of the world's urban population had sustainable access to an improved water source, and 81 per cent had access to improved sanitation, in 2002.[25] The global numbers are misleading, however – access to safe drinking water and decent toilets is not evenly distributed among populations within regions or even within cities, and gaining access may involve hardship or risk for residents.[26] Households without adequate water supply and sanitation suffer disproportionately from water-borne or water-related diseases; moreover, lack of these services within slums contributes to the degradation of water and land resources within cities.

Water and sanitation are intimately linked – where inadequate sanitation facilities exist, water contamination is common. This became startlingly clear when a cholera epidemic swept East Africa in 1997 and 1998, as a result of human waste contaminating water sources. The disease started in slums, where rainwater washed accumulated human waste into boreholes and other water sources and spread quickly throughout Kenya, Tanzania and Uganda.[27] More common than cholera is the incidence of diarrhoea, which contributes to rates of child mortality 10 to 20 times higher in areas lacking adequate water supply and sanitation than in cities with proper provision of services. The crisis is most acute in the cities of Africa and Asia. As many as 150 million urban residents in Africa lack adequate water supplies and an estimated 180 million people lack adequate sanitation; three-quarters of the global population without access to water supply, and more than half of the population without access to sanitation live in Asia.[28]

Access to improved water sources often changes over time. In East Africa, piped water systems in cities have degraded over the past 30 years – partly as a result of inadequate maintenance and urban population growth – leaving more households without reliable access and decreasing their overall water consumption.[29] Less water in households correlates with higher rates of illness, as it makes washing hands, cleaning cooking utensils and bathing difficult. Even where water is abundant, however, inadequate delivery systems and unsanitary conditions can lead to contamination and higher rates of illness.

In addition to degradation of urban water delivery systems, water scarcity results from over-exploitation of sources, which, in turn, contributes to environmental crises. Mexico City, for example, depends upon the Mexico Valley aquifer for 80 per cent of its water supply, but it has so depleted the aquifer that the land has shifted and the city is sinking.[30] In coastal areas, where most of the world's largest cities are located, pollution of water sources is posing major threats to human and ecosystem health. Lima, Peru, is one coastal city that is contributing to the global problem: it discharges 18,000 litres of wastewater per second into the Pacific Ocean.[31]

Renewable energy sources increase urban sustainability

Fossil fuels made cities what they are today, but dependence on them has led to high rates of air pollution and greenhouse gas emissions, poor health among urban residents and environmental degradation. Life depends upon stable ecosystems, and cities, as "eco-technical systems", have the potential to work in harmony with the natural world rather than to continue depleting its resources. Indeed, they must in order to survive over the long term. Switching to renewable energy sources – such as wind, solar, modern biomass, geothermal, and small hydro-electric systems – for the bulk of urban energy needs is vital to the sustainability of cities. It is also an essential aspect of meeting the Millennium Development Goals, from improving health and saving environmental resources to increasing global partnerships and reducing poverty. As developing countries adopt clean, efficient, reliable and renewable energy sources and upgrade existing urban systems, they offset many of the challenges to urbanization and make sustainable development possible.

Much of the current discussion about the need for reduced reliance on fossil fuels stems from 1997, when 84 countries signed the Kyoto Protocol at the Conference of the Parties associated with the UN Framework Convention on Climate Change. This recognition of the human impact on the world's climate and environmental resources spurred innovation around renewable energy technologies that curb the production of greenhouse gases and provide efficient, environmentally sound power. The use of renewable energy sources has increased since then, owing to major investments by governments and private industries around the world. The Renewable Energy Policy Network for the 21st Century (REN21) reports that at least 48 countries now have some type of renewable energy promotion policy, including 14 developing nations. In developing countries such as China, where electricity and industry are powered primarily by coal, promising renewable alternatives for meeting energy needs are emerging in the form of wind power and methane gas from decomposing solid waste. Renewable technologies now provide 160 gigawatts of electricity generating capacity – about 4 per cent of the world total – and global investment in renewable energy topped US $30 billion in 2004.

Government leadership in creating policies around renewable energy sources is important to their success in the market: renewables are used most in cities and countries with policy-bound targets for reducing carbon dioxide emissions and increasing reliance on renewable sources. Freiburg, Germany, was one of the first cities to adopt targets for reduced greenhouse gas emissions, in the late 1980s. Today, the city of 200,000 is a model of sustainable urban development, featuring city-financed solar building projects, subsidies for solar power in new construction, and an overall integrated approach to urban planning that considers the city's future energy consumption and sources. Freiburg has attracted renewable energy research institutes, companies, consultancies, solar engineers, and architectural firms that specialize in solar design. Other cities, including Portland, Oregon, USA, and Adelaide, Australia, have provided incentives and subsidies for "green building" projects that increase the energy efficiency of homes and commercial buildings and mandate the use of environmentally sound building materials.

One of the primary strategies for decreasing emissions, preventing climate change and improving resident health in cities is upgrading transport systems. Efficient mobility is essential to the economic success of cities, and mobility is facilitated by transport systems that are cost-effective, responsive to changing demands, environmentally sound, and accessible to all residents. Throughout Brazil, the practice of mixing regular gasoline with 26 per cent ethanol – derived from locally grown sugar cane – has led to reduced emissions and increased savings of fossil fuels for cities and citizens. Biofuels are gaining prominence worldwide: trains and buses in Sweden are running on methane produced by degrading animal waste; the entire Halifax, Nova Scotia, Canada, metro bus fleet runs on a mixture of 20 per cent biofuel and 80 per cent regular diesel; and zero-emissions buses powered by hydrogen fuel cells are currently in use throughout London and nine other European cities. In Helsinki, Finland, a compact urban design makes mass transit viable. Elsewhere in Europe and North America, transit-oriented development and "smart growth" – high-density development that facilitates pedestrian activity and provides easy access to commuter transit options – are helping to halt urban sprawl.

Sources: Girardet 2004; Martinot 2005a & 2005b; Balfour 2005; The Economist 2005; Franks 2005; Halifax Regional Municipality 2004; BBC News 2005, World Edition 2005a.

■ Poor solid waste management

Less commonly researched than sanitation provision is a related challenge to sustainability: the dearth of adequate solid waste management in many cities around the world. Municipalities in developing countries commonly spend 20 to 50 per cent of their available budgets on solid waste management, but many are only able to collect 30 to 60 per cent of the waste in their cities while serving just half of the population.

In Nairobi, Kenya, for example, only 25 per cent of the city's daily waste is collected.[32] Where sanitary landfills, recycling programmes and other properly managed means of solid waste disposal are not available, open dumping and burning are the norm. This leads to environmental and health hazards for urban residents – especially those who live closest to the dump sites. The Nairobi neighbourhood of Dandora, home to 250,000 people, has become the city's *de facto* dump: approximately 1,600 tons of solid waste is dumped every day on land formerly intended for housing. The sprawling waste has been blamed for numerous illnesses among the residents and has contaminated a nearby spring, upon which many residents depend for water when the municipal supplies stop working.[33]

Thin plastic carrier bags given out by supermarkets and local vendors comprise much of the solid waste in Dandora and similar neighbourhoods. Kenya produces 48 million of the bags each year, and businesses in Nairobi hand out more than 2 million of them annually. The bags can take up to 1,000 years to decompose; along with dumps and landfills, many bags end up in drains, sewers, riverbeds and the sea. Nobel Peace Prize laureate Wangari Maathai has also linked Kenya's plastic bag pollution to the incidence of malaria: discarded plastic bags trap rainwater, providing ideal breeding grounds for mosquitoes.[34]

Water Pollution, China ZHAO WEIMING/UNEP/STILL PICTURES

Water and sanitation are intimately linked – where inadequate sanitation facilities exist, water contamination is common.

■ Unsustainable practices most deeply affect the urban poor

It is clear that when city systems fail to manage basic urban issues – such as regulation of pollution-generating industries and provision of clean fuels, development of safe transport systems, provision of safe and adequate water and sanitation facilities and collection and proper disposal of solid waste – the most vulnerable residents suffer the greatest hardship. Per citizen, the urban poor in developing areas make vastly smaller resource demands, make much better use of those resources, and produce a much smaller pollution load than do their wealthier neighbours.[35] Yet, they endure the greatest environmental risk as a consequence of the consumption patterns of higher income groups and the production and distribution systems that serve them. The pollution and contamination produced by higher-income groups are immediately felt by lower-income groups who live and work in areas and industries that absorb it. The poor consequently bear heavy health burdens and other barriers to escaping poverty.

Low-quality, overcrowded housing and lack of basic services often lead to illness and absenteeism, which in turn affect the economic growth and sustainability of the urban system as a whole. Even as the consumption patterns of higher-income groups lead to suffering for the lowest-income residents, each depends on the other, for labour, employment and the availability of city services. Improving the urban environment in ways that remediate the harms experienced by the urban poor can reduce poverty and increase the overall economic, environmental and social sustainability of cities. Making progress toward sustainability, therefore, means designing and improving urban systems with people in mind – particularly those with the fewest resources.

Endnotes

1 The most recent UN-led international discussions on this topic were held at the World Summit on Sustainable Development in Johannesburg in 2002.
2 Girardet 2004.
3 The ecological footprint is a measure of how much biologically productive land and water area an individual, a city, a country or a region requires to produce the resources it consumes and the waste it generates.
4 Mougeot 2005b.
5 European Environment Agency 2004.
6 UN-HABITAT/DFID 2002.
7 Hinrichsen 2002.
8 These arguments are drawn from Satterthwaite 1999.
9 UN-HABITAT/DFID 2002.
10 This argument has been backed by various studies, including one on urban agriculture by Luc J.A. Mougeot of the International Development Research Centre (IDRC) in Canada.
11 United Nations Department of Economic and Social Affairs, Statistics Division, 2005.
12 Dooley 2002.
13 International Energy Agency 1996.
14 Figure from the World Health Organization, reported in Kirby 2004.
15 Watts 2005.
16 Karasov 2000.
17 Dooley 2004.
18 United Nations Department of Economic and Social Affairs, Statistics Division 2005.
19 Dooley 2004.
20 Dooley 2005.
21 World Health Organization 2004.
22 Dahl 2004.
23 World Health Organization 2004.
24 Dahl 2004.
25 Access to improved water sources refers to the percentage of the population who use any of the following types of water supply for drinking: household connection, public standpipe, borehole, protected dug well, protected spring, rainwater collection. Improved water sources do not include: unprotected well, unprotected spring, rivers or ponds, vendor-provided water, bottled water (due to limitations in the potential quantity, not quality, of the water), tanker truck water. Access to improved sanitation facilities refers to the percentage of the population with access to: facilities connected to a public sewer or a septic system, pour-flush latrines, simple pit or ventilated improved pit latrines. (United Nations Department of Economic and Social Affairs, Statistics Division 2005.)
26 See Part Two of this report for an in-depth discussion on water and sanitation issues in cities.
27 Ray 2003.
28 UNESCO 2003.
29 Thompson, et al. 2000.
30 UNESCO 2003.
31 Ibid.
32 BBC News World Edition, 2005b.
33 Kantai 2003.
34 BBC News World Edition, 2005b.
35 Atkinson 1996.

3.7 Double Jeopardy: The Impact of Conflict and Natural Disaster on Cities

Collapsed buildings after an earthquake in the Taiwan Province of China LO TSUNG HSIEN/UNEP/STILL PICTURES

Human conflicts and natural disasters impact cities differently – and often more deeply – than rural areas. Conflicts lead to the growth and proliferation of slums as displaced people seek refuge at the margins of urban areas; buildings and roads crumble and fall in the wake of major tremors, landslides and floods. The sheer concentration of people and infrastructure in cities often means greater loss of life when disaster strikes, and the social, political and structural capacity of cities to provide shelter for those in need is often limited. When conflicts or natural disasters hit, they can wreak havoc on urban economies, destroy communities and tear families apart. Such events perpetuate urban poverty, placing additional strains on people and places already burdened by lack of resources.

Conflicts generate slums

Conflicts and crises in war-torn countries often result in the mass exodus of rural communities to urban areas, where most end up in low-income, poorly serviced settlements, or slums. The continued threat of conflict in countries is, therefore, a significant contributing factor in the proliferation of slums in urban areas.[1] In Sudan, for instance, urban areas accommodated two-thirds of the more than 6 million internally displaced persons (IDPs) in the country in 1998; almost half of these IDPs moved to the capital city, Khartoum. Surveys indicate that the majority of IDPs in Khartoum are from Southern Sudan, the region most affected by a protracted civil war, and most reside in squatter settlements on the periphery of the city, with little access to basic services.[2]

The plight of displaced persons, whether they are settled in large groups in camps or merged into urban slums, therefore, needs special attention. The urban context presents unique issues and dynamics for people uprooted from their homes, separated from their families and community networks, and stripped of their livelihoods by conflict. Urban IDPs may come from different areas of the country, different post-conflict situations and possibly from different sides of the conflict, making resettlement in a diverse environment difficult.

In Azerbaijan, where conflicts with neighbouring Armenia have raged since 1988, the total number of internally displaced persons stands at nearly at 1 million.[3] Approximately 40 per cent of the country's displaced population lives in urban areas, which have proven unsuitable and unacceptable for long-term habitation, especially when employment opportunities are scarce. Almost 95 per cent of these urban IDPs state that they wish to return to their former homes – not because the material living conditions in rural settlements are better, but because they want to be able to continue engaging in their former source of livelihood and continue living in a familiar environment with their own community networks. Chronic insecurity or lack of rehabilitation of disaster-struck areas often inhibits the return process, leaving people stranded in substandard living conditions in urban areas.

In some countries, however, refugees and IDPs residing in urban areas often have no intention of returning to their place of origin, as the movement to cities is seen as an opportunity to escape the impoverishment of their rural homes. A review of rehabilitation and reconstruction programmes in Afghanistan conducted in the 1990s showed that a significant number of IDPs in cities such as Kabul and Mazar-i-Sharif used the opportunities presented by disruption to leave their rural homes for good and most had no intention of returning.[4] In some war-torn countries, IDPs literally have no place to return to as their homes and lands have been taken over forcefully by warring factions. This is the case in Somalia where years of lawlessness has led to illegal occupation of rural and urban land by armed militias.

City governments and municipalities have to deal with the burden of increased pressure on urban infrastructure caused by an influx of refugees and IDPs. Most IDPs and refugees are at risk and in need; they often live in densely populated squatter settlements on the periphery of cities, where widespread poverty and underdevelopment are prevalent. Local authorities face considerable additional pressure to absorb large numbers of refugees and internally displaced persons during the conflict; in many cases, they even have to deal with IDPs and refugees long after the crisis is over. For instance, preliminary reports indicate that despite the signing of a peace accord in January 2005, many of the IDPs in Sudan's urban areas are unwilling to return to their rural homes, either because assistance to reintegrate into their local communities is insufficient or local communities and administrations are acutely under-resourced to manage the return process.[5] Some IDPs have tried to return to their areas of origin only to find inadequate resources and services, forcing many to return to urban centers in an effort seek opportunities for a new start.

Low-income settlements are more vulnerable to natural disasters

Urban settlements are also prone to threats from natural and environmental hazards, and people living in poverty everywhere, especially in urban areas, are most at risk. Substandard housing and construction practices, lack of infrastructure, absence of secure tenure, inappropriate land use and increasingly degraded environments leave large sections of the poorest communities chronically vulnerable.

The world's largest cities are concentrated in developing countries, and many of them are in areas where earthquakes, floods, landslides and other disasters are most likely to happen. According to the UN's Bureau for Crisis Prevention and Recovery, some 75 per cent of the world's population lives in areas that were affected at least once by an earthquake, a tropical cyclone, floods, or drought between 1980 and 2000.[6]

At the same time, natural and man-made disasters are increasing in regularity, and perhaps more importantly, their adverse impacts on populations and human settlements are rising.

War-torn Bié, Angola EDUARDO LÓPEZ MORENO

Conflicts and crises in war-torn countries often result in the mass exodus of rural communities to urban areas, where most end up in low-income, poorly serviced settlements, or slums.

Calculations by the International Federation of the Red Cross and Red Crescent show that from 1994 to 1998, reported disasters averaged 428 per year. From 1999 to 2003, this figure shot up by two-thirds to an average of 707 natural disasters per year. The sharpest rise occurred in developing countries, which suffered an increase of 142 per cent.[7]

Poor people in developing countries are particularly vulnerable to disasters because of where they live; they are more likely to occupy dangerous floodplains, river banks, steep slopes and reclaimed land, and their housing is less likely to survive a major disaster. For instance, in Latin America, hundreds of low-income urban dwellers lost their lives when their precariously situated homes were swept away in floods and landslides during Hurricane Mitch in 1998. And investigation into the 2003 earthquake in Bam, Iran, found that most of the 40,000 people killed lived in housing that was built in the traditional mud-brick style without the necessary supportive structures to withstand tremors.[8]

In poor countries, the impact of disasters can be devastating on the economy; disasters not only wipe out decades of development in a matter of hours, but they also have a significant impact on a country's gross domestic product (GDP). Figures compiled by the World Bank show that between 1990 and 2000, natural disasters resulted in damages constituting between 2 and 15 per cent of the affected countries' GDP. While industrialized countries suffer higher losses in dollar terms – mainly because the cost of repairing or replacing destroyed infrastructure is higher – the overall impact of disasters on the economies of rich countries is negligible. According to the International Federation of the Red Cross and Red Crescent Societies, disasters in industrialized countries have inflicted an average damage of $318 million per event, compared with $28 million per event in developing countries. However, industrialized countries are able to quickly recover from the impact of disasters, mainly because of a surge in reconstruction activities and more public spending on rehabilitation of the affected areas. Moreover, rich countries are often more prepared to deal with the consequences of disaster, as they have more medical and emergency assistance services than lower-income nations. The prevalence of life and property insurance in the developed countries also means that affected populations suffer less personal financial loss than their counterparts in developing countries.

Disasters can paralyse developing countries, or even permanently destroy their social and economic assets. In Aceh, Indonesia, for instance, the total estimate of damage and losses from the December 2004 tsunami was $4.45 billion – nearly 97 per cent of the region's GDP.[9] Many developing countries also lack the health facilities to deal with large numbers of injured patients, resulting in a higher eventual death toll than in countries better equipped for disaster. The United Nations Office for the Coordination of Humanitarian Affairs (OCHA) estimates that in the 1990s, natural disasters killed almost seven times more people in developing countries per event than in industrialized countries; an average of 44 people per event died in industrialized countries compared with 300 people per event in developing counties.[10]

Surviving anarchy: Somalia's experience

Somalia UNICEF/HQ91-0914/ROGER LEMOYNE

Human settlements in Somalia have been severely affected by more than a decade of civil war, which has not only caused the destruction of infrastructure and services, but has also led to the breakdown of government institutions. The country had no functioning government for almost 14 years as warlords controlled most regions. In many cities, local and national government records have been ravaged and capacities have been greatly reduced or are non-existent. This has made it difficult to measure the country's level of human development or to make an assessment of needs vis-à-vis the Millennium Development Goals.

Although a new interim government was finally elected and installed at the end of 2004, the effects of the conflict will take years to erase, as much of the country has no fully functioning governance structures, either at the national or local government level. Nine months after the elections, for instance, the new government had still not moved to the capital Mogadishu – which is still considered to be too dangerous – and was operating from Jowhar, a town north-east of the former capital city. The elections themselves were conducted in neighbouring Kenya, which is host to thousands of Somali refugees and exiles.

Somalia is categorized as one of the 27 high-priority countries for the achievement of the Millennium Development Goals; available statistics indicate that the maternal mortality ratio per 100,000 births is 1,600 – much higher than that of neighbouring countries such as Kenya (590) and Eritrea (1000); primary education enrolment is only 13.6 per cent. The lack of a fully functioning government and the disruption caused by conflict makes the achievement of social and economic development precarious.

Despite the anarchic situation, since the early 1990s, isolated communities within Somalia have been able to achieve some level of stability and even manage to govern themselves despite the absence of a legitimate national government. In May 1991, five months after the state of Somalia collapsed and fell into civil war, the north-west region claimed independence and formed its own government with its own currency and institutions. Although it was not internationally recognized as an autonomous state, the Republic of Somaliland developed both formal and informal local governance structures, and even held its first local elections in 2002 and multi-party presidential elections in 2003.

In a country with no formal banking institutions, a widespread system of money vendors developed, allowing local and international agencies to make financial transactions. The relative stability in Somaliland ensured that the region invested in its own development and also attracted foreign capital and assistance. Compared to other regions in Somalia, Somaliland not only has more hospitals, but also has a vibrant private sector, which generates revenue for the region. The economy is also highly dependent on money sent home by members of the Somali diaspora. Although the economy of Somaliland is dominated by trade in livestock, the service sector, including mobile phone companies, has shown positive results in recent years. In the capital city, Hargeisa, revenue collection has become more efficient and the city of more than 400,000 inhabitants has become the centre of international aid agencies' operations as insecurity in other regions only allows limited operations.

However, like many war-torn cities, Hargeisa is becoming a destination for returnees and internally displaced persons (IDPs). The Hargeisa Municipality Statistical Abstract, published in 2003 by the Municipality of Hargeisa, shows that while the city has performed relatively well compared to other cities in the region and in the rest of Somalia, the spontaneous growth of settlements comprising returnees and IDPs is making it difficult for the municipality to provide basic services. In fact, in the city as a whole, only one out of seven dwellings has access to piped water and 20 per cent of households live in temporary, makeshift structures.

With an annual urban growth rate of 5.7 per cent, Somalia is one of the most rapidly urbanizing countries in Eastern Africa. The transition to democracy is likely to increase rates of urbanization, as returnees and IDPs make their way to cities. Land disputes are also likely to come to the fore, as years of lawlessness led to illegal occupation of land by militias.

Although Somaliland has distanced itself from Somalia's new transitional government – which it perceives as a threat to Somaliland's autonomy – this region of 3.5 million people has already begun working to ensure that urban governance structures are strengthened as national priorities are being redefined. The Municipality of Hargeisa is working with other local authorities to institute urban land reforms through the formulation of a City Charter on a range of issues, such as planning and taxation. Funded by UNDP and implemented by UN-HABITAT, the project's aim is to increase efficiency in urban planning, management and development, as well as service delivery and fiscal management.

UN-HABITAT is also working with the three regional authorities in Somalia, namely Somaliland, Puntland and South-Central Somalia, to improve the capacities of local authorities and civil society in basic leadership and urban planning – a need that was identified by the 2002 Somalia Urban Sector Profile Study. The European Commission, UNDP, UNICEF and the Governments of Italy and Japan are working with UN-HABITAT to implement projects in these areas. However, the success of these initiatives will largely depend on how effective the new government is in securing lasting peace in a country that has been managed through fear and violence for more than a decade.

Sources: UN-HABITAT 2005b; Hargeisa Municipality 2003; BBC News 2005a.

Earthquake victims in Kashmir UNHCR/V. TAN

The year 2005 was a particularly costly one in terms of lives lost and damage inflicted by natural disasters around the world. Nearly 125 million people were injured, lost their home, or required other immediate assistance as a result of disasters that year. More than 100,000 people were killed, in addition to the 230,000 who died in the tsunami at the end of 2004. Total economic damages in 2005 reached a record $200 billion, including $125 billion in losses from Hurricane Katrina alone. The single greatest human toll followed the October earthquake in Pakistan and India, the repercussions of which continued for months as affected families weathered out a difficult winter in makeshift shelters.[11]

Sustainable recovery from crisis

Disasters have serious consequences at every level, from far-reaching economic losses to personal hardship for individual families. The broad impacts of disasters exacerbate the fundamental challenges of crisis management and recovery processes: how to bridge the gaps that have repeatedly emerged between emergency recovery and sustainable development efforts, and how to provide all stakeholders with practical strategies to mitigate and recover from crises. The concept of sustainable recovery[12] does not entail an abrupt shift from relief to development, but rather an integrated approach in which those involved attend to basic needs while also supporting longer-term sustainable development.

Conflicts and disasters perpetuate poverty by placing an additional strain on already precarious social, environmental and economic conditions. Persistent urban poverty and lack of resources again increase vulnerability, weaken coping strategies and delay the recovery process. The urban poor are forced to accept a greater degree of risk because they lack the resources to live or work in safer environments. Urban poverty alleviation must therefore be central to any plan to effectively manage urban disasters and to sustain peace and stability. Other crucial pillars of sustainable recovery are good governance, public participation, inclusive decision-making, institutional development and empowerment of civil society. When governments adopt policies to make livelihoods more secure, institutions more responsive, public-private partnerships more effective, communities more safe and sustainable, and poverty less prevalent, personal and social protection are dramatically enhanced.

The need for durable settlement solutions for internally displaced populations is one of the key issues in post-crisis urbanization. In practical terms, this means either helping displaced people resettle in their areas of origin, or aiding their effective and sustainable social, economic, legal, and political integration into urban communities. Either way, the importance of supporting greater self-reliance among the displaced is apparent, in particular by ensuring their access to land, income-generating activities and skills development. The emergency phase after disaster or conflict tends to frame displaced populations as beneficiaries rather than partners in the process and agents of development. Economic recovery, for example, is recognized as one of the most difficult aspects of the post-crisis recovery process, yet many local resources can be tapped to assist with economic recovery, including the technical knowledge of skilled and semi-skilled local workers, the eagerness and assets of entire communities, and the resources of local authorities and the private sector.

Understanding urban vulnerability is the first step toward developing mitigation strategies that effectively improve resilience and reduce vulnerabilities of urban populations in the long term. The cornerstone of the implementation strategy is to build a "culture of prevention", or disaster mitigation, among the society at large. Disaster mitigation not only saves lives but also makes economic sense. The World Bank and the U.S. Geological Survey estimate that economic losses worldwide from natural disasters in the 1990s could have been reduced by $280 billion if $40 billion had been invested in preventive measures.[13] In China, the World Bank estimated that the $3.15 billion spent on flood control since the 1960s has averted losses of about $12 billion. Similarly, more federal funding for the levees in New Orleans might have reduced the scale of the tragedy when Hurricane Katrina struck in August 2005.

Medellin, Colombia, provides a good example of successful community-based disaster prevention. In the mid-1980s, following the destruction of the city of Armero by mudslides triggered by a volcanic eruption, the Colombian government established a National System for Disaster Prevention and Response. When a major landslide struck Medellin in 1987, the city and its inhabitants were able to mobilize resources to create a safer living environment, integrating risk management strategies with municipal physical, social and economic planning. Thanks to combined civic education and political and financial commitment, the landslides in Medellin decreased from 533 in 1993 to 191 in 1995.[14] Vulnerability reduction plans and disaster risk considerations are ideally integrated into sustainable development policies, planning and programming – in particular at local levels.

As the nature of disasters in cities becomes more multifaceted, so must the approach to their management. The impact of the recent Indian Ocean tsunami is a tragic reminder of the extreme vulnerability of the built environment to natural hazards. Natural disasters in and around cities are often anything but "natural", being triggered by deficient urban management practices, inadequate planning, excessive population densities, ecological imbalance, inadequate investments in infrastructure, and poorly prepared local governments. Furthermore, the increasing number of people displaced by crises and seeking refuge in cities is a call for attention all slum dwellers deserve: to improve their living conditions and address the urban context of poverty, as spelled out in the Millennium Development Goals.

Paradoxically, a crisis can also be an opportunity. During recovery from a disaster, communities have a unique opportunity to revisit past practices and rewrite policies to affect future development. In Rwanda, for instance, new land laws were instituted after the genocide in 1994 to give women and other vulnerable groups more rights to inherit and own land and property.

Endnotes

1 It is interesting to note that the region with the most conflicts - sub-Saharan Africa - also has the highest proportion of its urban residents living in slums.
2 Eltayeb 2003.
3 This comprises roughly 13 per cent of the country's population, which is one of the highest such rates in the world. See UNDP's Azerbaijan Human Development Report 2000.
4 This review was jointly conducted by UNCHS and UNDP in 1995.
5 See www.idpproject.org/sudan for more information.
6 United Nations Office for the Coordination of Humanitarian Affairs 2005a.
7 Ibid.
8 United Nations Office for the Coordination of Humanitarian Affairs 2005a.
9 UNEP 2005.
10 United Nations Office for the Coordination of Humanitarian Affairs 2005b.
11 Worldwatch Institute 2006a.
12 See UN-HABITAT Sustainable Recovery and Reconstruction framework, www.unhabitat.org.
13 Figures cited in *IRIN News*, United Nations Office for the Coordination of Humanitarian Affairs (OCHA) 2005b.
14 Medellin is Colombia's second largest city with 2 million inhabitants, close to 10 per cent of whom live in informal settlements on steep hillsides, vulnerable to floods and landslides. The landslide of 1987 killed more than 500 and left 3,500 inhabitants homeless. (Boulle & Palm 2004.)

Aftermath of Hurricane Katrina, New Orleans September 3, 2005 DIGITAL GLOBE

New Orleans: Poor residents suffer deepest impact of Hurricane Katrina

When Hurricane Katrina struck the Gulf Coast of the United States on 29 August 2005, the storm left more than one million people homeless and killed hundreds across three states. The city of New Orleans, in the southern state of Louisiana, suffered Katrina's greatest lasting impact.

Lashed by winds of more than 140 miles per hour and flooded by water overflowing the levees that kept the Mississippi River, the Gulf of Mexico and Lake Ponchartrain at bay, New Orleans lay almost entirely submerged and in ruins after the storm. Many of the city's 485,000 inhabitants fled before the storm via the well-planned interstate highway evacuation route. For tens of thousands of people without cars, cash or anywhere else to go, however, Mayor C. Ray Nagin's evacuation order had meant little. The refugees, disproportionately African American and living in poverty, took shelter under freeway bridges, in the city's sports arena, and in the nearby convention centre for days until help arrived. Reporters covering the scene likened it to something more akin to war-torn Somalia or post-tsunami Indonesia than a scene from one of the world's wealthiest nations. The disaster fast became a symbol of race and class division in the country.

Before Katrina pummeled its shore, the below-sea-level city of New Orleans was known for its European charm and grand-scale urban fêtes, including one of the largest Mardi Gras celebrations in the world. With the storm came a flood of facts about real-life conditions for average denizens of New Orleans, nearly one-quarter of whom were elderly or disabled, and more than 28 per cent of whom lived in poverty – double the national average of 12.4 per cent. Of those living in poverty, 84 per cent were African American and 43 per cent were children under the age of 5. Access to a car, the primary means out of the city during the crisis, was equally disproportionate: among African American households, 35 per cent did not have a car, while the same was true of only 15 per cent of white households.

As in many parts of the developing world, the poorest residents of New Orleans lived in the most hazardous areas of the city. Many of the city's lowest-income residents lived in the floodplains of the Lower Ninth Ward, a neighbourhood that sat below sea level and was inundated when the canals and levees failed. Although the Federal Emergency Management Agency had predicted that a hurricane would strike New Orleans since at least 2001, federal funds to reinforce the levees had been decreasing in recent years.

Speaking to a *New York Times* reporter, geographer Craig E. Colten of Louisiana State University said, "Out West, there is a saying that water flows to money. But in New Orleans, water flows away from money. Those with resources who control where the drainage goes have always chosen to live on the high ground. So the people in the low areas were hardest hit."

The Lower Ninth Ward neighbourhood – where more than 98 per cent of the residents were African American and more than a third lived in poverty – was built on a reclaimed cypress swamp, gradually drained and developed over the first half of the 20th century. The city's higher ground had been settled since the early 1700s, when French colonists fortified the swampland surrounded by large bodies of water and called it "leflontant" – the floating island. By 2005, the New Orleans metropolitan area was home to more than one million people. With population expansion came more reclamation; levees were built and water pumped away as settlements spread down from the high southern shore of Lake Ponchartrain to the low banks of the Mississippi. The lowest land was the only place European immigrants and African American families could afford to build homes in the early 1900s; the dirty, flood-prone parcels of land were adjacent to the city's commercial and industrial areas.

Poverty kept residents of the Lower Ninth in place over the years, unable either to move up and out or to renovate their increasingly run-down houses. In 2000, more than half of the neighbourhood's residents owned their homes and had occupied them for 10 years or longer; the opposite was true nationwide, with more than 60 per cent of American households having moved in the past 10 years. The houses in the neighbourhood held the history of New Orleans itself – 62 per cent were built before 1960, and only one-tenth were less than 20 years old. The age of the housing, along with its location, put residents of the Lower Ninth at risk. Having endured flooding before, the residents suffered another deep impact on their stability and their access to affordable housing with Hurricane Katrina. They may also be the last to benefit as the city is gradually rebuilt over the next several years.

Katrina was the first such storm to devastate a major urban centre in the United States. In other places similarly affected, the rebuilding of affordable housing has historically taken last place among the items on long-term plans, or has been left out altogether. Kobe, Japan, provides a case in point. When that city was destroyed by an earthquake in 1995, many residents lived in temporary housing for eight years, and areas of the city that had been affordable for families were rebuilt with housing beyond their financial reach.

But while the city of New Orleans will no doubt be eventually rebuilt, many fear its soul has been lost forever. As neighbouring cities and states struggled to cope with the hundreds of thousands of refugees fleeing the flooded city, authorities warned that draining the water from the city could take months, which would make it less likely that the refugees would return soon. According to one report, many of the more than 200,000 people who crossed into the neighbouring State of Texas in buses, planes and trains vowed never to return to New Orleans and its surrounding areas.

Sources: Applebome, et al. 2005; Greater New Orleans Community Data Center 2005; DeParle 2005; Leavitt 2000; Teather 2005; Gonzalez 2005; Walsh, et al. 2005; Luthra 2005.

3.8 Urban Insecurity: New Threats, Old Fears

Cities as targets

Prior to 11 September 2001,[1] urban violence and insecurity were considered peripheral to the development concerns of both rich and poor countries. In recent years, however, poverty, under-development and fragile states have created fertile conditions for the emergence of new threats, such as transnational crime and international terrorism, which are being played out in the world's cities.

Recent attacks on New York, Washington, Madrid, London, Nairobi, Dar es Salaam and Bali, among other cities, have demonstrated that urban insecurity is an emerging international issue that impedes economic growth and sustainability. Although cities have been the sites of warfare in both pre-modern and modern times, the near-invisible nature of modern-day attackers has made urban warfare and terrorism much more difficult to track and control, and is having a far more devastating financial, physical and psychological impact on cities than ever before. New York City is estimated to have lost $110 billion in infrastructure, buildings, jobs, and other assets as a result of the 11 September attacks.[2] The impact of the attacks has also extended outside the city's boundaries. The World Bank estimates that as a result of the terrorist attacks on New York and Washington, global gross domestic product (GDP) was reduced by 0.8 per cent, and some 10 million additional people were added to the world's poor.[3]

Unfortunately, the "war on terror" threatens to sideline the struggle against poverty. The security measures adopted by European and North American cities shortly after the recent attacks have greatly compounded terrorism's effects on poverty reduction. Increased security measures, such as metal detectors, street surveillance, stricter control of public areas and, in certain cases, curtailment of civil liberties, threaten the essence of cities.[4] Recent estimates indicate that the combined total expenditure on programmes for improving access to clean water, building sewage systems, reducing hunger, preventing soil erosion, eradicating illiteracy, immunizing children, providing reproductive health care to women, and fighting HIV/AIDS and malaria added up to a little more than half of the military budget appropriated for the Iraq war by the United States in 2004.[5]

Cities are not only targets of international terrorism, but also of localized ethnic and religious conflicts. According to the Worldwatch Institute, urban unrest is likely to increase in the largest cities of the developing world as more and more people from diverse ethnic and religious groups increasingly come into close contact with each other.[6] Although cities offer the opportunity for diverse interests to integrate, when resources are scarce, or when political interests collide, they can become the sites of warfare. In India, for instance, centuries-old animosities and grievances were played out in the country's biggest cities in 1992 when Hindu militants descended on the small town of Ayodhya, attacking security forces and destroying a 16th century mosque. Rather than spreading through the nearby countryside, the hatred exploded hundreds of kilometres away, in Mumbai, Calcutta, Ahmedabad, and New Delhi. In total, 95 per cent of those killed in the communal riots that ensued were city dwellers.[7] Cities such as Los Angeles, Belfast, Sarajevo and Mogadishu have all suffered from one form of urban warfare or another, creating "a new phase in the life of cities, where the concentration of ethnic populations, the availability of heavy weaponry, and the crowded conditions of civic life create futuristic forms of warfare ... and where a general desolation of the national and global landscape has transposed many bizarre racial, religious, and linguistic enmities into scenarios of unrelieved urban terror".[8]

Urban crime

Evidence shows that the probability of being a victim of crime and violence is substantially higher in urban areas than in rural areas. Approximately 60 per cent of urban dwellers in Europe and North America and 70 per cent of urban dwellers in Latin America and Africa have been victimized by crime over the past five years.[9]

Overall, recorded crime rates are stabilizing or even decreasing in some countries, but the risk of being a victim of a violent crime such as homicide, assault, rape, sexual abuse, or domestic violence has continued to rise worldwide. Globally, more than 1.6 million people die as a result of violence every year.[10] The increasing availability and use of firearms lends heavily to the increase in urban violence. On average, violence makes up at least 25 to 30 per cent of urban crime[11] and women, especially in developing countries, are twice as likely to be victims of violent aggression (including domestic violence) as men.[12] Increases in violence against women can be correlated with declining household economic security, but it is clear that poverty and unemployment on their own do not cause crime, violence and abuse. Rather, the costs of poverty and unemployment in the form of stress, loss of self-esteem and frustration appear to influence violent behaviour.[13]

World Trade Center site, New York © NATHAN GOODMAN. IMAGE FROM BIGSTOCKPHOTO.COM

Urban insecurity presents a major challenge to the social and economic development of cities because it compounds other factors, such as poverty and social exclusion, which already limit the quality of life for many. Violence and crime are no longer viewed exclusively as criminal problems but also as problems affecting the development of societies. Insecurity contributes to the isolation of groups and to the stigmatization of neighbourhoods, particularly those in which the poor and more vulnerable live. It creates conditions of fear, hinders mobility and may be a major stumbling block for participation, social cohesion and full citizenship. The most excluded groups – women, children, the elderly, widows, and people living with HIV/AIDS – are typically cut off from networks that provide access to power and resources, making them more vulnerable and increasing their risk of remaining poor or sinking further into poverty.[14] Communities where an increasing proportion of the population is excluded from society also suffer from higher levels of crime and violence than those that are connected to mainstream networks and power structures. Victimization surveys conducted by UN-HABITAT and its partners have also shown that people living in poverty are more likely to be victims of crime than higher-income residents.

Poverty is often cited as a cause of crime and violence, but increasing evidence suggests that poverty *per se* has little to do with crime and violence levels; rather, crime and violence occur more frequently in settings where there is an unequal distribution of scarce resources or power coupled with weak institutional controls. Crime is often linked to institutional weaknesses in society. Crime increases when the social control that operates through formal institutions – such as the police and judicial systems – and informal institutions, including civil society organizations and solidarity networks, breaks down or is weakened. Although there is no simple or direct causal relationship between inequality and violence, inequality does appear to exacerbate the likelihood of violent crime, especially when it coincides with other factors. For instance, Africa and Latin America – the regions with the highest levels of income inequality – exhibit high levels of homicide, which is often used as a proxy for the broader category of violent crime.[15] Cities where inequalities are most stark also appear more vulnerable to insecurity. In 2005, for instance, South Africa reported over 800 protests in the nation's slums, many of which turned violent.[16] Some believe that this can be explained by the theory of "relative deprivation", that is,

> Poverty, underdevelopment and fragile states have created fertile conditions for the emergence of new threats, such as transnational crime and international terrorism, which are being played out in the world's cities.

Manila FRIEDRICH STARK / STILL PICTURES

Worldwide, the majority of criminal offences are committed by youth between the ages of 12 and 25, and recently, youth delinquency has become increasingly violent.

inequality breeds social tension as those who are less well-off feel dispossessed when comparing themselves to others. This theory is based on the assumption that individuals or groups are more likely to engage in violence if they perceive a gap between what they have and what they believe they deserve. The consequences of relative deprivation seem to be playing out in the world's cities, which are sites of extreme inequality. This is not the case in rural areas, where levels of deprivation or prosperity are likely to be more evenly distributed.

Crime tends to impact people living in poverty more deeply and intensely than it does higher-income residents. Not only are low-income people often unable to protect themselves from crime, which can heighten their sense of helplessness and powerlessness, but they also lack adequate fall-back systems, such as insurance and savings, making recovery from the psychological and material impacts of crime difficult. Replacing stolen goods – such as the bicycle used to get to work – may be impossible, leading to further hardship. Inequality and exclusion exacerbate insecurity, which perpetuates the vicious cycle of poverty and vulnerability. Surveys conducted by UN-HABITAT in Nairobi, Johannesburg and other cities indicate that people living in poverty cite safety and security as a major concern – as important as hunger, unemployment and lack of safe drinking water. Supporting the physical security of the lowest-income urban residents is therefore crucial to reducing poverty.

Youth, unemployment and crime

Worldwide, the majority of criminal offences are committed by youth between the ages of 12 and 25, and recently, youth delinquency has become increasingly violent. Youth unemployment is a significant contributing factor, given the high proportion of young people, high rates of population growth and slow economic growth in many cities. Some studies have also suggested a link between excessively high urban growth rates and violence and conflict in cities.[17] According to Population Action International, countries with rapid rates of urban population growth – greater than 4 per cent per year – were roughly twice as likely to experience civil conflict during the 1990s.[18]

Unemployment tends to be two or three times higher for young people than for the general population, and the lack of work opportunities may increase frustration, especially if young people's expectations have been raised through expansions in education. Estimations for Africa reveal that more than 8 million people enter the labour market each year for whom jobs will have to be found.[19] In developed countries, youth unemployment is usually twice the rate of adult unemployment; in developing countries it is often much higher. According to the International Labour Organization (ILO), an estimated 88 million young people between the ages of 15 and 24 were without work in 2003, accounting for nearly half the world's jobless. In the developing world – home to 85 per cent of youth – unemployment in this group is particularly high.

The incapacity of a country to integrate a young labour force into the formal economy has a profound impact on the country as a whole, ranging from the rapid growth of the informal economy to increased national instability. But while the informal sector offers a solution to urban unemployment, it is characterized by low salaries, dangerous work and job insecurity, all of which make it harder for youth to escape poverty.

Long-term unemployment among youth is known to be associated with negative consequences such as ill health, involvement in crime and delinquency and substance abuse.[20] In this context, the boundary between what is legal or lawful and what is illegal and illicit becomes ambiguous. Disenchanted urban youth are among the first recruits to organized criminal gangs and violent rebel groups. However, urban conflict and unrest is not simply confined to the poor. Studies have shown that the risks of instability among youth may increase when skilled members of higher income and social groups are marginalized due to lack of opportunities or when the salary or benefits they receive are not commensurate with their socioeconomic background or educational achievements, and hence what they feel they are entitled to earn.[21] Unemployed youth seek alternative models of success and peer recognition, which sometimes implies illicit and criminal activities but may also lead to violent behaviour.[22]

Fortress cities and the architecture of fear

High levels of urban crime and violence also impact the social fabric of entire cities; they instil fear and suspicion in the lives of urban residents, often leading to residential fortification among the rich, who build higher walls around their homes and spend more on private security, in effect "locking themselves" in enclaves that are physically separated from the rest of the city.[23] The fear of crime has led to increased fragmentation and polarization of urban communities, characterized by enforced segregation through gated communities, stigmatization and exclusion.[24] Insecurity has resulted in the abandonment and stigmatization of certain neighbourhoods and the development of an *architecture of fear* and the gradual establishment of so-called "fortress cities" where response to crime has led to spatial transformation that has changed parts of cities into protected enclaves and "no-go areas" separated by high walls, gates, electronic surveillance cameras and private security guards. As one commentator put it, "Creating fortified environments may reduce the opportunities for crime but may raise levels of fear."[25] The result is a fragmented urban environment that may contribute to the fear of crime outside protected areas, which could make cities more vulnerable in the long term.

In cities of the developed world, new forms of international terrorism that target public infrastructure, such as underground train networks, have promoted a culture of fear among urban residents. In some countries, such as the United Kingdom, the threat posed by international terrorism has resulted in stringent immigration policies and stricter policing, which threaten to polarize urban communities even further. There are fears that the threats posed by international terrorism may also lead to new forms of xenophobia in European and North American cities.

Dealing with perceptions of crime, particularly anxiety and fear of crime, is as important as reducing crime levels. Fear of crime affects quality of life and has negative economic and political consequences. It can also affect people's willingness to trust, interact and cooperate with the authorities, particularly the police, but also with local government crime prevention practitioners.[26] Fear of crime does not affect everyone to the same extent. The most vulnerable in society, such as women, the elderly and the poor, fear crime the most and have the most difficulty recovering from it.[27]

Crime makes cities less competitive

Crime and lawlessness impede growth and development, discouraging foreign investment and domestic economic activity. Urban insecurity impacts productivity in several ways. In many cities, employees resist working or leaving work after dark when the streets are more insecure. Employers and investors are less likely to invest in cities where their assets are likely to be destroyed or stolen. This, in turn, limits the assets and livelihood sources of the poor. Crime and the fear of crime curtail urban investment. Both individual improvements in standard of living – as minor as acquiring a radio or painting a room – and entrepreneurial investments in buildings and services are hindered by the likelihood of crime and violence. In 2001, 61 per cent of surveyed firms in Kenya reported experiencing criminal victimization. In such an environment, businesses are forced to divert resources away from productive

LUIS FELIPE CABRALES BARAJAS

GATED COMMUNITIES ARE NOT THE SOLUTION TO URBAN INSECURITY*

How can a distinction be made between legitimate diversity and illegitimate inequality?
- Norbert Lechner

Starting in the 1950s, many North American families, guided by a pioneering spirit, decided to move their homes to fortified suburbs, an attractive housing product promising a healthy environment, public safety and lifestyle benefits. That urban process rapidly became widespread during the 1970s, when it began to extend worldwide.

Because of the range of urban models and diversity of cultural settings into which these urban spaces have been incorporated, there is no standardized vocabulary to describe them. For example, in Spanish-speaking countries they are called *barrios cerrados, fraccionamientos cerrados* or *urbanizaciones privadas*; in Portuguese, they are known as *condomínios fechados* and in English the terms *walled communities* and *enclosed neighbourhoods* are used, although the most universally recognized term is *gated communities*.
Implicit in the notion of a gated community is the decision to create private urban spaces which are set apart from the rest of the city with the aim of providing an escape from undesirable social disorders. They have precedents in the 19th century, linked with the idea of garden cities and the preference for moving to the suburbs, to which later were added the rules of modern urban planning, aimed at dividing up the urban space into single-use areas. Another type of urban planning that aims to create protected surroundings is the design of cul-de-sacs or no-through roads in areas that were previously open to traffic.

As a phenomenon that is part of the globalization process, the way gated communities develop in different parts of the world tends to reflect local economic conditions. In some countries, governments in favour of keeping a tight hold on urban planning support social policies based on redistribution, as is the case in some European countries, which would explain why the trend is hardly noticeable – or even non-existent – in those countries. The same cannot be said for Latin America, where social divides coupled with the official permissiveness of neo-liberal urban planning have been conducive to the development of gated communities. Frequently, this has occurred in breach of urban regulations that have proved incapable of controlling new processes and has led to the privatization of streets and community areas that were traditionally open to the public.

The image evoked by gated communities is rooted in ideological principles and urban models that can be replicated: large or small housing estates typically aimed at the middle and upper classes and usually surrounded by a wall or fence. Those designed for the elite have sophisticated electronically controlled security systems and police surveillance, club houses, plenty of green spaces and sports grounds, and sometimes include a golf course.

Luxury enclosed estates are usually based on architectural and urban designs for low-density and low-rise housing, but they also occur in a high-rise format. Swanky towers have been introduced in areas with high environmental value in cities, as well as in suburbs such as Santa Fé in Mexico City and coastal areas, such as Palm Islands in Dubai or Miami in Florida.

Dr. Luis Felipe Cabrales Barajas is a senior professor in the Department of Geography at the University of Guadalajara, Mexico and is editor of the book, ***Latin America: Open Countries, Closed Cities***.

These trends have been possible thanks to a combination of various factors, such as growing insecurity and the deterioration of public areas, although reasons such as the desire for an exclusive address and the guarantee of a high social status – either real or perceived – also play a role.

All this has generated an ideological debate which has played into the hands of real estate agents and private security companies. Set against government inability to provide an effective guarantee of such universal rights as public security or the provision of public areas, these concepts are becoming commodities that are obviously accessible only to a minority.

The real estate market both derives benefits from and promotes paranoia about insecurity and environmental degradation, which encourages the proliferation of enclosed neighbourhoods. The standard approach is to mark out – both physically and symbolically – the boundaries of the residential estate, adding value to it by providing public areas and launching a good marketing campaign. This makes it possible to push up the prices of the products – both houses and land – and at the same time is conducive to speculative activities aimed at converting the estates into lucrative capital havens. Notwithstanding the high prices they pay, buyers are confident that their properties will retain their economic value, provided that negative externalities are not allowed to filter in.

The gated community shares many of the characteristics of postmodernism: the privatization of urban services, the deregulation of public utilities, individualistic practices, selective socialization, the rejection of the best urban traditions and placing emphasis on the use of private vehicles. If the value of urban development systems is assessed from a very broad perspective, however, private – and, in particular, low-density – urban planning runs counter to the fundamental principles of sustainable development. Gated communities are major consumers of land space and they conduct activities that constitute a wasteful use of resources, including water and electricity. For example, in the metropolitan area of Buenos Aires, enclosed urban estates occupy approximately 30,000 hectares yet house only 1 per cent of the city's population; and in the case of Guadalajara in Mexico, enclosed communities occupy 10 per cent of the city's land space yet house only 2 per cent of the population.

From being an exceptional approach to housing until the 1970s, gated communities have gradually become more widespread, and are now an increasingly common model across the world and are a sign of changing times for urban planning. The most conspicuous cases of urban planning geared towards ensuring safety are possibly found in the southern hemisphere, in polarized societies where the most affluent inhabitants want to avoid contact with the rest of the city. The private cities of Dainfern, Nordelta and AlphaVille, in South Africa, Argentina and Brazil, respectively, are good examples of this.

Dainfern, which is located in Johannesburg, has an area of 320 hectares and 1,208 houses with surface areas ranging from 450 to 1,600 square metres. It has high-quality facilities on offer, including an on-site university that carries the same name as the residential estate. The security measures are operated through a network of 57 cameras and the estate has a three-metre high electric perimeter fence (www.dainfern.com). Nordelta – a sort of city-within-a-city, known in Spanish as a "ciudad-pueblo" – located in the suburbs of Buenos Aires, covers an area of 1,600 hectares and has nine private neighbourhoods, three educational centres, a medical centre and artificial lakes (www.nordelta.com). AlphaVille is an urban development located in São Paulo that started in 1975 as an industrial zone and gradually became a residential area. It covers a surface area of 500 hectares and accommodates 50,000 residents as well as extensive commercial areas and various facilities (www.alphaville.com.br).

Even if it is conceded that gated communities are a legitimate option for the people who live there, it is important to recognize that they pose new problems or heighten previously existing ones. Is it beneficial to humankind that this model of settlements should continue to spread? Are there sound arguments in favour of their growing popularity? Do they really provide a solution to urban problems or do they simply cover them up and masquerade them as something else? However naïve or controversial they might be, the possible answers to these questions will be useful for the purposes of picturing future scenarios and directing new urban policy strategies that foster social cohesion.

Given that issues such as the impact of gated communities on social segregation, urban fragmentation, vehicular mobility and the consumption of natural resources must be analysed, local governments should discuss the appropriateness of these communities, reaffirming the democratic principles to be applied in efforts to achieve cities that are more habitable and a world that is less exclusive in the future.

Sources: Améndola, et al. 2000; Garay 2000; Glasze et al. 2005.

** Translated from Spanish*

Urban crime trends

The analysis of international crime trends is often made difficult by the limited information available, especially as regards time series. Most frequently, statements about crime levels are not based on any statistical evidence but are drawn exclusively from media reports and local perceptions. Data is scattered among different sources, is hardly comparable (especially when referring to police statistics) and is rarely available at the city level – particularly in developing countries. Yet, large urban agglomerations around the world share certain characteristics that make comparison of crime trends across cities a meaningful exercise, often more informative and useful than those across countries.

Between 1989 and 2000, the International Crime Victim Survey (ICVS) collected data on victimization experiences of citizens from large cities and urban areas in more than 70 countries around the world. Crime situations surveyed by the ICVS were quite general, but indicated an array of possible victimization experiences that are more likely to occur in urban contexts than in rural areas.

Data collected between 1992 and 2000 in 33 cities and urban areas (populations of more than 100,000) reveal an overall trend towards a *decrease* of victimization experienced by citizens. The average percentage of respondents who experienced any crime in the year preceding the survey went down from 32 per cent to 29 per cent. A sharp decrease was observed in 20 out of 33 cities and urban areas, while 4 cities were stable and only 9 showed a marked increase. However, in only 6 out of 20 cities and urban areas was a major decline in actual victimization matched by citizens feeling more safe. Citizens continued to feel less safe in Buenos Aires, Bogotá and Warsaw, and in the large cities of the United Kingdom, the Netherlands and Switzerland, even though victimization rates dropped. On average, more than half of the citizens surveyed in the 33 cities and urban areas felt either "very safe" or "fairly safe" (57 per cent). This percentage remained unchanged between 1992 and 2000.

Declines in victimization rates were primarily attributable to decreased frequency of house burglaries and ordinary thefts. These crimes are among the most common, and their reduction has a significant impact on citizens' quality of life. On average, burglary rates went down from 4 to 3 per cent, while theft rates decreased from 9 to 8 per cent. Indeed, house burglaries decreased in most urban areas of developed countries (the only exceptions were the cities in Australia and Scotland), as well as Kiev, Tirana, Sofia, Tallin, Warsaw, and Riga. A decrease in burglary rates was also observed in cities in Asia, Africa and Latin America (with the exceptions of Bogotá and Johannesburg).

Trends in car theft, robbery and personal assault only showed a slight decrease in average victimization rates. Car theft decreased in approximately half of the observed sites, including most capital cities in developing countries, while it increased in the capitals of the new European Union Member States and in south-east Europe. The average proportion of the surveyed population that experienced car theft remained stable over the study period, at 1.5 per cent.

Victimisation by robbery, which on average affected approximately 2.5 per cent of the respondents, increased in Buenos Aires, Johannesburg and Ulaan Baatar, as well as in the capital cities of the new European Union Member States, the United States, Australia, the United Kingdom, France, Finland and Sweden.

New York. CELIA ANDERSSON

Finally, victimization rates for personal assaults were, on average, stable at 4.5 per cent. Personal assaults decreased in cities in developing countries, North America and South-Eastern Europe while increasing in cities in Central and Western Europe and Australia.

Levels of personal assault and violence may be closely related to homicide levels. Although the latter are not measured by the ICVS, homicide rates recorded in police statistics largely match the assault trends described above. Police data from a group of 35 large cities, mostly in Europe and North America, indicate a decrease in homicide rates in the vast majority of cities (21) between 1998 and 2002. Homicide rates in 10 cities with populations of more than one million people (Chicago, Dallas, Houston, Las Vegas, Los Angeles, New York, Philadelphia, Phoenix, San Antonio and San Diego) continued to decrease between 2001 and 2002 from an average of 13.2 to 12.5 per 100,000 people (with homicide numbers falling in all 10 cities). Nevertheless, the positive trend of the largest cities was not matched by some smaller cities such as Washington, Atlanta, Detroit, and Richmond, where homicide rates were already high (above 30 per 100,000 people) and further increased between 2001 and 2002.

Homicide rates in the 8 cities with the lowest rates in 1998 (less than 2 per 100,000 people) increased the most sharply to 2002, while in "high-risk" cities, the rate remained relatively stable. This was the case, for example, in Cape Town and Rio de Janeiro. It should be noted, however, that homicide rates in these cities were above 40 per 100,000 people, with some cities, such as Medellin in Colombia and East London in South Africa, showing homicide rates above 100 per 100,000 people. In such extreme cases, where homicide represents the leading cause of death for juveniles, it is even more important to collect regular information in order to monitor the impact of any initiative aimed at crime reduction and crime prevention.

The available data series, although very limited, provides some reassuring signals on decreasing crime levels in large cities, especially in developing countries. However, significant gaps still need to be filled, since accurate and timely information is essential to building proper strategies to combat urban crime and insecurity.

Source:United Nations Office on Drugs and Crime (UNODC). Based on data from: United Nations Interregional Crime and Justice Institute, UNICRI (International Crime Victim Survey 1992/96 and 2000); Home Office of the UK, Crime in England and Wales 2003/2004: Supplementary Volume 1: Homicide and Gun Crime; US Government, Bureau of Justice Statistics, Sourcebook of Criminal Justice Statistics 2002.

uses, leading to reduced competitiveness and fewer investments – or even disinvestments. Crime therefore has a significant impact on economic development and investment.

While crime has many dimensions, strategies to tackle the problem are often too narrow, focusing on tougher penalties and law enforcement. A more effective way to control crime is to combine enforcement with prevention – developing positive strategies to tackle the underlying causes of crime. In Bogotá, Colombia, a multiple action plan for citizen education focused on improving law, culture and morale; this contributed to a 30 per cent decline in homicide rates between 1995 and 1997. The programme targeted strategic areas such as crime monitoring, police and judiciary action, education, public services, youth programmes and restrictions on bearing arms.[28]

Urban violence, poverty and ineffective governance are inextricably linked and mutually reinforcing. Unless urban violence – and its manifestation in exploitation of the poor – are addressed as part of poverty reduction and governance improvements, programmes to improve the lives of urban dwellers will have limited impact.[29] Good governance both supports and is supported by safe cities where inhabitants are free from fear. Poor governance increases the risk of insecurity. Where safety is improved, interaction between the people and public institutions becomes possible, creating an enabling environment for economic growth and participation.

Expanding the definition of human security

The current concerns over crime and violence in cities should not obscure the fact that in most developing country cities, the poor are contending with other forms of insecurity that threaten their lives and livelihoods. The security of the poor, in particular, is affected by their health status, which influences both their ability to work and their access to health care. The HIV/AIDS pandemic has particular implications for urban security as it means loss of household income, growth in the phenomenon of orphaned street children, and disintegration of the family unit. Many urban poor families also face the constant threat of eviction. Insecurity is exacerbated by insecure tenure with respect to both housing and land.

As economist and Nobel Laureate Amartya Sen has noted, "The demands of human security include a balanced view of tragedies that are the result of terrible omissions as well as commissions."[30] He defines an adequate concept of human security in the contemporary world as one that includes the following: a clear focus on human lives, as opposed to the technocratic notion of national security in the military context; an appreciation of the role of society and of social arrangements in making human lives more secure; and a fuller understanding of human rights, which must not only include political freedoms, but also rights to food, medical attention and basic education.

Endnotes

1 Although the attacks on New York and Washington are the most widely known, several other cities, such as Nairobi and Dar es Salaam had been targets of terrorism prior to 11 September 2001.
2 Cohen 2002.
3 Department for International Development 2005.
4 Marcuse 2001.
5 Worldwatch Institute 2005. Estimates based primarily on United Nations-generated data.
6 Ibid.
7 Worldwatch Institute 2005.
8 Appadurai 1996.
9 Results from International Crime Victim Surveys conducted by UNICRI. Comparative statistics available online: www.unicri.it.
10 WHO 2002.
11 Results from International Crime Victim Surveys conducted by UNICRI. Comparative statistics available online: www.unicri.it.
12 Vanderschueren 2000.
13 Moser & Rodgers 2004.
14 Narayan 2000.
15 United Nations 2005a.
16 Reported in the New York Times, 25 December 2005.
17 Worldwatch Institute 2005.
18 Ibid.
19 Economic Commission for Africa 2002.
20 O'Higgins 2002.
21 United Nations 2005a.
22 Vanderschueren & Vezina 2003.
23 Moser 2005.
24 Caldeira, 1996, writes, "the talk of crime is the principal discourse in São Paulo everyday life". The perception that the institutions of order, particularly the police, are also violent has magnified the fear and led to the growth of private guard systems.
25 Landman 2003.
26 Robertshaw, et al. 2001.
27 These trends have been supported by victim surveys in South Africa and abroad.
28 Kathuria & Oberai 2004.
29 Dwyer 2005.
30 Amartya Sen, quoted in CNN's Principal Voices series in 2005: www.time.com/principalvoices.

STEPHEN GRAHAM
THE URBANIZATION OF POLITICAL VIOLENCE

Since the dawn of urban and military history, cities, warfare and organized political violence have always helped to constitute each other. As symbolic targets, urban centres have few equals. Contemporary cities are actually made up of almost infinite concentrations of sites, assets and spaces that can either be improvised as weapons projecting political violence themselves, or attacked as "soft targets" (entities that are not fully militarized or equipped to fight back) by terrorists, insurgents and state militaries alike. Urbanites, moreover, are especially vulnerable to the disruptions caused by political violence. This is because they rely on extensive concentrations of technical and social infrastructure or capital to survive, feed themselves, access water and energy, avoid disease, remove wastes, and so on. Disrupting these systems through political violence, deliberately or unintentionally, leaves many urbanites with few alternatives.

The post-Cold War period has seen a dramatic reduction in the number of state-versus-state conflicts. Meanwhile, wars pitching state military or paramilitary forces against non-state insurgent, terrorist, or organized crime groups have proliferated. This trend has been associated with a dramatic urbanization of political violence around the world. Like other facets of global social change, political violence is, in a sense, being urbanized. More than ever, geopolitical concerns increasingly merge into and irredeemably zero in on the very local sites and symbols of city life. Not surprisingly, this change has been associated with a major change in the balance of civilian rather than military casualties through political violence. Between 1989 and 1998, for example, approximately 4 million people were killed in violent conflicts around the world; an estimated 90 per cent of these were civilians – primarily women and children.

In an increasingly urbanized world, insurgent and guerrilla groups, rather than seeking shelter within rural proletarian groups, are colonizing the world's burgeoning urban spaces. At the same time – after centuries when cities were seen as sites to be either avoided or "rubbleized", state military doctrine, particularly in the West, now sees urban sites as the *de facto* terrain for current and future struggle. Whilst they still occur, attempts at the complete annihilation of cities are now unusual, but the targeting of urban soft targets as a means to coerce and win victory over a political enemy is now axiomatic to terrorists, insurgents and state militaries alike.

Bank Underground Station, London PHILIPPE HAYS/STILL PICTURES

The methods and styles of this targeting could not be more varied. On the one hand, non-state insurgent and terrorist groups increasingly exploit the embedded assets of cities as weapons bringing instantaneous death, terror and mediated violence. In the absence of sophisticated military hardware, the very bodies of volunteers are often mobilized to project violence directly against the sites and symbols of the modern city.

Such projections of violence are becoming more spectacular and sophisticated as the infrastructural and technological fabric of global cities intensifies in reach and complexity: airplanes became cruise missiles of mass murder in New York and Washington; mobile phones were used to trigger subway bombs in Madrid; and London's underground trains and buses provided the intensely crowded and enclosed spaces necessary for suicide bombers' actions. And in a widening range of suicide and car bomb attacks in Indonesia, Iraq, Israel, Jordan, Kenya, Lebanon and Morocco, among other countries, the unavoidably crowded spaces of urban everyday life are instantly being transformed into soft targets by terrorists.

Whilst not reaching the levels of the total urban annihilation that characterized 20th century warfare, targeting of cities raises concern because large numbers of innocent civilians are often killed as "collateral damage". Cities are unavoidably crowded and it is virtually impossible to distinguish between insurgents and wider civilian populations, even with high-tech tar-

Stephen Graham, Professor of Human Geography at Durham University, is the editor of ***Cities, War and Terrorism: Towards an Urban Geopolitics***, which was published by Blackwell in 2004.

geting. The worry here is that urban assaults merely radicalize the civilian populations on the receiving end of violence, adding legitimacy to retaliatory attacks by terrorists. Moreover, in cities, combatants can target water, sanitation, electricity, and food distribution systems through biological agents and anti-infrastructure weapons such as missiles and bulldozers, waging a "war on public health". When this happens – as in Iraq after the 1991 Gulf War – far more civilians eventually die from preventable diseases than from the immediate effects of bombs and missiles.

The deepening sense of urban exposure and vulnerability, as transnational flows and networks erupt on city streets in violent acts of terror, has provoked widespread search for technical or architectural solutions to terrorist attacks. This is especially the case in rich, Northern cities. However, the almost infinite complexity and necessary openness of globalized cities means that new solutions are unlikely to emerge beyond the highly restricted environments of airline systems.

State military action against cities, on the other hand, is very much defined by the technological capability of the military involved. In Africa, Latin America and Asia, a wide range of state anti-insurgency campaigns mobilize relatively unsophisticated technologies in their targeting of urban insurgencies, their host populations, or increasingly militarized gang and organized crime networks. Western and Israeli militaries, meanwhile, are being remodelled to adapt to the new demands of urban warfare as part of their "war on terror".

As levels of urbanization around the planet continue to intensify, all projections point to the deepening urbanization of organized political violence in the future. As the sites, symbols and embedded assets of cities become both weapons and targets in increasingly mediated conflict, a vital challenge for all concerned with the widest aspects of human security is to resist the temptation to try and fortify cities against the putative risks in a narrow technical or architectural sense. Such a "fortress city" approach to "homeland security", whilst lucrative to burgeoning military and security sectors, is a red herring because it is largely ineffective against determined attackers who can simply select the next unprotected, soft target out of the millions of options on offer in contemporary cities. Moreover, such an approach also risks undermining the interchange, openness, flow and density that sustain cities in the first place.

The challenge, rather, it is to work at all scales of governance and conflict mediation to try and ensure that the grievances, injustices, extreme ideologies and hatreds that fuel political violence against cities and urbanites are, as far as possible, ameliorated. This must be done to the extent that the murderous assaults on urban soft targets, by terrorists, insurgents, and state militaries alike, are prevented or are rendered politically or ideologically illegitimate.

Such a challenge is daunting. This is especially so as urban research, policy and activism have tended to neglect the urbanization of political violence thus far, leaving the subject to international relations specialists. But, in an increasingly urbanized world dominated by intensifying resource conflicts, global warming, proliferating refugee, water and food crises – sometimes precipitated by aggressive nation states and transnational terrorist groups – the process fuelling the urbanization of political violence seems set to accelerate further. Through the rest of the 21st century, these challenges are likely to become even more critical. The time for a specifically urban treatment of geopolitics, which concentrates on how local urban sites and infrastructure are enrolled into global networks of political violence, is upon us.

4

Part Four

Policies and Practices that have Worked

Using slum estimates produced in previous parts of the Report, this Part provides a policy assessment of slum upgrading and prevention policies in more than 100 countries across the world. An analysis of the policy responses implemented by these countries shows that the most successful ones share similar attributes. Considering that policy outcomes in reducing shelter deprivations have been rather bleak, this Part highlights some necessary bold actions needed to scale up improvements today and prevent slums tomorrow. It concludes by outlining key issues related to development assistance provided by the international community.

Road Construction, Bogotá, Colombia MARK EDWARDS/STILL PICTURES

4.1 Milestones in the Evolution of Human Settlements Policies 1976-2006

"Human settlements are linked so closely to existence itself, represent such a concrete and widespread reality, are so complex and demanding, so laden with questions of rights and desires, with needs and aspirations, so racked with injustices and deficiencies, that the subject cannot be approached with the leisurely detachment of the solitary theoretician."
– Opening statement of Canadian Prime Minister Pierre Trudeau at Habitat: United Nations Conference on Human Settlements, Vancouver, 31 May 1976

1976: Habitat conference brings human settlements issues to the fore

When the first United Nations Conference on Human Settlements (Habitat I) took place in Vancouver, Canada, in 1976, the world was largely agrarian, with two-thirds of world's three billion people residing in rural areas. Urbanization, as a phenomenon, was beginning to be recognized, but more as a "problem" rather than as a positive force for economic, social and cultural development. Cities were viewed as generators of pollution, sites of unsustainable growth and a drain on national resources. These negative sentiments were echoed by various leaders attending Habitat I who blamed "uncontrolled urbanization" for overcrowding, pollution and psychological tensions and lamented "the disintegration of rural life and the disappearance of farm lands through the spread of cities and their satellites".

The Vancouver Conference emphasized rural-urban disparities and called on governments to improve the rural habitat by adopting policies that promoted a more equitable distribution of the benefits of development between urban and rural areas. This led to a general "rural bias" among development agencies, which focused their efforts on issues such as agricultural productivity and provision of basic services to rural areas. The thrust of the recommendations in the Vancouver Declaration was also more towards provision of public goods and housing, a policy that would change dramatically in the coming years. Nonetheless, the Vancouver Conference was the first milestone in the "habitat" agenda; it recognized that human settlements – both rural and urban – were a new category of analysis and international policy intervention. Perhaps the most significant aspect of this new habitat agenda was the recommendation to adopt an integrated – as opposed to a sectoral – approach to housing, infrastructure and basic services. The search for a "harmonious integration of components" was an important outcome of Habitat I.

1978: UN establishes focal point for human settlements development

The importance accorded to human settlements at the Vancouver Conference was endorsed by the United Nations in 1978 when it created a special agency to serve as the focal point for human settlements action and coordination within the United Nations system. The United Nations Centre for Human Settlements (UNCHS) was tasked with the responsi-

The Habitat Conference, Vancouver, 1976 UNITED NATIONS.

FIGURE 4.1.1 SLUM POPULATION AND URBAN POPULATION GROWTH IN THE WORLD (1976-2006)

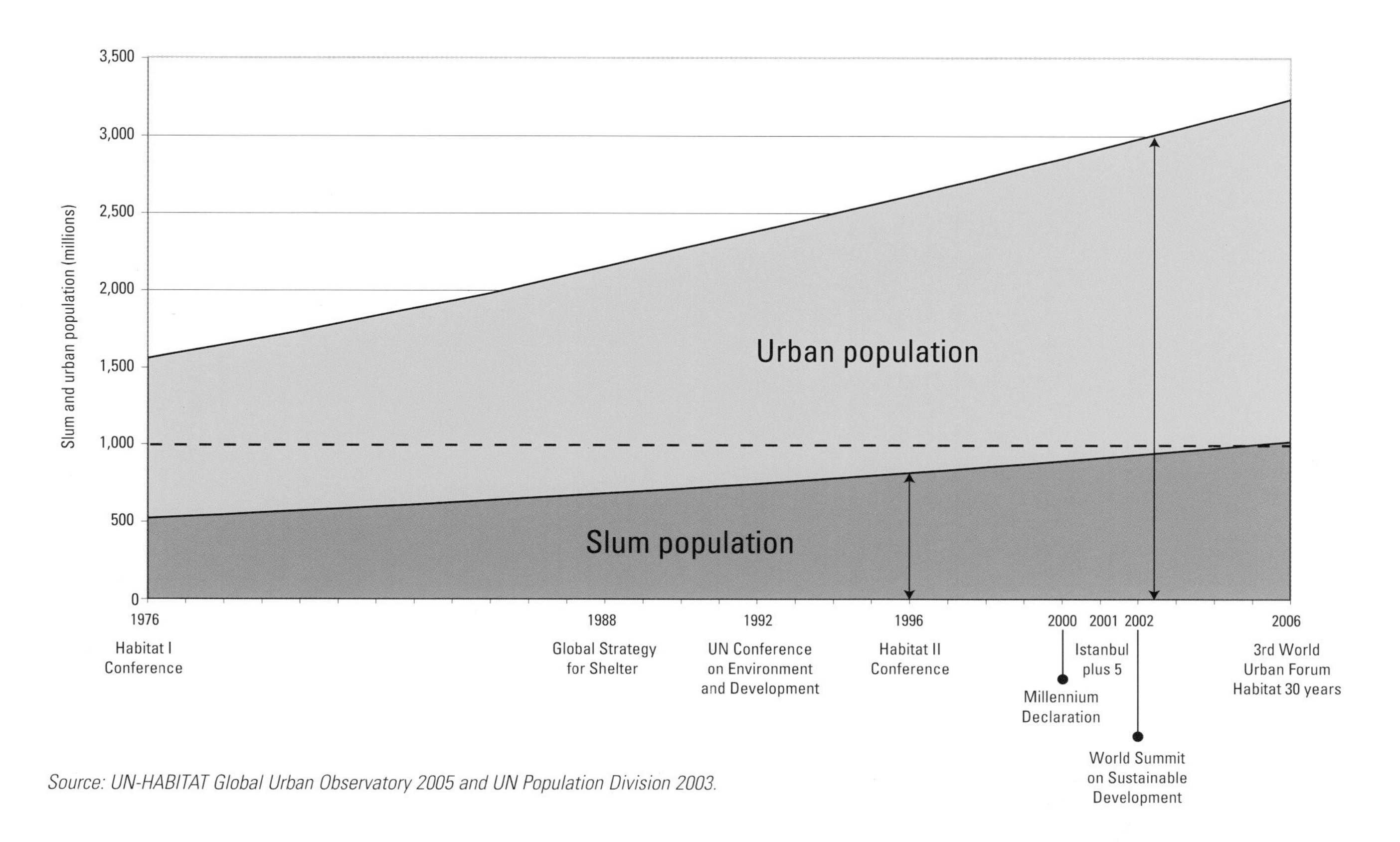

Source: UN-HABITAT Global Urban Observatory 2005 and UN Population Division 2003.

bility of ensuring that the shelter dimension was reflected in UN programmes and projects. In its early years, the organization (which later became the United Nations Human Settlements Programme or UN-HABITAT), worked with other UN agencies to formulate shelter policies and implement housing programmes in both rural and urban areas, particularly in least developed countries and those ravaged by conflict or disaster.

1988: Global Shelter Strategy transforms role of the State from provider to enabler

In the 1980s, scarce public funds and increasing urban populations were putting a strain on governments in developing countries, which soon came to realize that provision of public housing was neither affordable nor sustainable in the long term. This realization led to the adoption in 1988 of the Global Strategy for Shelter (GSS) to the Year 2000, which advocated an "enabling strategy" that shifted the role of governments from provider to "facilitator". Governments were expected to remove obstacles and constraints that blocked people's access to housing and land, such as inflexible housing finance systems and inappropriate planning regulations, while people were expected to build and finance their own housing. The GSS also accorded a fundamental role to the private sector in shelter delivery; it was based on a sectoral approach that aimed to introduce innovations in building technology, new construction methods and affordable building materials. During this time, assistance was focused mainly on the promotion of low-cost self-help housing, also known as the "bricks and mortar" approach. In theory, this approach offered many advantages: it allowed greater flexibility in building, and the possibility of community development and construction took place incrementally over time, giving beneficiaries the option of pacing the construction according to their household earnings. In practice, however, the approach had major shortcomings: it was entirely dependent on the supply of public land (or acquired private land), and for many years remained project-oriented and limited to some small-scale demonstration projects.

The GSS also came at a time when the idea of public provision of services or the "welfare state" was losing legitimacy on the global stage. Structural adjustment programmes (SAPs) that drastically reduced the role of the State in socio-economic development were being adopted by many countries in the developing world, which resulted in deregulation and privatization of essential services. Unfortunately, public expenditure cuts in health, housing and education resulted in serious housing and basic services deficits that also increased levels of urban poverty in many parts of the world.

1992: Sustainability emerges as a development challenge

The 1980s also saw a shift towards democratization and decentralization, which focused on strengthening the capacity of local governments to improve urban management. By the early 1990s, other non-state actors, such as civil society organizations and the private sector, were clamouring for a greater say in public affairs. These actors played an increasingly important role in the development of international policies on human rights and the environment.

Meanwhile, a 1987 report by the World Commission on Environment and Development concluded that human activity was the leading cause of environmental degradation and pollution. This led many governments to rethink their strategies on environmental management. By the time the United Nations Conference on the Environment and Development (UNCED) took place in Rio de Janeiro in 1992, the concept of "sustainability" (the idea that development should meet the needs of present and future generations without harming the environment) was gaining ground. However, while the environmental movement focused its energies on addressing issues such as biodiversity, global warming and desertification, it wasn't until the World Summit on Sustainable Development (WSSD) in Johannesburg in 2002, that "sustainable urbanization" emerged as a new multidimensional concept that covered not only the impact of cities on the environment, but also their potential to manage the urban environment in a way that benefits urban residents both socially and economically. The move towards sustainability spawned a variety of projects aimed at improving access to basic services, such as water, sanitation and waste disposal, among others, which were also the main areas of focus of new joint initiatives, such as the Sustainable Cities Programme and Localizing Agenda 21.

1996: Globalization and urbanization shape the Habitat Agenda

By the early 1990s, a series of United Nations conferences focused the world's attention on the challenges facing an increasingly globalizing world. The second United Nations Conference on Human Settlements (Habitat II), the last of this series of conferences, took place in Istanbul, Turkey, in June 1996. By this time, 45 per cent of the world's population was already living in urban areas, with demographers predicting a major shift in the world's population from rural to urban in the early years of the new millennium.

When world leaders met in Istanbul, globalization and urbanization were becoming powerful forces of economic growth, as well as exclusion. The rural idealism of the 1970s had given way to the harsh urban reality of the 1990s where cities were seen as engines of economic growth, innovation and creativity, but also as sites of extreme poverty, exclusion and environmental degradation. The world's population had doubled from 3 to 6 billion. Urban populations were growing at a rate 2.5 times faster than rural populations, with the result that almost half the world's population was already living in urban areas. The role of the State was diminishing as cities became major centres of trade and finance, negotiating directly with each other. Information and communication technologies were changing the ways in which people, cities and countries were communicating with each other. National borders were becoming less significant and local authorities were becoming key players in national and international negotiations. Civil society organizations were also playing a more active role and demanding to be heard at UN Conferences, including Habitat II.

Participation and partnerships were seen as the guiding principles to "sustainable human settlements in an urbanizing world" and "adequate shelter for all" – two of the main goals of the Habitat Agenda and the Istanbul Declaration on Human Settlements, which were adopted by the 171 governments at the Habitat II Conference. Unlike the Vancouver Conference, the Istanbul Conference affirmed the positive role of cities in a globalizing world, and focused the world's attention on urbanization as an emerging issue in the 21st century.

With the massive mandate set out in the Habitat Agenda, UN-HABITAT struggled alone among United Nations agencies to prevent and ameliorate problems stemming from rapid urbanization and deteriorating living conditions. Although it tried to address this shortcoming by forging partnerships with other international agencies, such as the World Bank, the impact of its programmes was not being felt among the most vulnerable populations in cities – the urban poor. Part of the problem was that while the Habitat Agenda offered the most comprehensive guide to addressing these issues, it did not focus the organization's work in one particular area, which led to

FIGURE 4.1.2 GDP PER CAPITA BY INCOME GROUPS 1975-2004

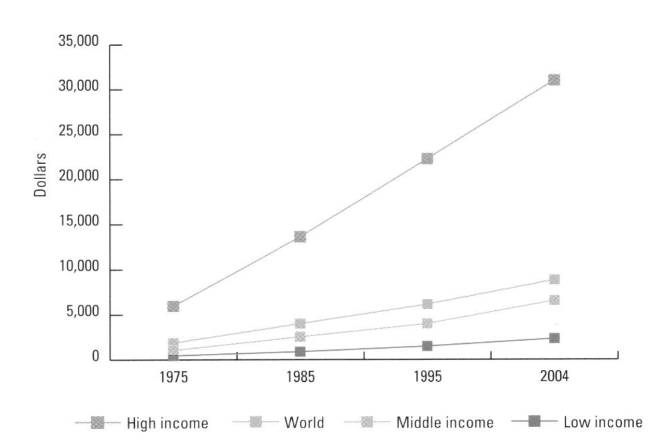

Source: World Bank, World Development Indicators database.
Note: GDP per capita based on purchasing power parity (PPP). Data is in current international dollars. Income grouping based on World Bank's definition:
High income countries: those in which 2004 GNI per capita was $10,066 or more.
Middle income countries: those in which 2004 GNI per capita was between $825 and $10,065.
Low-income countries: those in which 2004 GNI per capita was $825 or less.

duplication with other agencies. Also, without an agreed prioritization and integration of recommendations with time-bound targets, Member States of the United Nations were left with a mission that encompassed all objectives, but no central objective. This made it difficult to monitor and review progress in implementing the Habitat Agenda at the national, regional and international levels.

2000: World leaders set the "cities without slums" target

In September 2000, the United Nations convened the Millennium Summit where world leaders were expected to set priorities and targets towards poverty reduction. This resulted in the Millennium Declaration from which the Millennium Development Goals and time-bound targets were derived. The Millennium Declaration commits governments to addressing key development issues and sets broad goals in order to eradicate poverty by the year 2015. It encompasses key goals related to poverty reduction, health, gender equality, education and environmental sustainability.

The Millennium Declaration was adopted at a time when almost half the world lived in urban areas, and almost one-third of this urban population lived in slums. Governments were increasingly recognizing that slums could not just be considered an unfortunate consequence of urbanization, but needed to be treated as a major development challenge through coordinated policies and action at the global, national and city levels.

UN-HABITAT is mandated to monitor progress towards achieving Millennium Development Goal 7, target 11 – *"By 2020, to have achieved a significant improvement in the lives of slum dwellers"*. Known as the "Cities without Slums" target,[1] this clear, time-bound target enables the organization and its partners to re-orient their policies towards slum improvement and prevention. Slum upgrading is now a very important area of focus of UN-HABITAT's, work, with increasing emphasis being placed on policy and operational support to scaling up of projects and programmes. Monitoring progress towards achievement of this target at the national, regional and global levels also comprises a significant part of UN-HABITAT's work in this area.

Despite the grand proclamations and commitments by governments at the Millennium Summit, the reality on the ground suggests that poverty is rising in cities and that slums are emerging as a dominant type of settlement in many cities in the developing world. UN-HABITAT estimates indicate that almost a billion people already live in slum-conditions around the world and that slums are growing dramatically in the world's poorer cities, particularly in sub-Saharan Africa and Asia.

Participants attending the 3rd World Urban Forum to celebrate the 30th anniversary of the first Habitat Conference in Vancouver in June 2006 will have to contend with this reality: the situation of slum dwellers is likely to worsen if policies and actions are not put in place to alleviate, prevent or improve the housing conditions of the world's urban poor. If no action is taken, the number of slum dwellers is expected to rise to 1.4 billion by 2020. It is hoped that 30 years after the "habitat" agenda made its way into the development agenda, words will finally lead to the kind of concerted action needed to stem the tide of rising poverty, exclusion and deteriorating living conditions in cities.

Burdur, Turkey DARCY VARNEY

Endnotes

1 Millennium Declaration, paragraph 19.

Vancouver: The world's most liveable city combines multiculturalism with environmental sustainability

"Vancouver is home to a multitude of cultures and languages from around the world. The City of Vancouver values this diversity, and considers it a source of our strength, vitality and prosperity." — City of Vancouver 2005

In October 2005, the Economist Intelligence Unit (EIU) voted Vancouver, British Columbia, the world's most liveable city. Vancouver topped the EIU's list of 127 cities around the world, having earned the honour for its well-established infrastructure, cultural and environmental richness, low crime rate, and easy access to goods and services.

While many point to Vancouver for its environmental consciousness and physical beauty, it is, in effect, the city's cultural diversity and resources that work most effectively toward strengthening its environmental sustainability and liveability. Vancouver's population is a mosaic of cultures and ethnicities: 38 per cent of the metropolitan region's total population is foreign born, and two-thirds of the adult population is directly connected to immigration, being immigrants themselves or children of foreign-born parents. British Columbia's largest city-region has become a gateway to opportunity for people across the Asia-Pacific region, an exemplar of Canada's effort to build a cooperative multicultural society and a leader in integrated planning for liveability and sustainability.

Vancouver is a city of 550,000 residents set within a region – the Greater Vancouver Regional District – that hosts a total population of approximately 2 million. In 2001, Canada's last census year, half of the City of Vancouver's residents were "visible minorities" (non-Caucasian people who are not Aboriginal in origin), and 46 per cent were immigrants. The largest proportion of the immigrant population comes from Asia – particularly China, India, the Philippines, and South Korea. Vancouver is home to 14 per cent of the population of British Columbia but 24.5 per cent of its total immigrants. Less than half the city's population speaks English as a first language.

Multiculturalism has served as a codified Canadian value and policy framework since 1971, when the country expanded its bicultural policy – which recognized the equality of the English and French cultures in the development of the nation – to include respect for and protection of the full spectrum of the country's citizens. The passage of the Canadian Multiculturalism Act in 1988 further provided a comprehensive legislative structure for the country's emerging multicultural identity. As one of Canada's most diverse urban centres, the City of Vancouver has initiated many strategies and partnerships designed to promote social inclusion, increase civic engagement, recognize the many cultural and linguistic groups in and around Vancouver, and address the needs of Vancouver residents from a range of backgrounds.

One of the major building blocks upon which multicultural programmes and services are built is the Civic Policy on Multicultural Relations. Adopted in 1988 as a municipal extension of the national legislation, the policy recognizes ethnic, cultural and racial diversity as a source of strength for the city and resolves that all residents have the right to freedom from prejudice as well as the right to access civic services regardless of background or language. A variety of institutional supports within the city exist to help implement the policy. These include a Social Planning Department that participates in the overall planning of the city by reviewing city developments and providing advice on how to address specific cultural and social needs.

The Social Planning Department was instrumental in helping diverse cultural communities and demographic groups participate in Vancouver's component of the Greater Vancouver regional planning process. The six-year participatory process resulted in an award-winning Liveable Region Strategic Plan, Greater Vancouver's official regional growth strategy, adopted in 1996.

The City of Vancouver carried out its part of the regional process between 1992 and 1995, when approximately 20,000 residents participated in the CityPlan initiative. As in the overall regional vision, the city's residents expressed a strong commitment to connecting social and environmental principles for achieving sustainability. They requested more efficient public transport, more and safer bicycle routes through the city and more pedestrian-friendly streets. Residents also prioritized the following: development of distinctive neighbourhoods featuring diverse public spaces, affordable housing and access to services; a strong sense of community and increased public safety; and economic development that keeps jobs in the city. The city is also investing millions of dollars to reduce homelessness and drug addiction within the Aboriginal population, which for many years was left out of Canada's development agenda.

Sources: Mendes & Holden 2005; Leaf 2005; UN-HABITAT 2002a; City of Vancouver 2001; City of Vancouver 2004; Statistics Canada 2001; Economist Intelligence Unit 2005.

ELISABETH GATEAU

LOCALIZING THE MILLENNIUM DEVELOPMENT GOALS

Vancouver GOVERNMENT OF CANADA

At the Millennium+5 Summit in New York, the United Nations General Assembly explicitly recognized the important role of local authorities in contributing to the achievement of Millennium Development Goals. The outcome document also opened up new possibilities for direct dialogue between the General Assembly and the world organization of local governments, and put local authorities first in the list of major actors that work with the United Nations.

As the United Nations Secretary-General stated when he met with a delegation of mayors: "A state which treats local authorities as partners, and allows public tasks to be carried out by those closest to the citizens, will be stronger, not weaker. Weak cities will almost certainly act as a brake on national development, whereas strong local democracy can be a key factor in enabling a country to thrive."

Despite these milestone achievements, there is still a great distance to cover if we are to bridge the local versus global knowledge gap. The new global realities of today's fast-moving cities and diverse towns and villages require that decisions be taken at the level of government closest to the citizens. Decentralization is essential for a democratic system of governance and it is the key to basic service provision for, by and with the community.

We must strengthen local governance, through enhanced citizen participation and effective partnerships with all local stakeholders, if we are to succeed in achieving the Millennium Development Goals. For this reason, 2007 will see the launch of a United Cities and Local Governments (UCLG) Global Report examining progress in decentralization and local democracy across the globe.

Elisabeth Gateau is the Secretary-General of United Cities and Local Governments (UCLG), an international organization

4.2 Countries Taking Slums Seriously

Commitment from the top[1] obviously has something to do with why some countries have been more successful than others in managing slum growth. It is widely accepted that political will in responding to the reality of slums is pivotal in mobilizing commitment to help the urban poor to gain access to adequate shelter, livelihoods and services. The experiences of 23 countries analysed by UN-HABITAT indicates that political support for slum upgrading, slum prevention and urban poverty reduction in general varies significantly over time among countries and within cities. Some national and local governments, particularly during the last decade, chose to largely ignore their burgeoning slums or confine actions to symbolic gestures, often preferring to continue with practices of slum clearance and mass evictions. By contrast, other governments have taken the responsibility squarely on their shoulders: making commitments backed by bold policy reforms; scaling up upgrading programmes; and adopting urban planning measures and equitable economic policies to prevent future slum growth. However, most governments seem to lie somewhere between action and inaction, sometimes experimenting with new forms of planning and financing more in line with the needs of the urban poor, trying to push through much-needed reforms, but ultimately holding back on political commitment to make a significant impact. This part of the Report analyses how some governments have demonstrated clear commitment and leadership in dealing with slums.

Where political commitment to slum prevention and improvement has made a real difference

Brazil, Egypt, Mexico, South Africa, Thailand and Tunisia stand out as countries that demonstrated consistent political commitment over the years to large-scale slum upgrading and service provision for the urban poor. These countries are the same ones that have been most successful in reducing or stabilizing slum growth rates in the last 15 years, as revealed by the scorecard results reported in Chapter 1.3. Egypt and Tunisia recorded falling annual slum growth rates of 1.6 per cent and 5.4 per cent per year, respectively, while in Mexico and South Africa slum growth fell to only 0.5 per cent and 0.2 per cent, respectively.

Other countries, while often showing considerable political determination, were ranked slightly lower according to the criteria set to analyse leadership. In Indonesia and the Philippines, the political upheavals of the last two decades have somewhat undermined slum improvement efforts. In India, despite major slum upgrading efforts at national and State levels, the overriding concern seems to be with creating an urban environment conducive to attracting foreign investment and selling India as an attractive partner in the liberalized marketplace. While there are progressive projects and policies being implemented in the larger cities, there is a parallel and continuing trend towards evictions without adequate or appropriate relocation. Other countries, such as Ghana and Morocco, also show more modest support for upgrading, having recently stepped up actions to tackle slum growth. More moderate political support in these countries over the last 15 years appears to have held back governments' performance in achieving the kind of turnaround in slum numbers seen in the top-performing countries, where political commitment was consistently stronger. These "moderate" performing countries experienced sustained slum growth rates ranging from 1.4 per cent to 2 per cent per annum.

However, it would be unfair to say that all the countries that are struggling to cope with high slum growth rates in this category have shied away from committing to change. In sub-Saharan Africa, Burkina Faso, Senegal and Tanzania have in recent years shown promising signs of growing political support for slum upgrading and prevention that include reforms in policies governing land and housing. For example, Tanzania has embarked on land reforms and has shown a more tolerant and responsive attitude to its growing informal settlements. Yet translating new-found support into significant improvements on the ground will no doubt take time.

So there appears to be a pattern: the stronger the political commitment by national and local governments to slum improvement, the more significant the progress in reducing the growth of slums. Several questions remain, however. Exactly how have political leaders and top public servants shown – or not shown, as the case may be – their commitment to slum upgrading and preventive measures, and how does this make an impact on the living standards of slum residents? And, perhaps the most critical question is: what motivates them to take affirmative action? While the focus of this chapter is on political willingness by national governments to act decisively, it is recognized that political leadership at the city level is also vitally important in driving slum improvement programmes and policies. In some countries, cities, under dynamic leaders, have gone ahead with innovative citywide upgrading programmes despite a lack of political and institutional support from the centre. But for the most part, national governments have the power to make or break slum policies and programmes – it is their political support, legislation, reforms and macro-economic policies that create the overall environment and framework through which programmes and policies can operate locally, as well as on a countrywide scale.

■ Signs of commitment

In some countries, *political pronouncements* have often acted as a major driving force behind change, sending out explicit signals for policy reform and unleashing innovation and initiative in the delivery of shelter and services to the poor. This commitment may be enshrined in the highest legal instruments, such as the South African constitution and the Brazilian city statutes, both of which are based on the principle of equity. In other cases, statements by heads of state, such as the Royal Letter in Morocco, or the public commitment of the Cambodian Prime Minister, have set the benchmarks for urban poverty reduction and slum prevention.[2] Presidential decrees have also helped to set in motion pro-poor programmes and budgets. In Mexico, a presidential executive order launched the "Habitat Programme" in 2003 as part of the president's agenda to address poverty, health and education in Mexico's larger cities.

However, executive proclamations only make a difference when they are backed by *long-term strategies with realistic national targets for slum improvement, adequate budgetary allocations, and a programme of policy, legal and regulatory reforms* to meet the targets. Morocco provides a good example of a country which, through its ambitious *villes sans bidonvilles* programme for in-situ slum upgrading and greenfield development, set the goal of becoming a "slum-free" country by 2010. One hundred thousand units of social housing are to be delivered per year through significant budgetary allocations. A few countries in Latin America, including Brazil, Chile and Mexico, have made concerted efforts to develop long-term action plans for slum upgrading and urban poverty reduction by setting clear targets and establishing institutions to ensure that these plans are implemented. In Brazil, following the introduction of the "Slum Action Plan" by the Secretariat for Housing and Urban Development, the President established a housing fund of $1.6 billion for financing new housing construction and upgrading *favelas.* The action plan is starting to make an impact with the upgrading of 30 slums and the improvement or construction of 31,000 housing units is already underway.[3] In Colombia, the National Council of Economic and Social Policies (CONPES) has made a public commitment to improve the living conditions of slum dwellers. The national goal is to reduce the proportion of people living in slums from 16 per cent to 4 per cent by 2020.

Consistency in political commitment is crucial in mobilizing long-term support for slum upgrading. The Government of Tunisia's consistent support for upgrading over the past three decades has all but eliminated the national backlog, which now stands at some 24,000 units.[4] In 1997, the Government of Chile inaugurated the Chile Barrio Programme to achieve a substantial improvement in the housing and employment of families living in informal settlements. This carried the extensive regularization process that started in the 1980s further. The foundations for these initiatives were laid in the late 1970s when Chile implemented a comprehensive national housing programme. In recent years, the barrio programme has utilized 70 per cent of the budget of the Ministry of Housing and targeted 30 per cent of the poorest population.[5] Since 2003, the Government of Thailand has been implementing programmes for the construction of homes for one million low-income households in partnership with commercial and public banks. This commitment is a continuation of almost three decades of attention to low-income housing, going back to 1977 when the National Housing Authority (NHA) was given the responsibility for it and carried on as the umbrella body for slum upgrading. Such long-term support for low-cost shelter and slum upgrading has undoubtedly contributed to Thailand's extraordinary success in improving living conditions in slums – the slum growth rate has fallen by an average 18.8 per cent per year since 1990.

Hammamet, Tunisia JOCHEN TACK/STILL PICTURES

Another indication of strong political commitment for slum improvement is the *inclusion of upgrading and urban poverty reduction policies in the national development agenda.* This has led to prioritizing the urban sector and slum upgrading within national budgets and macro-economic frameworks. As part of Jordan's National Strategy for Eliminating Poverty and Unemployment, adopted in 1998, a major new upgrading pro-

Costing the slum target*

In 2003, the UN Millennium Project began a costing process for the various targets that had been established in the Millennium Declaration, and laid down principles and a general methodology for costing individual goals. For this purpose, they proposed a modified form of Millennium Development Goal 7, target 11, namely, "By 2020, improve substantially the lives of at least 100 million slum dwellers while deterring new slum formation". In practice, a costing was sought for two "technologies of intervention":

a) for 100 million people in existing settlements, a typical slum upgrading package was to be provided in line with recent programmes;

b) for the 700 million potential new slum dwellers for whom new construction on vacant land was to be undertaken, to a modest level of quality.

First, a general methodology was developed that could be used for all kinds of interventions and to test all sorts of options. With staff from UN-HABITAT and the Millennium Project, a very specific set of interventions and standards was established in line with industry standards and best practices in constructing human settlements at modest cost. A number of recent surveys of upgrading and construction in different parts of the world were reviewed to establish the average cost of each type of upgrading activity, which ranged from streets, lighting and kerbing, through basic community physical and social infrastructure, to the purchase of land and the construction of basic dwellings. The assumptions used were in strict accordance with those used for other Millennium Development Goals, and were also very similar to the costing benchmark rules used by private sector engineering firms in tendering for projects.

The average global costs for these activities were found to be on average about $1800 per person assisted, for both new sites and upgrading, with governments paying about $1090 and the beneficiaries meeting the remaining costs. Overall, the total ground costs to governments and donors of meeting the modified target 11 and assisting 800 million people was about $830 billion over 17 years. The construction programmes for the participating developing countries would be similar in size (relatively speaking) to those in which China and South Africa have typically engaged.

** Estimate based on work undertaken by Joe Flood in collaboration with the UN Millennium Project.*

gramme was launched focused on improving living and environmental conditions, as well as promoting employment opportunities and vocational training. As a consequence, a new upgrading programme was launched and resulted in 29 new sites being upgraded, which helped to improve living conditions for some 327,000 residents.[6]

The experiences of these and other countries show that a government's positive stand on slums can set off a chain reaction of new strategies, policy reforms, laws, institutional development, and scaling up of existing programmes which, over time, improves access to services, shelter and employment among the urban poor. Of course, the impact of these interventions depends on many other factors, including the capacity of local and national governments, economic conditions, the willingness of donors, or simply the scale of the slum problem. But one thing is for sure; not much moves without a committed, energized national government. In other words, countries that have the political will also tend to elicit the capacity and resources needed for slum upgrading and prevention.

At the other end of the spectrum, a number of countries have shown weak political support for slum upgrading and urban poverty reduction. In some cases, there may be declared political intention but there is limited evidence to suggest follow-up support through clear policies, programmes and allocation of public funds; implementation is also usually weak. Some countries appear not to prioritize shelter and services for the urban poor at all. Afghanistan, Bangladesh, Ethiopia and Haiti are among the surveyed countries that fall into this overall category. In Haiti, for example, a national strategic plan on "Urban Development and Slum Upgrading" exists, and the Prime Minister has constituted a "National Commission on Social Integration" targeting slum dwellers, but not much progress appears to have been made in implementation. Similarly, Afghanistan has yet to fully develop and implement its new urban strategy. The National Urban Programme is the Third Pillar of the Afghanistan Development Forum 2005 and is intended to focus on upgrading the living conditions of households in under-serviced informal settlements. Understandably, countries such as Afghanistan that are emerging from conflict have been less able to mobilize the institutions and political support required for slum upgrading, as the focus is usually on building the capacity of institutions and governance structures that were destroyed or failed to perform during the conflict. In both Afghanistan and Somalia, for instance, one of the top priorities of the new governments is to build the capacities of institutions and to rebuild destroyed infrastructure, rather than improve the lives of slum dwellers *per se.*

A test of political leadership: Recognizing the right to the city

As the experiences of a number of countries show, government commitment that stays strong enough and long enough can help to reduce the numbers of slum dwellers. Other countries, where political leaders once looked the other way, are also getting more serious about slums. Much depends on the political juncture at which countries find themselves. Democratization and decentralization hold the promise of shifting the balance of power, giving the city's poor a chance to bargain for a better life, but this remains a far distant goal in many countries. Meanwhile, the fact that centralized, and by most measures "undemocratic", governments have also managed to make significant inroads into pro-poor shelter and services suggests that a top-down approach that is focused, efficient and is backed by resources can also bring about positive change. Another force for change has been international aid. Donors continue to exert influence on countries, bringing pressure to bear on governments through aid for slum upgrading and urban poverty reduction – the issue is whether bilateral and multilateral development partners themselves are serious or not about stepping up aid for slums and how this might further shape government attitudes.

City planning is about balancing and reconciling conflicting interests. And, in the long run, cities are planned, built and managed to reflect the values of those who hold power to make public decisions. Slum dwellers have for too long been forgotten in this equation. The present system – including the way official local plans are made, land is allocated, and administrative rules and regulations are set – is usually stacked against them and instead favours better-off communities and bigger businesses. But with a greater willingness to improve slums over the last few years, governments, it could be said, are taking a more enlightened view towards the urban poor and increasingly recognizing the "right to the city" of squatters and slum dwellers in official circles.

But despite this sunnier mood, planners and politicians are still caught between two worlds as they continually try to juggle the interests of the poor and disadvantaged with those of the urban middle class and elite. This tension is continually being played out in the struggle for land, services and jobs – and more often than not it is the poor who are left without. City politics, in many places, continues to be dominated by practices of political patronage – handing out favours and services to certain communities in return for their political loyalty – rather than on the basis of more objective criteria of need and entitlement.

Even in countries ranked highly in this Report for their political determination to provide for the urban poor, there is evidence to suggest a complete about-face by the authorities at particular points in time. More than anything this shows that commitment to slum upgrading can be a fickle thing. A well-meaning government that today shows firm support through new pro-poor legislation, a major land titling push for the urban poor, or a reinvigorated national upgrading programme, could tomorrow authorize forced evictions of the very same communities it sought to help.

Take the major cities of Rio de Janeiro, Jakarta and Mumbai. The cities have shown, to some degree, political commitment by adopting progressive, citywide policies and programmes – helped significantly by reformed policy and enabling legislation at the national level. In Rio de Janeiro, the Programa Favela-Bairro resulted in an investment of more than $600 million and has improved access to basic infrastructure, health and education for nearly 500,000 people in the city.[7] In Mumbai, initiatives such as the Slum Redevelopment Scheme and Slum Sanitation Project, built on partnerships among the government, non-governmental organizations and community groups, have also made a difference to the living conditions of slum dwellers in the city. In Jakarta, the Kampung Improvement Programme (KIP) has been hailed as Jakarta Administration's "best practice" response to slums in the city.[8]

And yet a closer look at these cities reveals that many of their slum communities have recently gone through or are facing the threat of eviction by the same authorities. In Rio, hundreds of poor families are facing eviction as 14 shantytowns, the majority in upper-class neighbourhoods, have recently been earmarked for removal by the city's public prosecutor on the grounds of preserving the environment, boosting tourism and economic development, and diffusing urban violence.[9] In Jakarta, criticisms from some human rights groups have drawn attention to the administration's track record of evicting the urban poor. According to some sources, between 2000 and 2005, the city evicted 63,000 people and a further 1.5 million people are under threat in the wake of several new development projects.[10] Mumbai, above all, witnessed one of the most spectacular shifts in policy as 90,000 shanties fell under bulldozers in late 2004 and early 2005, all in the name of becoming a "world-class" city.

Leaders in office today face a daunting task to respond to the pressures around them from different interests. In committing to change, many of these leaders are helping to kick-start a new generation of policy reforms and large-scale programmes targeting the urban poor. This has taken political courage because it means convincing those who hold power to share their wealth, land and other resources with those less powerful in society.[11] At times, the balancing act becomes a hard one to maintain. In the face of mounting pressures to make their cities more competitive or to respond to the demand for high- and middle-income housing or commercial interests, authorities can often slip back to policies and measures that once again exclude the city's poor – and, in the most extreme cases, revert to some of the worst types of evictions and demolitions experienced in their history.

Endnotes

1 This statement refers to political leadership at city level, but perhaps more importantly at the level of central government.
2 Tebbal 2005.
3 UN Millennium Project 2005b.
4 Cities Alliance 2003.
5 Government of Chile 2004.
6 Hiasat 2005.
7 Inter-American Development Bank 2004.
8 Darrundono 2005.
9 Philips 2005.
10 Urban Poor Linkage (UPLINK) in Indonesia: www.uplink.or.id.
11 Biau 2005.

4.3 Pro-poor Reforms on Slum Upgrading and Prevention

UN-HABITAT's policy analyses show that countries performing well in managing slum growth have strategically targeted investments, legislation and pro-poor policy reforms in tackling basic shelter deprivations: the absence of secure tenure; overcrowded houses; poor durability of house construction; and the lack of safe drinking water and adequate sanitation. Such reforms have gone a long way towards enabling central government bodies, local authorities and urban poor communities themselves to improve people's access to land, housing and basic services. In this Chapter, we examine some of the policy reforms that have made a positive impact on the lives of slum dwellers in various regions. It should be noted, however, that many of these policy reforms and programmes have yet to reach an appropriate scale to deal with current deficits and future needs in housing and basic service delivery.

Increasingly, a number of countries are offering pragmatically designed tenure rights based on a spectrum of formal and informal legal arrangements, ranging from formal titling to customary rights of tenure. In parallel, these countries are trying to ensure an adequate supply of well-located, affordable serviced land that can increase the housing supply in the future and prevent the growth of new slums. They are attempting to make the land market work better by removing or reducing the legal and administrative rules and regulations of both central and local governments; this includes establishing a simpler land registration process, increased flexibility in approved building materials and standards, and reduced minimum plot sizes and infrastructure standards. Through land-use planning and zoning, more progressive local authorities are taking action to increase land supply for future low-income housing and economic activities. Better performing countries are also increasing investments in water and sanitation and establishing pro-poor policies and reforms in the sectors, allowing basic services to be provided at rates affordable to the poor through appropriate design and innovative structures of tariffs and subsidies.

Latin America and the Caribbean

Latin America and the Caribbean stands out as a region that has adopted various pro-poor policies and instruments to integrate the urban poor into the legal and social fabric of cities. Several countries in Latin America and the Caribbean have moved towards progressive national housing programmes and policies and land reforms resulting in a marked improvement in the provision of adequate low-cost housing with secure tenure. Governments across the region are increasingly prioritizing land allocation to meet the needs of low-income families and people living in informal settlements with a long-term view that considers future growth and slum prevention. Some of the best examples of national policies and reforms come from Brazil, Chile and Colombia.

For example, Brazil has been a leader in establishing innovative enabling instruments that have helped to improve the land development and housing rights of the urban poor. The country's recently-adopted City Statute provides the legal and guiding framework for municipalities to implement land management and regularization processes. Several cities have already successfully used the provisions of the City Statute to regularize informal settlements and provide secure tenure to the urban poor. They have sought to address land speculation by including specific measures for the compulsory use of non-built underutilized or non-utilized land, and in case of non-compliance, they envisage progressive imposition of property taxes. In 1996, Brazil also established an innovative planning and zoning instrument called ZEIS (Zone of Special Social Interest). A ZEIS is local authority demarcated area which allows the application of flexible standards to promote upgrading.

Profavela is another piece of pro-poor legislation that is helping low-income communities in Brazil to gain access to secure tenure. The Profavela federal law has been adopted at the local level by the city of Belo Horizonte and is currently enabling slum dwellers to negotiate with public authorities and service providers to establish a plan for the improvement and regularization of their settlements. Belo Horizonte has 177 slums and 63 public housing projects, totalling 240 low-income neighbourhoods housing approximately 500,000 people. The legislation enables the local authority to suspend and relax, on a temporary basis, relevant laws, by-laws and standards to facilitate improvements and land tenure regularization. Once the plan for improvement and regularization is approved, it becomes a legal instrument that further enables public intervention, and empowers slum dwellers to invest in improvements and to comply with agreed codes and standards, and ultimately, to gain legal recognition and title deeds.[1]

However, despite a long history and commitment to improving the lives of slum dwellers, Brazil has been unsuccessful in improving the lives of the poorest of the urban poor; inequality and chronic poverty are on the rise, and perceptions about *favela* (slum) dwellers have not changed. One recent study in Rio de Janeiro found that there is still a pervasive stigma against *favela* dwellers in the job market; in fact, living in a *favela* seems to be bigger barrier to gaining employment than being dark skinned or female.[2] Clearly, there is a need to change people's perceptions about slum dwellers and institute reforms that go

Improving urban planning and monitoring in the city of Aleppo

In Syria, the planning process has traditionally been guided by central planning authorities reinforcing hierarchical procedures for reporting and approval for projects. The country's approach to urban management, which has consisted of comprehensive master plans and regulations aimed at directing growth and organizing service delivery at a macro-scale, does not allow flexibility for municipalities to decide on a wide range of functions. Most municipalities also lack the capacity to deal with the considerable overloads in their daily activities.

However, recently a number of municipalities have been able to overcome some of these challenges. The renowned old city of Aleppo, famous for its history and monuments, is a good case in point. The municipality has been able to introduce a new set of codes and regulations in order to improve its built environment and the living conditions of its residents. The municipality has also initiated a decentralized system of monitoring urban space, dividing the city into nine sectors, each with its own monitoring unit and maintenance group. This has allowed a closer relation between the municipality and residents, in addition to improving the city's urban management scheme. In this regard, the municipality has set up its own "Local Urban Observatory" as a focal point for developing an information network to support planning decisions. The observatory is helping the city to collect data in order to create a more accurate profile of the current urban situation, to identify major challenges and areas of intervention, and track progress towards sustainable urban development.

The city of Aleppo has also implemented an upgrading scheme for its old city core that has given it widespread visibility and recognition for its planning practices. The new planning scheme, which aims to upgrade the historic core of the city and preserving its architectural heritage, conceived of this task within a broader structure of urban management, including land use regulation, housing, technical infrastructure, traffic, and others. Through an action areas approach, the project engaged various local stakeholders such as non-governmental organizations, citizen groups, and other state agencies and was hence able to extend its partnership network even broader. The project led to substantial improvements in the living conditions of the old city's inhabitants and at reducing neighbourhood degradation. This plan provided an important departure from earlier centralized master planning experiences towards a more flexible development plan in priority action areas. The municipality has also been able to involve several stakeholders and generate innovative partnership with local and international actors that allowed it to compensate for lack of know-how as well as the scarcity of its resources.

Source: ESCWA 2001, UN-HABITAT 2004.
Economic and Social Commission for Western Asia, Sustainable Urban Development: A regional perspective on good urban governance, United Nations, 2001
World Urban Forum 2004, Barcelona, Presentation on Aleppo Local Urban Observatory, Mayor Mann Chibli.

beyond slum upgrading and regularization in order to safeguard poor people's livelihoods.

The Government of Brazil is also implementing progressive reforms in the provision of water and sanitation. The National Sanitation Policy emphasizes environmental sanitation as a social right and the adoption of new regulation and inspection guidelines that establishes the rights and obligations of all providers and users of services. The government has also significantly raised investments in water supply and sanitation systems.[3] It is estimated that these investments are likely to benefit 9 million families across the country. Furthermore, the government is stepping up its actions to improve the capacity of sanitation operators.[4]

In Colombia, the new constitution in 1991 transformed the way land was utilized in urban areas. The Law on Spatial Planning that followed is based on the overarching principle of protecting the public over individual interest – this has led to a more rational use of land, greater equity in the provision of basic infrastructure and services, the protection of the environment and the preservation of cultural heritage. In cities such as Bogotá and Medellin, it has enabled the adoption of innovative practices in the integration of low-income settlements, as well as a more efficient network of roads, infrastructure and transport. A key success factor has been the instrumental role of the Ministry of Economic Development in nationwide campaigns to familiarize the public with the law.

Other countries have experimented with large-scale government subsidies to potential home-buyers and developers. For example, Chile reformed its housing policy in 2001 to increase subsidies in an attempt to reach the poorest 20 per cent of the population and to meet the rising costs of social housing; this policy has been credited with reducing poverty levels in urban areas. In parallel, the country has also instituted a national land tenure and titling program called "*Póngale titulo a sus sueños*".

Asia

The performance of Asian countries in carrying forward major pro-poor reforms and programmes in land and housing provision has also been generally good. In India, national policy guidelines on housing are being finalized, drawing on the Urban Land Ceilings and Regularization Act from the 1970s, which allowed municipalities to set aside land for the shelter needs of the urban poor. Individual states and cities have undertaken innovations such as the use of transferable development rights to free up land for low-income housing. In 1996, India's Slum Rehabilitation Act allowed state authorities to offer land development rights to slum and pavement dwellers.

India is also implementing reforms that go beyond the housing sector, but which have the potential to significantly improve the lives of slum dwellers. India's five-year development plan aims to promote universal coverage of water supply and sanitation. In pursuit of this goal, several central government-sponsored schemes and programmes have been implemented. Under this programme, by the end of the first quarter of 2005, a total of 5 million sanitation units were constructed. Similarly, the National Slum Development Programme (NSDP) looks specifically into upgrading of urban slums by combining physical infrastructure with social services, including water supply, community latrines, storm water drainage, community bathrooms, sewers and other amenities. The government has also recently launched a new programme – the Jawaharlal Nehru National Urban Renewal Mission (JNNURM) – that aims to bring about mandatory reforms both at State and municipal levels to improve basic service provision and secure tenure in urban poor neighbourhoods. The programme, the single largest national government initiative in the urban sector, was launched in December 2005 and is to be implemented over a period of 7 years. The programme's special components include water supply and sanitation, sewerage and solid waste management, construction, and improvement of drains and storm water drainage. However, because of massive backlogs in housing and basic services, and because of high levels of urbanization, these important programmes may prove insufficient in the face of the huge challenges that they need to address, particularly with regard to annual slum growth rates that are estimated to be 1.72 per cent per annum.

In Sri Lanka, the provision of housing has been considered a major priority by successive governments since independence in 1948. In 1977, the government ventured into significant and ambitious attempts to increase the housing stock in the country and increase home ownership. A distinct and separate authority, the National Housing Development Authority, was established to implement and promote mass housing programmes such as the One Hundred Thousand Housing Programme and the One Million Housing Programme, and provide long-term subsidized loans for new developments and upgrading activities. In 1994, it focused its attention on high density housing in urban areas. In view of increasing land scarcity, the government decided to stop slum upgrading and initiate a programme to relocate slum and shanty dwellers in apartments built by the government. The Indian Ocean tsunami of December 2004 had a deep impact on housing in the country's coastal towns, but the government's commitment to rebuild houses and communities has remained unaltered, which bodes well for the island nation's prospects for recovery.

Cities in Thailand have adopted different kinds of innovative land-use mechanisms. Local context-specific solutions are designed with close guidance from government, community groups and NGOs. In the 1980s the ministry of finance reorganized the Government Housing Bank, which is now a leading institution in Thailand's housing finance system. It has improved housing affordability in the country and enabled large sections of the population to buy houses. In the past seven years, the economy of Thailand has grown at a rate of more than 8 per cent per annum. Since public sector housing has become very common, the private sector has had led to lower the costs of housing, which is making housing affordable to the majority. The private housing market has also developed its operations effectively and offers the lowest lending rates in the market, challenging other financial institutions to lower their interest rates in order to compete for business.

Bangkok, Thailand ©PHIL DATE. IMAGE FROM BIGSTOCKPHOTO.COM

Sub-Saharan Africa

Africa's policy experiences with improving tenure security and access to land have been mixed, and fall way short of the progress made in Latin America and Asia. Post-independence land reform was postponed in countries undergoing political upheaval, military rule and experiments with different forms of government. The emphasis was on rural development, and it is only now that governments in the region are beginning to address land and housing for the poor in urban areas. Many of the region's problems stem from the fact that many countries have inherited tenure and administration systems that are not appropriate or relevant to the needs of modern African cities. Formal systems of land registration and administration have been unable to cope with rapid urban growth, with the result that between 50 per cent and 70 per cent of all urban land in the region is delivered through informal systems.[4] Governance failures, lack of economic development and industrialization in some countries, rising urban poverty levels and high proportions of slum populations in cities have all compounded problems associated with securing tenure and gaining access to land and housing in urban areas.

The recording and registration of urban land could be a first step in the process, but there is also considerable potential for developing systems that create an interface between informal and formal systems. In some countries, this interface is occurring at an incremental level, but still not on a scale that can address the roots of the problem, which are structural, legal and economic. Both Ghana and Tanzania are embarking on the process on a pilot basis. In Ghana, following a long period of military rule and a slow shift back to civilian rule, the housing markets have become severely crippled. A substantial portion of all land and housing in the country is informal. Land titling and registration are major bottlenecks in slum upgrading and work on this is beginning in two of the bigger cities, Accra and Tema. In Tanzania, the 1995 land policy recommends registration and recording of all urban land and states that "existing squatter areas ...will be upgraded". The 1999 Land Act initiated the process of registering all properties in unplanned areas. So far, 3,000 titles have been issued to people who hold properties and land in the slum areas of Dar es Salaam. The main objective of the project is to enhance land tenure by issuing residential licences to slum dwellers. The Tanzanian government also embarked on another project in 2002 to allocate 20,000 serviced plots to residents who would have otherwise built housing in unplanned areas.[5]

Reforms in Burkina Faso and Senegal, on the other hand, appear to have a longer history and run deeper. Measures that Burkina Faso took from the early 1980s to address informal settlements were very much a result of a central government decision to undertake large-scale physical and tenure upgrading of all irregular settlements in the capital city. Although this was done through the nullification of individual titles (an exercise that was reversed later), a culture of improvement and acceptance of informal dwellings did set in. After 1991, new reviews of land legislation brought back the land titles abolished by the upgrading process of 1984 –1990. *De facto* security of tenure was enforced and further reforms have oriented land management towards a market approach. In 1997, the National Company for Urban Land was set up for the provision and sale of serviced land. Despite these reforms, both annual slum growth rates and slum prevalence are still very high (4 per cent and 76 per cent, respectively). In 1986, the Government of Senegal attempted to implement a tenure regularization programme at the national level. Physical upgrading and tenure regularization were carried out in parallel. In effect, this established a tradition of tolerance for informal settlements and led to low rates of eviction.

In 1994, after the first open democratic elections, the Government of South Africa promised to build one million houses a year. In order to reduce the housing backlog, the government established a number of social housing institutions accompanied by a people-driven housing process, but housing provision has not lived up to the promise. While the supply of housing increased between 1994 and 1998, it actually decreased between 1999 and 2004. Although considerable work is being done to improve security of tenure, scarcity of urban land has forced the government to place urban poor households on the outskirts or outside the main commercial centres, which impacts their ability to earn a living. In many areas, racially-segregated distribution of land in the apartheid era still dictates urban landholding patterns. While there is impressive expansion of housing stock, urbanization is also escalating, creating new situations of unmet demand. The need to build skills at the local and central levels to deal effectively with informal settlements upgrading and tenure provision remains urgent. South Africa has found that despite sincere intentions to fast-track social housing and upgrading, the absence of adequately qualified technical personnel creates a major constraint.

Reforms in other sectors are beginning to make an impact. South Africa stands out as a country that has made universal access to water and sanitation a high priority. In order to redress the imbalance in access to water and sanitation brought about by past apartheid laws, the government has been thoroughly reviewing its national policies and legal frameworks so as to ensure that all new legislation conforms to the principles of equity, fairness and sustainability. A new water law provides every household with 200 litres of free water per day, which has substantially increased coverage in urban areas.

In other parts of the continent, there are growing signs of more progressive water and sanitation policies by governments. For instance, the three East African countries of Uganda, Tanzania and Kenya are undergoing radical water sector reforms. The key objectives of the reforms are to promote good governance and improve the performance of the sector. The reform process has followed a similar route in the three countries: separation of water services and water resource management activities; separation of asset-holding and development, service provision and regulatory activities; decentralization and devolution of responsibilities to the lowest practical level (subsidiarity); greater transparency through increased civic engagement; reduction of political interference at all levels through the introduction of commercialized operations, and public-private partnerships, where appropriate.

Endnotes

1. UN-HABITAT Best Practices Database: www.bestpractices.org.
2. Perlman 2005.
3. Government of Brazil 2005.
4. Rakodi 2005.
5. Based on unpublished briefs prepared by the Tanzanian Ministry of Lands and Human Settlements.

4.4 Governing from the Bottom, Governing from the Top, Connecting the Two

Local governance takes centre stage

The concepts of good urban governance and the "inclusive city" have gained considerable currency in recent years. In particular, UN-HABITAT has been a major campaigner for inclusive, participatory decision-making in cities and devolution of power from central to local governments – two of the cornerstones of good urban governance.[1]

In practice, improvements in urban and local governance have taken different forms in different countries and regions of the world. The major transformation in local governance over the last decade or so has been through the process of ***decentralization***. The conceptual basis for decentralization is provided by the notion of subsidiarity, which implies devolution of responsibilities to the "lowest appropriate level". The expanded responsibilities of local governments as described above are a result of the devolution of both power and responsibilities from national to local governments. They are primarily a consequence of administrative decentralization, wherein decision-making authority and responsibilities are transferred to sub-national governments. Administrative decentralization is often preceded or accompanied by political decentralization. Financial decentralization, on the other hand, is the most complex step in the decentralization process. It is often the most contentious as well, as national and local governments struggle to retain and wrest control of local budgets. Decentralization exists in its most advanced form when elected local governments are empowered and capable of setting development priorities, making major development and expenditure decisions, and determining and collecting local revenue. The other critical trend in local governance in recent years, as a spin-off to decentralization, is the growing trend towards ***direct, broad-based participation of communities in decision-making*** as way of improving responsiveness of local policies and initiatives to citizens' priorities and needs.

While many developing countries are now preoccupied with carrying forward reforms in decentralization and trying to give communities a greater voice in local planning, it is worth considering for a moment the state of play – *who's actually doing well in making important strides towards good local governance? And is any of this making a difference in the lives of slum dwellers?*

Empowering cities: A look at country performance in building local governance

Where local governance is becoming a reality

By most criteria, several Latin American countries, including Brazil, Colombia and Mexico, take the lead in the area of improved local governance. The process of "re-democratization" in the late 1980s and early 1990s in Latin America resulted in the adoption of progressive policies aimed at reinforcing local government and promoting inclusion by allowing grassroots movements to take part in decisions at the local level. Numerous initiatives, such as participatory budgeting, participatory planning, popular movements for access to land and housing, and empowering women with a greater voice and choice in local governance have emerged from this region. Brazil was among the first countries to introduce "participatory budgeting", an innovative mechanism in representative democracy that allows community-led city councils to decide on health, education and other policies and on the allocation of municipality budgets. In Belo Horizonte, for instance, up to half the local resources for investment were allocated using this method in 1999. In other cities, participatory budgeting has resulted in better and more social services. Since the city of Porto Alegre adopted participatory budgeting in 1989, for instance, the number of public schools has risen from 29 to 84 and the proportion of the population with access to the municipal sewer network has grown from 46 per cent to 84 per cent.

Participatory budgeting has been praised, both nationally and internationally, as a shining example of good governance. By 2002, over 140 of 5,571 municipalities in the country had adopted participatory budgeting as a policy. Although each city adopts different formats to define investment criteria and to select community representatives (who are generally from low-income districts), the process has resulted in more active participation of civil society in municipal decision-making.[2] The revised 1988 Constitution also decentralized resources to the sub-national and local government levels, making it possible for local governments to institute various pro-poor policies aimed at integrating low-income communities into the fabric of urban society. Through initiatives like the City Statutes and participatory budgeting, Brazil has made major advances in developing a participatory and sustainable way of city planning and management, and has empowered the country's 5,000 or so municipalities.

Asia, as a whole, is not far behind Latin America in efforts to improve local governance. In Southern Asian countries, such as India, Nepal and Sri Lanka, municipal authorities are semi-autonomous bodies with substantial independence from central government, and have significant responsibilities for the provision of public services.[3] South-Eastern Asian countries are also starting to make progress on decentralization. After shifting from military rule to a democratic system, Indonesia began a decentralization process in 2001 in an effort to give more political and financial clout to local governments. In the Philippines, the adoption of the Local Government Code in 1991 devolved to municipal authorities the responsibility for basic services such as health, primary education, public works and housing, and helped increase the financial resources of local government.

Asia is also home to a strong civil society, focusing in particular on the rights of the poor and slum dwellers, but civil society action has had a mixed record with regard to its impact on the ground. In India, while a vibrant, organized civil society liaises with government on innovative shelter solutions and service provision to the poor, it has not been able to halt an ongoing trend in some large cities of evictions without adequate or appropriate relocation; more often than not, partisan politics or commercial interests determine whether or not an eviction is to take place. In other countries, community-level participation has made a real change in people's lives. The strength of Sri Lanka's slum upgrading approach, for instance, lies in the tradition of keen community participation. For example, programmes and initiatives that began as far back as the 1970s, such as the Urban Settlements Improvement Programme and the Urban Basic Services Programme, created local community level institutions that continue to exist today. But there are as yet inadequate channels for community-based processes to link to government decision-making.

In Eastern Asia, Thailand provides a shining example of participatory governance that has resulted in successful slum upgrading efforts. In 1992, the central government initiated the largest community-driven programme for assistance to the urban poor in any developing country through its Urban Community Development Office (UCDO). UCDO supported community organizations with loans, small grants and technical support and encouraged them to form networks to negotiate collectively with the city and provincial authorities. In 2000, UCDO merged with a Rural Development Fund to form the Community Development Institute or CODI and it continues to be a parastatal. CODI facilitates active dialogue among communities in informal settlements and municipalities, non-governmental organizations, and the private sector. By promoting such partnerships at the city level between municipalities, the agency helps to make sure that slum upgrading projects are well designed and, as far as possible, sustainable.

In sub-Saharan Africa, very few countries have attained a significant degree of devolution. Most decentralization initiatives are relatively recent, and many are poorly implemented due to resource constraints and weak institutional capacity. In Cameroon and Senegal, laws on decentralization were passed in 1996; Tanzania followed suit in 1999 and South Africa in 2000.[4] However, some of the other African countries that have attempted to devolve power away from the centre have ended up giving significant authority to districts and regions, not to local governments. In some countries, including Burkina Faso and Senegal, decentralization policies have been developed but implementation has been slow. For example, Burkina Faso adopted a new constitution in 1991 that devolved administrative powers and the authority to tax to the local level. In 1995, a new decentralization law was passed that gave municipalities a large array of responsibilities, including land management. But in fact, this promise did not materialize as the central government has shown to be reluctant to hand over land management responsibilities to municipalities. In countries where there is a degree of decentralization, such as South Africa and Tanzania, municipalities have often been stopped in their tracks by a lack of funds, inadequate technical capacity, insufficient administrative resources and ambiguous regulatory guidelines on how to implement legal frameworks at decentralized levels. In fact, wherever there has been progress on pro-poor reform, it can be attributed to clear direction and support from national government institutions.

Decentralized versus centralized governance: What works best in improving slums?

Where bottom-up local governance works

Are countries that are getting better at governing their cities from the bottom up also improving the lives of slum dwellers? In Latin America, it appears that those countries where decentralization and people's participation is strongest, such as Brazil, Colombia and Mexico, have performed well in stabilizing slum growth rates since 1990. Brazil and Mexico recorded 0.3 per cent and 0.5 per cent annual slum growth rates, respectively, while Colombia registered a slightly higher growth rate of 1.1 per cent. On the other hand, Asian countries appear to have struggled somewhat more than their Latin American counterparts in translating policies into significant improvements in the lives of slum dwellers, despite implementing wide reforms in decentralization. For instance, India, Nepal and the Philippines registered slum growth rates of 1.7 per cent, 4.8 per cent and 1.9 per cent per year, respectively, since 1990. Thailand, however, is one major exception in the region where the number of slum dwellers has fallen by a remarkable 18.8 per cent per year and where, as shown earlier, there is strong track record in community-driven upgrading with extensive government support. Sri Lanka, too, has performed very well, recording a decline of 3.7 per cent in the annual slum growth rate since the start of the 1990s.

Sub-Saharan Africa lags behind Latin America and Asia in efforts to improve local governance at the local level and it is

TONY HILL

CIVIL SOCIETY AND THE URBAN AGENDA

Since the fall of the Berlin Wall in 1989 and the subsequent expansion of more democratic forms of governance around the world, civil society, non-governmental organizations (NGOs) and citizen's groups of all kinds have emerged in great number everywhere and have shown themselves to be a vital force in tackling some of the world's most pressing problems. Whether it is a mass lobby for a better deal on aid, trade and debt for developing countries, or the provision of services and material and moral support for the poorest communities and people, or self-organized citizen groups demanding their basic human rights, civil society has emerged as a key driver of progressive social, economic and political change in all regions of the world. As UN Secretary-General, Kofi Annan, observed in 2004, *"The partnership between the UN and civil society is...not an option; it is a necessity."*

There is a growing recognition that the battle to achieve the Millennium Development Goals, to eradicate poverty, to achieve gender equity and human rights for all, and to move towards environmentally sound patterns of production and consumption, will increasingly take place in the world's cities. Cities are prolific users of natural resources and generators of waste, pollution and the greenhouse gases that cause climate change; and with one in every three urban dwellers living in a slum, cities concentrate and manifest extensive poverty and exclusion in some of its most shocking forms. With demographers projecting that 60 per cent of the world's population will live in cities by 2030, up from around 30 per cent in the 1950s and 50 per cent today, it is not hard to imagine the momentous challenges that have to be faced in securing clean water supplies, waste and pollution management, decent housing, employment, urban transport and so on, all within a framework of law and respect for citizens' human rights.

In response to these daunting challenges, new directions in urban governance, policymaking and action are beginning to emerge, based upon greater recognition of the legitimate claims of different stakeholders involved in urban issues and problems. This movement is bringing together central governments, local authorities and municipalities, and civil society organizations and groups in joint efforts to address the most pressing problems. This is manifest in the tremendous upsurge of different forms of international, regional, national and local alliances, coalitions and partnerships focused on city issues around the world over the past 15 years. At the same time, there has been growing understanding and acceptance that civil society advocacy work on urban issues is a legitimate part of good governance and democratic politics, and can lead to more just, effective and efficient outcomes. Of course, this is not the case everywhere, and even where these new forms of participatory politics are taking root, these are still early days with much more work to be done if the dynamics and destinies of the world's cities are to be truly taken in hand. Yet the momentum is growing and will surely prove unstoppable in the long run.

As was underlined by the UN Secretary-General's High Level Panel on UN-Civil Society Relations, the United Nations and its agencies, funds and programmes, such as UN-HABITAT, are at their most effective in promoting positive change around the world when they reach out to the diverse range of real actors on the ground and use their convening power to bring these actors together to negotiate and agree on the way forward. In many respects, the second United Nations Conference on Human Settlements, held in Istanbul in 1996, broke new ground in the vision and practice of partnership. In addressing the mounting challenges related to the huge population shifts in the world's cities, UN-HABITAT will need to build upon and develop this legacy and deepen its relations and cooperation with a wide array of governmental and civil society organizations everywhere that are vital to its mission.

Vancouver, 1976, UNITED NATIONS

Tony Hill is the Coordinator of the United Nations Non-Governmental Liaison Service (NGLS).

also the region where there has been the greatest upsurge in slum growth rates since the start of the 1990s. In the countries such as Burkina Faso, Ethiopia, Senegal and Tanzania, where decentralization initiatives are underway, the number of slum dwellers has risen from between 4 per cent and 6.2 per cent per year on average. However, other countries are performing better, including Ghana, with a slum growth rate of 1.8 per cent per year. South Africa, in particular, stands out in its efforts to keep slum growth rates down to only 0.2 per cent annually.

In this group of "reforming" countries, there does appear to be some association; as a general rule, the more established the local governance practices in a country, the more able a country appears to be in managing its slum growth rates. However, the relationship between good local governance and its effect on reducing slum growth is often far from clear-cut. For one thing, the move towards more decentralization and local democratization does not seem to automatically result in improvements in the lives of the urban poor, especially over the short-term. With the possible exception of South Africa, decentralization can, and often has, placed power in the hands of local elites, particularly those that played an established role under earlier, centralized systems, and has simply worsened inequalities. Secondly, in most parts of the developing world, especially sub-Saharan Africa, decentralization is a relatively recent process and if experience of other regions is anything to go by, it will take decades for decentralization to make an impact. And thirdly, governance alone cannot explain why slum growth rates have gone up in some countries and down in others; economic development, levels of urbanization, good and bad policies, all contribute to a country's overall performance.

It is nonetheless pertinent to ask: why does good local governance make a difference in slum growth rates in some places? Better local governance is starting to contribute to the success of slum upgrading operations and their scaling up to citywide and nationwide scales through various means. First, decentralization brings with it new incentives for municipal governments to participate in upgrading existing slums and related poverty-reduction schemes – including the design, implementation and financing of basic infrastructure and services in poor neighbourhoods and informal settlements. Second, municipal reforms and capacity building, in some places, have led to the improved operation of local authorities through changes in systems and administrative procedures, streamlining of functions, and reform of municipal financial systems. Enhancing the rule of law, efficiency in service delivery, and fiscal transparency and accountability, have been the major objectives of these processes. For example, a key success factor behind the Favela Bairro programmes in Rio de Janeiro is that it is financed and executed entirely by the municipality that has built up its own capacity and institutional structures for efficient service delivery. A technical committee drives investment decisions while a coordination committee makes sure that all municipal departments are on board for implementation. Furthermore, the programme uses modern management techniques that ensures smooth running of between 40 to 50 neighbourhood projects at the same time all over the city. The programme is also strong on ensuring community involvement in decision-making and in the operation of some services such as garbage collection and reforestation[5]. Third, new waves of democratization and decentralization have also led to a marked improvement in the organizational capacities of civil society groups and opened up opportunities for slum dwellers and other marginalized group, to get involved in planning and project design and implementation. With growing political maturity of grassroots organizations, the urban poor are in a better position to negotiate with local authorities for services and land rights.

Local governance works, but in many countries it works best with strong support from the centre. Countries that have performed well in decentralizing and strengthening local governance have done so with strong commitment and support from central government. National reforms and legislation – for example, covering decentralization, fiscal transfers, municipal elections, community participation, and spatial planning – create the enabling environment for city administrations to carry out their functions, including slum upgrading and prevention, more effectively. Despite handing over power, governments continue to play a significant role in taking decisions from the top that invariably affect the provision of shelter and services for the poor at the local level. For example, while many Latin American countries have demonstrated a good track record in decentralization, reforms still often tend to emphasize de-concentration rather than real redistribution of power. Decision-making power, in many ways, remains centralised. In addition, despite new systems of political representation and participation of civil society, countries in the region often display centralized structures that serve to strengthen the position of the ruling political party, reflecting the continuing centralist character of Latin American political culture.[6] For instance, in Mexico, the central government, through the Ministry of Social Development (SEDESOL), is implementing a large-scale national urban poverty reduction programme – "Habitat Programme" – involving significant transfers of resources to the local level. However, the resources and activities are, in effect, managed by local branches of the central administration or by local structures reinforced by the central government through so-called "local development agencies". Despite working in close collaboration with the municipality and other local stakeholders, control lies principally in the hands of the central government. The ministry can indeed play a very positive role in coordinating resources and delivering services for the poor, implementing redistributive policies that aim at bridging the gap between regions and cities. What is important is to ensure that bottom-up approaches to governance connect with top-down systems of decision-making.

Highly centralized systems of governance cannot benefit the urban poor if they are being run by regressive, anti-poor political leaders and inequitable policies.

Where top-down centralized governance also works

Moving to the other end of the spectrum of countries, an interesting finding has emerged: among those countries that have highly centralized structures of governance some also appear to be performing well on the slum target. By taking a tough policy stand on improving slum and housing conditions, governments in these countries have been able to set up the institutional arrangements, allocate important budgets, and execute projects to effectively meet their targets and commitments. And they have done so with limited involvement of local authorities or communities in decision-making processes. This is particularly the case in the countries of Northern Africa, such as Egypt, Morocco and Tunisia, where local governments have no real political or administrative power, and where most decision-making is centralized. These are also the countries that are experiencing low or negative slum growth rates and where central governments continue to exert enormous control over allocation of resources and decision-making. Cuba also performed well, recording a slum growth rate of 0.7 per cent per annum since 1990, making it one of the best performing countries in the Latin American and the Caribbean region.

What is clear, however, is that highly centralized systems of governance cannot benefit the urban poor if they are being run by regressive, anti-poor political leaders and inequitable policies. In all centralized countries that have performed well on reducing slum growth rates, benevolent, progressive leadership appears to be key to the success of slum upgrading programmes. For instance, Morocco's centralist tradition has benefited from the new monarch's pro-poor stance towards upgrading. In Cuba, the island's greatest achievements have been based on the government's consistent commitment to the principles of universality and equitable access that have defined social and economic policy since the 1959 revolution. On the other hand, the Government of Zimbabwe, through its nationwide eviction programme in 2005, demonstrates how top-down rule can adversely affect the urban poor.

Why has centralized governance, contrary to current thinking, also produced meaningful results in these places? It appears that command and control from the centre has often given cohesiveness to the design and implementation of slum upgrading projects. For instance, in Tunisia, a large number of institutions are involved in slum upgrading, including the Urban Upgrading and Renovation Agency (ARRU), the Housing Bank, the Solidarity Bank, the municipalities and some non-governmental organizations and community groups. The entire operation is managed by ARRU, which acts on behalf of the municipality. Through this approach, the programme has succeeded in upgrading more than 250,000 housing units (18 per cent of the urban housing stock) benefiting about 1.5 million people. In Morocco, slum upgrading is also driven from the centre by the Ministry of Housing among others, and has produced good results – between 1993 and 1999, 82 slum upgrading projects were implemented reaching nearly 99,000 households[7]. *As these examples show, top-down planning and implementation by strong central institutions can play a key role in the success of slum upgrading projects by providing clear purpose and direction, effective coordination, and institutional capacity to achieve results.*

While a direct, top-down approach to slum upgrading and shelter provision has brought benefits in these countries, there is a significant downside. Without meaningful decentralization and participation, it becomes harder to motivate municipal governments, civil society and citizens to take more control over the processes that affect their material well-being and contribute to development. In addition, participatory processes make possible collective learning, promote social and institutional innovation and facilitate different forms of inclusion. Centralized interventions can be extremely effective in redressing inequalities, but they come with a social cost. In Cuba, for example, the idea that the government is in a better position to determine how housing construction resources are to be used has stifled community and individual initiatives to improve housing, particularly when these initiatives fall outside national priorities.

However, in some cases, governments are starting to realize that top-down decision-making can only go so far. For instance, the Government of Morocco is now trying to introduce greater participation in the planning and implementation of slum upgrading projects through a new concept called "Social Project Control" and laws on promoting participation. In Cuba, self-help community-driven construction – through government-sponsored "microbrigades" – has been broadly implemented in the last two decades, although the impact of these initiatives is not yet fully felt in larger cities. In Egypt, too, there are signs of change. The entire regulatory and policy framework, including the constitution, is being examined and revised in a massive effort to broaden democratic processes within the country.

Endnotes

1. Taylor 2000.
2. Souza 2002.
3. Pieterse 2000.
4. Halfani 2004.
5. Inter-American Development Bank 2004.
6. Pieterse 2000.
7. This information was gathered during UN-HABITAT's research on slum policies in 23 countries. Sources included telephone interviews and various country reports.

ANTOINE HEUTY AND SANJAY G. REDDY

SLUM IMPROVEMENT: TECHNICAL FIXES ARE NOT ALWAYS THE SOLUTION

The ultimate costs of achieving the Millennium Development Goals in cities may well be either significantly underestimated or significantly overestimated because most cost estimates are based on unjustified assumptions and weak data. For example, existing models assume that the unit costs of required interventions to achieve a Goal are fixed, even as the Goal is progressively attained. However, there are strong reasons to believe that decreasing or increasing marginal costs (economies and diseconomies of scale) may play an important role in determining costs. For instance, in poor urban areas, those to whom coverage of relevant services (such as piped water, electricity or telephones) must be extended may be those who are most difficult to reach, for geographical or social reasons. In this case, the initial cost of delivering services may be high, as it may depend on the installation of infrastructure needed to make service delivery possible. On the other hand, it can become progressively easier and cheaper to provide more of the same services when the delivery network and infrastructure already exists or has previously been installed. Which of these is the case may depend on local circumstances.

The UN Millennium Project's estimates of the cost of achieving Millennium Development Goal 7, target 11 (developed in a background paper prepared for UN-HABITAT and the Millennium Project's "Task Force on Improving the Lives of Slum Dwellers") are based on an approach that assumes away such increases or decreases in marginal costs and also fails to address differences in unit costs between and within countries (although it does distinguish between investments in existing and in new settlements). It also fails to take account of interdependencies in the attainment of different Goals: there may be economies or diseconomies of scope that operate between distinct kinds of services; the cost of expanding services of one kind may depend on the extent to which services of another kind have already been expanded. For example, it may be less costly to bring about improvements in child health if children attend school. On the other hand, increases in child survival will increase the number of children for whom schools must be provided, and thereby increase the cost of achieving school enrolment objectives. Interactions of this kind among distinct social development objectives are numerous, but are largely neglected in current work on the costs of achieving the Millennium Development Goals. Finally, the data required to assess the baseline scenario of the Goals and to monitor their progress over time are at present severely deficient. As a result, it is often not possible meaningfully to judge either the extent of progress required or the costs of achieving the required level of progress. The baseline data may also understate the scale and depth of poverty, especially in urban areas. The Millennium Project's estimate of the total cost of meeting target 11 for developing countries ($113 billion for slum upgrading for 100 million existing slum dwellers, and $1176 billion for new sites construction for 700 million potential future slum dwellers) may end up being quite inaccurate because it is premised on a model in which these considerations hardly figure.

Our work (see "The Cost of Achieving the Millennium Development Goals: What's Wrong with Existing Analytical Models" on www.millenniumdevelopmentgoals.org) has demonstrated that the costs of ultimately achieving the Goals are potentially significantly affected by the assumptions that are made regarding unit costs and the nature and extent of economies of scope and scale, within plausible ranges of variation.

Although we already know about *some* interventions that are likely to be effective in enhancing human well-being, the solutions to a great many other problems in urban areas are unknown. Technical fixes may not exist for the most important problems in urban areas, which may require institutional innovations, political will and more innovative governance structures (such as participatory budgeting). The use of technocratic models is ultimately unlikely to provide an adequate basis for improving the living conditions of the poor in cities.

Even the most carefully constructed future scenarios are ultimately unlikely to prove accurate, especially when projected far into the future. Unpredicted shocks, whether at national, regional or global levels, are sure to eventually undermine the accuracy of such forecasts. Examples of significant shocks of this nature that have arisen in the past or may occur in the future include new diseases (such as HIV/AIDS), disruptive large-scale climatic events (such as the December 2004 Indian Ocean tsunami or the recent hurricanes in Louisiana in the United States), and civil and regional conflicts. Of course, those who frame analytical models do not claim that the strategies that they recommend can be applied without regard to changes in the circumstances of their application. However, these models do not take note of the likelihood that such changes in circumstance will arise.

The unreliability of the informational base and the undue restrictiveness of current approaches for planning to achieve the Millennium Development Goals means that these approaches cannot serve adequately to guide policymakers in the poorly charted, uncertain and changing environment that they actually face. The weaknesses of technocratic predictive models can be mitigated but not overcome. The potential damage from the use of incorrect predictive models in decision-making is likely to be greater when such models are applied to guide long-term decision-making – whether over the level of resources to be raised or over how those resources will be deployed. Inaccurate predictive models can eventually cause significant misallocation of resources and errors in policy choice, which may make it difficult or impossible to achieve the Goals.

An alternative approach to strategic planning should establish an institutional framework for continuous informed policy choice by representative decision-makers. The alternative approach to achieving the Millennium Development Goals can be implemented through a process of periodic peer and partner review, through which countries, regions and cities periodically formulate and review one another's plans, with the broad involvement of citizens and the support of experts. Participatory planning and budgeting that promote context-sensitive judgment and learning from experience can be important elements of such a strategy.

Antoine Heuty is Public Finance Economist (Poverty Group) at the United Nations Development Programme (UNDP).
Sanjay G. Reddy is Professor of Economics at Barnard College and in the School of International and Public Affairs, Columbia University.

4.5 Time for Bold Action: Scaling Up Improvements Today, Preventing Slums Tomorrow

Countries changing course in the new millennium

With some notable exceptions, the 1990s presented a bleak picture of policy outcomes in reducing shelter deprivations and improving the lives of the urban poor. In the absence of more effective, ambitious policies and programmes, city conditions will deteriorate rapidly in most of the "off track" and "at risk" countries described in Chapter 1.3. The challenge is daunting, yet a few developing countries have made remarkable leaps in improving the lives of slum dwellers over the last decade and a half. These are not seismic shifts by any stretch of the imagination; rather change has been driven by patient, consistent policies and leadership over time. Poor-performing countries, if they choose to, can take inspiration and hope from fast-track countries. Their governments can make a decision to change course today by making a serious commitment to slum improvement and implementing a bold action plan to meet target 11. Alternatively, they can decide to stay on the same path towards 2020 and watch the numbers of people living in slums grow.

It is not too late – countries off track can get back on track with the determination and foresight to introduce long-term planning and reforms for achieving the 'cities without slums' target. The first step for these countries is to take target 11 and the other Millennium Development Goals and targets seriously and to mobilize political will behind the Goals. As the report of the Millennium Project to the UN Secretary-General, *Investing in Development: A Practical Plan to Achieve the Millennium Development Goals* emphasizes:

"To enable all countries to achieve the Millennium Development Goals, the world must treat them not as abstract ambitions but as practical policy objectives. The Goals are essential for transparency and accountability, so it is important that they be taken literally since the pressures in development policy push overwhelmingly for lower rather than higher expectation. National governments and international donors, not wanting to be held accountable for their role in poverty reduction, will always want to water down the Goals – particularly if achieving them requires increased budgetary commitments or major policy changes. In many countries the Goals are deemed 'unrealistic' because they would require dramatic progress."

There are signs to indicate that, at least in some countries, governments are starting to take target 11 more seriously. Countries that struggled during the last decade, such as Mauritania and Senegal, for instance, are now showing political determination to make slum upgrading a core business and are moving towards longer-term, scaled-up slum policies. Other well performing countries from the nineties are stepping up their actions. For instance, Brazil, Mexico, South Africa and Thailand are carrying forward comprehensive, national upgrading policies to deal with regularizing and improving conditions in existing slums, as well as trying to plan ahead to avoid future slum growth. Yet, most countries that performed poorly during the 1990s continue to lag behind in making the political commitment and reforms needed – our projections indicate a worsening situation in all of these places. Can the governments of poor-performing countries find a compelling enough reason to act now and with the ambition needed to achieve target 11?

If the answer is yes, they would be well served taking on board the major lesson learned by successful countries: *success is driven by following a two-pronged strategy for, one, scaling up improvements in existing slums and, two, planning well ahead to provide better, alternative solutions to avoid the spread of future slums.* Countries such as Brazil, Mexico, Colombia, South Africa, Thailand and Tunisia, have all managed to successfully scale-up programmes for slum and informal settlements upgrading and urban poverty reduction to countrywide levels. In doing so, these programmes have resulted in a measurable impact on national indicators of slum growth. One-off, local projects, however successful, are usually incapable of making such a mark without widespread replication and scaling up. Pilot projects provide valuable test cases and, when they work, demonstrate the technical and financial feasibility of providing better housing and services to the urban poor. Many of today's successful national slum upgrading projects, such as Indonesia's Kampong Improvement Programme, began life on a modest scale, covering a few neighbourhoods or a single city and, with a proven track record, were expanded to national level. Equally importantly, these countries have realized that the magnitude of deficiencies in basic infrastructure, service and shelter provision for slum communities today will fade into insignificance compared to conditions in the next five, ten or fifteen years. Governments have, therefore, taken a much longer view on the expansion of slums and have begun to focus their energies on measures that effectively meet future need by developing plans that can effectively halt the growth of new slums and promote more sustainable cities. This has meant creating a planning system that makes land and infrastructure available and affordable to low-income housing.

Towards national and local strategies to achieve Millennium Development Goal 7, target 11

Those countries that are serious about achieving target 11 should make sure they have a long-term, national strategic plan for scaling up remedial and preventive measures to meet the

China UNEP/STILL PICTURES

basic needs of slum dwellers and future low-income populations. The national plan would provide a blueprint for action that signals the government's commitment to improving the lives of slum dwellers and sets out clear, time-bound targets and new policy vehicles for achieving widespread governance and sectoral reforms. The plan should include appropriate budgets and expenditure frameworks for achieving its stated objectives and targets. Such a planning process would offer countries an opportunity to formulate a countrywide policy for urban poverty reduction that is aligned systematically with all of the Millennium Development Goals and targets. Brazil, for example, has taken steps to shape new national policies responding to target 11 and the achievement of other Goals and targets in its cities. The Ministry of Cities, established in 2003, is currently responsible for formulating and implementing the new National Urban Development Policy. The overarching policy framework is based on the principle of universal access to adequate housing, urban land, safe drinking water, sanitation and mobility with safety.[1]

Likewise, municipal authorities should also be in the forefront of translating the Millennium Development Goals, particularly Goal 7, target 11, into their own city-level goals and targets, and subsequently adopt citywide strategies for achieving them. The goals may be global in character but they must be implemented locally, at city and community levels, where people live and shelter and services are required. However, many local governments are barely aware of target 11 and the other Millennium Development Goal targets and even if they are, they often have no incentive or commitment to meeting these targets. Thus the first task before local governments is to build awareness about the Goals, why they are important, and what they mean to a slum dweller in the city.

Local governments are no strangers to setting targets – for housing, infrastructure, services, health, education and other sectors. The key to achieving the Millennium Development Goals is: (a) to benchmark the targets against the Goals; and (b) to ensure that the targets are bold enough to deal with current shortfalls and are established in consultation with national governments and local stakeholders. This would require many rounds of consultation, discussion and explanation, sharing of experiences and best practices. Such participatory processes can build awareness among local authorities and stakeholders, and demonstrate how the Goals are linked to their own objectives and priorities. Local authorities should also try to produce better information to give as accurate a picture as possible of the situation – be it the number of people living in slums, the numbers without clean water and sanitation or the numbers of children dropping out of school. This kind of detailed information helps planners and policy-makers to make more informed decisions and keep track of change.

Strategies and action plans to achieve the Goals must cut across sectors and institutions. A common folly is to establish a stand-alone local authority project or department for poverty reduction that is separate and distinct from the project or department that monitors the achievement of the Goals at the national level. This separation defeats the very purpose of the Goals, particularly Goal 7, target 11. The Millennium Development Goals and targets must be built into all development activities and projects, and resources must clearly be allocated for these. Often, intervention in just one sector, such as improving sanitation or regularizing tenure, can have a huge impact on the quality of living conditions in slums. Targeted interventions, aimed at the most vulnerable urban populations, can sometimes be more effective than physically upgrading slums, which may not be feasible or viable in the short-term. In the regions that suffer from one major shelter deprivation, for instance, intervention in just that sector could drastically reduce the number of slum dwellers. For instance, in both Tanzania and Uganda, where over 80 per cent of the urban population suffers from lack of proper sanitation, investment in sanitation in slum areas could reduce the proportion of slum dwellers from more than 90 per cent to 40 per cent of the urban population in both countries, assuming that the other shelter needs, such as water or sufficient living area remain the same. Egypt is one country that managed to dramatically decrease slum incidence by investing heavily in water and sanitation.

Mobilizing financing for pro-poor urban strategies and programmes

Most of the investments needed to achieve target 11 and the other Millennium Development Goals and targets in cities will have to come from domestic sources. With limited budget support from the centre, cities need to turn to public sector borrowing in the domestic financial markets to fund major investments in infrastructure and services. However, most developing cities have sourced debt financing mainly from government financial institutions or on the basis of government guarantees. In this regard, national policies and regulatory reforms can play a key part in removing distortions in the market and attracting private capital to finance public infrastructure[2]. Another crucial measure to narrow the gap between municipal financial resources and expenditure is to enhance the revenue base of the local authority, for example, through increasing the efficiency of property tax collection or rationalizing water rates and ensuring that revenue collected is devoted to slum upgrading and prevention.

In addition, slum communities can make a major contribution to upgrading through their own savings and by leveraging various sources of local funding. Innovative mechanisms should be looked at to consider how best to improve access to credit among the urban poor. For instance, housing microfinance has been quite successful in reaching low-income groups. Municipal *subsidies* are also being explored in different countries. However, what is perhaps most needed is long-term finance for low-income shelter. This is a gap that the private sector, non-governmental organizations and donors are currently trying to fill by supporting demonstration pilot projects in which local authorities and civil society recipients can borrow funds for shelter development and slum upgrading.

Leveraging Local Resources for Slum Upgrading

International assistance towards housing and basic services in developing countries has not been sufficient to address the significant shortfall in these areas. It is estimated that combined public and private investment and official development assistance meets only 5 per cent to 10 per cent of the financing required for slum upgrading in sub-Saharan Africa, Southern Asia and South-Eastern Asia. In order to deal directly with this "finance gap" in slum upgrading, investments must be predominantly domestic, community-driven and market-based.

Any international donor interventions must therefore be catalytic and should "leverage in" local and other resources. Many developing countries have a large amount of resources that could potentially be directed towards slum improvement – the challenge is that such resources need to be harnessed, prioritized and restructured. The real barrier is the lack of political will, accountability and institutional capacity, rather than financial affordability.

The international donor community has an important role in helping to develop sustainable financing mechanisms for slum upgrading and prevention that should be built on the following three pillars:

(a) Harnessing and enhancing individual and community resources;
(b) Strengthening and reallocating finances of city and national governments to meet the needs of the urban poor, and the introduction of appropriate financial and non-financial public policy instruments; and,
(c) Promotion of access to domestic capital markets.

There are increasing levels of community mobilization and savings in slums. Simply put, slum dwellers are taking matters into their own hands in the absence of affordable housing and related urban infrastructure – and the absence of public and private resources to finance such improvements. The result is a proliferation of daily savings associations, work-based savings and credit schemes, revolving loan funds, and microfinance lending. While social lending arrangements of this kind vary from slum to slum, city to city and country to country, they share in common powerful mechanisms for both mobilizing savings and undertaking community-based initiatives to improve housing and infrastructure. The experiences of community-led and government-enabled programmes in countries such as India, Morocco, Tunisia and Uganda demonstrate the principles of a workable community-driven approach.

There are also cases of countries, such as India, Indonesia, South Africa and Sri Lanka that have restructured public finances and market-based financing for urban upgrading. However, the experiences so far have largely remained as "isolated islands of innovation" in many countries and there is an urgent need to replicate and scale up actions. The process of scaling up will require massive policy and regulatory reforms that promote community driven approaches, healthy local government financial capacity and domestic capital markets.

The liberalization of the domestic financial service industry is a key trend that the international community should seek to capture and harness for the purposes of financing slum upgrading. While Brazil, India, Mexico and South Africa have gone through significant liberalization of the banking sector and opening of domestic capital markets, a similar, largely unnoticed trend is also unfolding in lower-income countries in sub-Saharan Africa, Southern Asia and South-Eastern Asia. Pension funds, insurance companies and private investors in these countries maintain enormous stocks of domestic capital (estimated annual value of domestic capital in Nairobi, Kenya, for instance, is $1 billion) and are increasingly trading on local stock exchanges.

Taking upgrading projects to scale also requires access to multiple forms of investment and the use of several kinds of corresponding financial instruments and products. In some cases, "credit enhancement" may be needed to attract domestic capital. There are a variety of institutions currently providing different forms of credit enhancements for projects that seek to access capital markets, including, for example, the International Finance Corporation (IFC), GuarantCo, United States Agency for International Development (USAID) Development Credit Authority (DCA) Facility, and the Emerging Africa Infrastructure Fund. Domestic guarantee facilities are also gradually emerging in a few countries in order to attract private capital, including Colombia, India and South Africa.

One of the constraints in attracting private capital is lack of adequate "bankable" projects that addresses the risks and concerns of communities, governments and the private sector. This is the reason why UN-HABITAT, in association with several donor agencies and development partners, set up the Slum Upgrading Facility (SUF). The central objective of SUF is to assist developing countries to mobilize domestic capital for slum and urban upgrading activities. A major focus of SUF will be to package the different forms of investment and to structure the projects so that these can attract not one but multiple forms of financing. This process of rendering projects "bankable" will involve facilitating partnerships and strengthening capacity at country level among development partners and the domestic financial service industry. It will also include linking these local actors with key international financial institutions, donor facilities, and regional development banks and funds that will be in a position to "credit enhance" domestic financial instruments through risk reduction and risk sharing and, by doing so, enhance the mobilization of domestic capital into slum upgrading projects.

Endnotes

1 Government of Brazil 2005.
2 Cities Alliance 2004.

4.6 Is the International Community Ready to Keep the Promise?

In order to improve the living conditions of people in slums, responsibility for change ultimately lies with governments. Some countries are starting to show real determination in taking on the target of improving the lives of slum dwellers and are making it a reality within their cities and towns. Yet too many governments remain in a state of inaction. Time has come for governments to place the urban agenda much higher on the list of national developments priorities – by identifying local resources, mobilizing domestic capital and developing mechanisms to attract external funds for innovative solutions that would maximize slum upgrading and prevention programmes.

Lack of investment in slums bears enormous social and economic costs, which add to the burden of cities and governments. Development assistance in improving the capacity of governments – institutional reforms, better local governance, improved urban planning and management and providing affordable land and housing solutions to the urban poor – can go a long way towards creating a pro-urban environment in countries of the developing world that currently do not address slum upgrading or prevention as part of their overall poverty reduction strategies.

In addition, carrying forward commitments made by rich countries in recent years, particularly during international conferences, it is possible to arrive at a "new deal" as part of the Millennium Development Goals. The international system – the United Nations, bilateral donors, the World Bank, and the regional development banks – has proven to be an important source of financing for poverty reduction in several developing countries around the world. For many countries in Africa and in least developed countries, aid is still the largest source of external financing and, it is argued, is critical to the achievement of the Millennium Development Goals and targets.[1] With this in mind, making the international system work better for poor countries, especially by raising the amount and quality of aid, is another major target that the world has set for itself and expressed in Goal 8: to develop a global partnership for development.

A question of money, a question of donor attitude

Donor financing has played an important role in supporting slum upgrading over the last few decades. International financial lending institutions, regional development banks and bilateral donors have provided consistent support to slum projects in this period. In many countries, especially in Africa, international agencies have provided the bulk of financing for slum upgrading primarily through investments and loans. The volume of aid for the urban sector has consistently risen among some international actors. For example, the Inter-American Development Bank, operating in Latin America – the most urbanized of developing regions – has seen its portfolio of urban loans grow in volume and complexity; more than $25 billion of loans has gone towards urban projects in the last 40 years, representing nearly 15 per cent of the total lending by the Bank.[2]

The "Cities without Slums" target[3] has helped to generate a renewed interest in slum improvement among many of the donor agencies. The target has sent out a clear signal to donors and governments alike to re-orient their policies towards urban poverty reduction. Multilateral and bilateral development agencies are taking up this challenge and starting to streamline their assistance to respond directly to the slum target. To some extent, this explicit support for the slum target has the potential of translating into higher levels of development assistance targeted specifically at slum upgrading and slum prevention.

Within this framework, UN-HABITAT has transformed its work programme in line with target 11 as well as other Millennium Development Goals and targets, including those on water and sanitation. Slum upgrading is now an important area of focus for the organization, with increasing emphasis being placed on policy and operational support to the following areas: scaling up of slum upgrading projects and programmes; campaigns on secure tenure and urban governance; urban water supply and sanitation; and pro-poor planning and management. Monitoring progress towards achievement of the slum target at the national level is also an important part of UN-HABITAT's work. The agency is also leading a major new initiative – the Slum Upgrading Facility (SUF) – designed to assist local partners to mobilize local domestic capital for slum upgrading, low-income housing and related infrastructure. The Cities Alliance, a joint initiative of UN-HABITAT and the World Bank that brings together a global coalition of cities and their development partners, has also played a catalytic role in coordinating and mobilizing broad-based international support for scaling up slum upgrading activities.

Despite this overall improvement, development assistance to alleviate urban poverty and improve slums remains woefully inadequate. Although investments have been made in related sectors, such as health, water and sanitation, these are not targeted specifically at slums, particularly in developing countries. The cost of meeting the slum target alone – that is improving the lives of at least 100 million slum dwellers – has been estimat-

Kibera, Nairobi SEAN SPRAGUE/STILL PICTURES

Rio de Janeiro from Corcovado Peak SEAN SPRAGUE/STILL PICTURES

ed at $67 billion[4]. Furthermore, providing decent housing and basic services for the additional 400 million people who are expected to join the ranks of slum dwellers during by 2020 is estimated to cost a total $300 billion or over $20 billion per year. Successful models have demonstrated that, when appropriately supported by local and central governments, local residents can provide about 80 per cent of the required resources. This would leave 20 per cent to be provided by international aid, that is roughly US$5 billion a year. Yet, according to one estimate, total urban sector assistance to developing countries is just $2 billion a year, a fraction of what is needed to meet the slum target and cope with future growth[5], that is, if all funds are used for this purpose.

Although urban poverty is beginning to be recognized as an issue that deserves attention, it is apparent that international assistance to the urban sector has not been able to match the scale of the problem. Associated with low levels of resource mobilization, the real question is that urbanization has not been fully understood by developing agencies both in terms of positive outcomes and negative externalities. The international policy environment needs to rethink the urban agenda and place it among national priorities. This is the only way resources will flow to address slum upgrading and prevent slum formation. There are, however, some positive signs, particularly among regional development banks; some governments are also revisiting, or at least debating about[6], their aid programmes with the aim of integrating urban challenges in their development assistance priority list.[7]

Sharpening the focus of aid

Another trend worth noting is that donors are becoming more strategic in the way they support urban poverty reduction. As this chapter has highlighted, some countries have achieved dramatic improvements in the living conditions of the urban poor by hitting the most critical policy levers for scaling up and intensifying slum upgrading and prevention measures, including building much-needed political leadership, enhancing local and central governance, and making bold reforms in land, housing, basic infrastructure, financing and planning. International aid, in recent years, is more targeted towards trying to meet these larger policy objectives. In this way, multilateral and bilateral development partners have gone for a more "programmatic" approach to urban development projects, focusing, for example, on institutional development, municipal management, provision of security of tenure, and reform of central-local fiscal relations. This marks a broad swing away from targeted area investments that proved inefficient because of counterproductive policies, especially concerning land regulation, and weak local institutions that lacked the mandate and resources to deliver services on the ground[8].

Yet, there is plenty of room for improving the strategic focus of donor interventions in slum upgrading and slum prevention. Firstly, despite increasing efforts to target policy failures at the national level and reinforce country leadership behind slum improvement, more could be done in this area. There is sometimes a tendency among governments and donors to focus more on the local scene rather than dealing head on with some of the basic weaknesses in the broader, national policy environment. Tackling the root causes – weak institutions, stifling legal and regulatory systems – by putting in place key reforms is often a much more difficult task. In doing so, the international community is no longer providing simply technical solutions but challenging governments to make political choices. It takes courage for governments to institute national reforms and programmes to benefit slum dwellers because it means sharing power, wealth and land in a more equitable way. The evidence suggests that countries, such as Brazil, South Africa and Sri Lanka that have taken this path have made major breakthroughs in reducing slum growth rates.

Furthermore, while there is a deliberate attempt by donors to "move purposefully to promote initiatives owned, generated and designed by cities"[9] in order to counter the dominant power of central government that is often perceived to block city development, the evidence presented here suggests that the State, when it wants to, can play an extremely positive role in coordinating and delivering slum upgrading programmes. This perhaps calls for a further consideration of balancing support for both top-down and bottom-up governance approaches. Poverty Reduction Strategy Papers (PRSPs) are a good channel to do so, but the urban chapter needs to be prioritized, confronting, somehow, the anti-urban attitude that is often found on both sides of the table (governments and donors). Evidence shows that countries such as Jordan and Tunisia that have successfully linked slum improvement initiatives to wider poverty reduction strategies have managed to secure a bigger share of the national budget for slums and found it easier to push through reforms in key sectors. In most of the other countries, governments and donors have not performed well in linking slum upgrading and urban poverty reduction to the broader, national development agenda. High slum growth rates in these countries prove this point.

It is time for governments and donors to take a more proactive stand in promoting Millennium Development Goal 7, target 11 in their PRSPs – this may be one of the most critical entry points for rapid scaling up of slum upgrading and slum prevention measures. Yes, governments and donors, given their limited resources, will need to balance competing priorities. But the Goals and targets point them to a very clear set of priorities and outcomes, improving health, education, the environment, gender disparities and urban poverty issues. Hence, there is a clear opportunity to generate momentum behind target 11 and provide a commitment to long-term planning for slum growth and prevention and connecting this with reforms in land, water and decentralization policies.

But this takes the will of governments and donors to act – and this is, for the most part, sorely lacking. The prioritization of sectors and budget allocations by donors and recipient government are often based on political decisions and are usually the result of competing interests among different agencies and sectors. This is unfortunate, since such a process can ignore real need and fail to consider past performance. When the issue of urban poverty and slums are already well below the radar screen of donors and governments, it makes it all the harder to ensure a place at the negotiating table. To gain more visibility, donors may consider new tactics, for example, when it comes to the development and execution of urban projects, donors traditionally deal with the line ministries such as planning, housing, environment and local government. Perhaps, they could also try to influence other parts of the central government, including the ministries of finance, health and education that have more clout in the PRSP process. This Report has shown the very significant links between shelter deprivations in slums and health, education and employment – it makes little sense to put "slums" in a box without taking into account the critical linkages with these other key sectors.

Endnotes

1 United Nations 2002.
2 Inter-American Development Bank 2004.
3 Millennium Declaration, paragraph 19.
4 UN Millennium Project 2005a.
5 Cohen 2004.
6 For instance, the February 2006 discussion held at the United Kingdom House of Commons about urban poverty issues.
7 Mauritania is one example.
8 Kessides 1997.
9 Cities Alliance 2004.

Bibliography

A

Addis Ababa City Council (2004). Household survey results shared with UN-HABITAT as part of the Monitoring Urban Inequities Programme.

African Population and Health Research Centre (2002, April). *Population and Health Dynamics in Nairobi's Informal Settlements: Report of the Nairobi Cross-Sectional Slums Survey 2000.* Nairobi.

Améndola, G. & Garay, A. (2000). "La visión urbanística". In *La Fragmentación Física de Nuestras Ciudades.* Municipalidad de Malvinas Argentinas.

Appadurai, A. (1996). *Modernity at Large: Cultural Dimensions of Globalization.* Minneapolis: University of Minnesota Press.

Applebome, P., Drew, C., Longman, J. & Revkin, A. (2005, 4 September). "A delicate balance is undone in a flash, and a battered city waits". *The New York Times.*

Arputham, J. (2001). "Whose city is it anyway?" *Our Planet.* UNEP.

Asian Coalition for Housing Rights (2001, October). "Building an urban poor people's movement in Phnom Penh, Cambodia". *Environment and Urbanization,* 13(2).

Atkinson, A. (1996). "Sustainable cities: dilemmas and options". *City,* 3/4, 5-11.

Augustinas, C. (2003). "Land in an urbanizing world". *Habitat Debate,* 9(4).

Australian Bureau of Statistics (2004, September). "The health and welfare of Australia's aboriginal peoples". *Shelter WA Newsletter.*

Ayieko, F. (2006, 30 January-5 February). "Form a Building Authority, say experts". *The East African.*

B

BBC News (2005a, 15 July). "Regions and territories: Somaliland". Retrieved from http://news.bbc.co.uk.

BBC News (2005b, 30 August). "Immigrants die in new Paris fire". Retrieved 30 October 2005 from http://news.bbc.co.uk.

BBC News (2005c, 15 November). "Chirac in new pledge to end riots". Retrieved 16 November 2005 from http://news.bbc.co.uk.

BBC News, World Edition (2005a, 14 January). "Green hydrogen buses are a 'hit'". Retrieved 26 October 2005 from http://news.bbc.co.uk.

BBC News, World Edition (2005b, 23 February). "Kenya urged to ban plastic bags". Retrieved 16 November 2005 from http://www.bbc.co.uk.

Bartram, J., Lewis, K., Lenton, R., & Wright, A. (2005). "Focusing on improved water and sanitation for health". *The Lancet,* 365, 810-812.

Bazoglu, N. & Moreno, E. L. (2005). *UN-HABITAT Methodology Paper to Globally Monitor Secure Tenure.* UN-HABITAT, unpublished document.

Bennhold, K. (2005, 30 August). "Seven die in Paris fire, the 2nd in 4 days". *The New York Times.*

Biau, D. (2005). "Spatial inequalities – the need for affirmative action". *Habitat Debate,* 11(2).

Bidinger, P.D., Nag, B. & Babu, P. (1986). "Nutritional and health consequences of seasonal fluctuations in household food availability". *Food Nutrition Bulletin,* 8, 36-59.

Billig, P., Bendahmane, D. & Swindale, A. (1999). *Water and Sanitation Indicators Measurement Guide.* Washington, D.C.: Food and Nutrition Technical Assistance Project.

Bindra, S. (2005, 24 April). "Give shade to jua kali businesses". *Sunday Nation.* [Nairobi, Kenya].

Boulle, P. & Palm, E. (2004). *Strategies to Reduce Urban Risk in the New Millennium.* Geneva: SDR Secretariat, Office for the Coordination of Humanitarian Affairs. Retrieved November 2005 from www.nidm.net.

Burra, S. (2005, April). "Towards a pro-poor framework for slum upgrading in Mumbai, India". *Environment and Urbanization,* 17(1).

Bwayo, J., Plummer, F., Omari, M., Mutere, A., Moses, S., Ndinya-Achola, J., Velentgas, P., & Kreiss, J. (1994, 27 June). "Human immunodeficiency virus infection in long-distance truck drivers in East Africa". *Archives of Internal Medicine,* 154(12).

C

Caldeira, P. R. (1996). *Building Up Walls: The New Pattern of Spatial Segregation in São Paulo.* UNESCO Report ISSJ 147/1996.

Canada Mortgage and Housing Corporation (2005, 1 April). *Canada's UN-Habitat Urban Indicators: Monitoring "The Habitat Agenda" and "The Millennium Development Goals".* CMHC File 0981-178. Report prepared for UN-HABITAT.

Carael, M. & Allen, S. (1995). "Women's vulnerability to HIV/STD in sub-Saharan Africa: an increasing evidence". In P. Makinwa & A. M. Jensen (Eds.), *Women's Position and Demographic Change in Sub-Saharan Africa.* Westport, CT: Greenwood Press.

The Center for International Earth Science Information Network (2005, 8 August). *Earth Institute News.* Columbia University. Retrieved September 2005 from http://www.ciesin.org.

Central Bureau of Statistics (Kenya) (2000). *Economic Survey.* Government of Kenya.

Centre on Housing Rights and Evictions (COHRE) (2003, 23 June). "Seven million left homeless as forced evictions double in two years". [Media Release]. Retrieved 15 February 2006 from http://www.cohre.org.

Centre on Housing Rights and Evictions (COHRE) (2004, December). *Evictions Monitor,* 1(2). Retrieved 15 February 2006 from http://www.cohre.org.

Centre on Housing Rights and Evictions (COHRE) (2005, 29 November). "China named a 2005 Housing Rights Violator for widespread forced evictions". [Media Release]. Retrieved 15 February 2006 from http://www.cohre.org.

Chambers of Justice, Care Kenya & Cradle (2005). *The Defilement Index.*

Cherp, A. (1999). *Water, Environment and Children in Central Asia.* Paper prepared for UNICEF. Retrieved December 2005 from http://www.ceu.hu/envsci/aleg/projects/water.pdf.

Chibli, M. (2004). *Aleppo Local Urban Observatory.* Presentation made at the second World Urban Forum, 13-17 September, Barcelona.

Cities Alliance (2003). *2003 Annual Report.* Retrieved from http://www.citiesalliance.org.

Cities Alliance (2004). *2004 Annual Report.* Retrieved from http://www.citiesalliance.org.

City of Vancouver (2001). *CityPlan: Directions for Vancouver.* Retrieved 17 November 2005 from http://vancouver.ca/commsvcs/planning/cityplan/dfvs.htm.

City of Vancouver (2004). *City of Vancouver Planning Department Information Sheet: CityFacts 2004.* Retrieved 17 November 2005 from http://vancouver.ca/commsvcs/planning.

Clichevsky, N. (2003). ECLAC, "Pobreza y acceso al suelo urbano. Algunas interrogantes sobre las políticas de regularización en América Latina". *Medio Ambiente y Desarrollo,* Número 75. [Chile.]

Coaffee, J. (2004). "Recasting the 'ring of steel': designing out terrorism in the city of London". In S. Graham (Ed.), *Cities, War and Terrorism.* Malden, Oxford and Carlton: Blackwell Publishers.

Cohen, M. (2002). *Spiderman, the Twin Towers and Saskia Sassen.* Unpublished paper presented in Barcelona.

Cohen, M. A. (2004). *Cities and the Wealth of Nations.* Paper presented at a parallel event held at the UNCTAD XI-URBIS meeting, June, São Paulo.

Cornia, A. G. & Menchini, G. (2005). *The Pace and Distribution of Health Improvements During the Last 40 Years: Some Preliminary Results.* Paper presented at the Forum on Human Development, 17 January, Paris.

Crombie, I. K, Irvine, L., Elliott, L. & Wallace, H. (2005). *Closing the Health Inequalities Gap: An International Perspective.* Geneva: WHO Europe.

Dahl, R. (2004). "Vehicular manslaughter: the global epidemic of traffic deaths". *Environmental Health Perspectives,* 112(11). Retrieved 10 September 2005 from http://ehp.niehs.nih.gov.

Daily Nation [Nairobi, Kenya]. (2005, 2 December). "AIDS infection rate declines".

D

Darrundono, P. (2005). "The first large-scale upgrading programme: an Indonesia success story". *Habitat Debate,* 11(3).

Davis, K. (1951). *The Population of India and Pakistan.* Princeton: Princeton University Press.

Davis, M. (2004, March-April). "Planet of slums". *New Left Review.*

DeParle, J. (2005, 4 September). "What happens to a race deferred". *The New York Times.*

De Soto, H. (2000). *The Mystery of Capital.* London: Black Swan.

Department for International Development (DFID) (2005, March). *Fighting Poverty to Build a Safer World: A Strategy for Security and Development.* London: DFID.

Dodoo, F. N., Sloan, M. & Zulu, E. M. (2003). "Space, context and hardship: socializing children into sexual activity in Kenyan slums". In S. Agyei-Mensah & J. B. Casterline (Eds). *Reproduction and Social Context in Sub-Saharan Africa.* Westport, CT: Greenwood Press.

Dooley, E. (2002). "Canada restricts cars". *Environmental Health Perspectives,* 110(7). Retrieved 10 September 2005 from http://ehp.niehs.nih.gov.

Dooley, E. (2004). "Death toll of biomass burning". *Environmental Health Perspectives,* 112(6). Retrieved 10 September 2005 from http://ehp.niehs.nih.gov.

Dooley, E. (2005). "Africa afire". *Environmental Health Perspectives,* 113(9). Retrieved 10 September 2005 from http://ehp.niehs.nih.gov.

Dwyer, G. (2005). *Violence and the Poor.* Asian Development Bank. Retrieved December 2005 from http://www.adb.org.

E

Eaton, L., Flisher, A. J., & Aaro, L. E. (2003). "Unsafe sexual behaviour in South African youth". *Social Science Medicine,* 56, 149-165.

Economic Commission for Africa (2002). *Youth and Employment in Africa.* Paper presented at the Youth Employment Summit, 7-11 September, Alexandria, Egypt.

Economic Commission for Latin America and the Caribbean (ECLAC) (2005). *El Rostro de la Pobreza en las Ciudades de América Latina y el Caribe.* Report co-published with UN-HABITAT.

Economic and Social Commission for Asia and the Pacific (ESCAP) (2003). *Asia-Pacific Economies: Resilience in Challenging Times.* Economic and Social Survey of Asia and the Pacific, 2003. New York: UN ESCAP.

Economic and Social Commission for Western Asia (ESCWA) (2001). *Sustainable Urban Development: A Regional Perspective on Good Urban Governance.* New York: United Nations.

The Economist (2002, 5 September). "Driven to Alcohol." Retrieved 26 October 2005 from http://www.economist.com.

Economist Intelligence Unit (2005, 3 October). "Vancouver tops liveability ranking according to a new survey by the Economist Intelligence Unit." [Media Release] Retrieved 10 November 2005 from http://store.eiu.com.

Economy, E. (2005, 5 December). "The lessons of Harbin". *TIME.*

Ellaway, A. & Macintyre, S. (1998). "Does housing tenure predict health in the UK because it exposes people to different levels of housing related hazards in the home or its surroundings?" *Health & Place,* 4(2), 141-150.

Eltayeb, G. E. (2003). "The case of Khartoum, Sudan". In *Understanding Slums: Case Studies for the Global Report on Human Settlements 2003: The Challenge of Slums.* Development Planning Unit, University College London/UN-HABITAT.

Energy for Sustainable Development Limited (2003). *Fuel Substitution: Poverty Impacts on Biomass Fuel Suppliers.* Final Technical Report to the UK Department for International Development, KaR Project R8019.

Esray, S. (1996). "Water, waste and well-being: a multi-country study". *American Journal of Epidemiology,* 143(6).

European Environment Agency. (2004). *Air Pollution in Europe 1990-2000.* Topic report No. 4/2003. Copenhagen: EEA.

Evans, B. (2005). *Securing Sanitation – The Compelling Case to Address the Crisis.* A report commissioned by the Government of Norway as input to the Commission on Sustainable Development and its 2004-2005 focus on water, sanitation and related issues. Stockholm. Retrieved December 2005 from http://www.siwi.org.

F

Fafo Institute for Applied International Studies (2003, May). "Living in Port-au-Prince". A presentation prepared for the Habitat Urban Indicators Workshop, Nairobi.

Faroohar, R. (2005, 21 November). "It's about jobs". *Newsweek.*

Fernandes, E. & Varley, A. (Eds.) (1998). *Illegal Cities: Law and Urban Change in Developing Countries.* London and New York: Zed Books Ltd.

Food and Agriculture Organization of the United Nations (2002). "FAO methodology for estimating the prevalence of undernourishment". In *Proceedings of the International Scientific Symposium on Measurement and Assessment of Food Deprivation and Under-Nutrition.* Rome: FAO.

Ford, P. (2005, 2 September). "Fires reveal housing crunch". *Christian Science Monitor.* Retrieved 5 September 2005 from http://www.csmonitor.com.

Forum on the Global South (2003). *Anti Poverty or Anti Poor? The Millennium Development Goals and the Eradication of Extreme Poverty and Hunger.* A study commissioned by UN Economic and Social Commission for Asia and the Pacific, Bangkok.

Fouchard, L. (2003). "The case of Ibadan, Nigeria". In *Understanding Slums: Case Studies for the Global Report on Human Settlements 2003: The Challenge of Slums.* Development Planning Unit, University College London/UN-HABITAT.

Franks, T. (2005, 24 October). "Cows make fuel for biogas train". BBC News, UK Edition. Retrieved 26 October 2005 from http://news.bbc.co.uk.

Fry, S., Cousins, B., & Olivola, K. (2002). *Health of Children Living in Urban Slums in Asia and the Near East: Review of Existing Literature and Data.* Environmental Health Project, Activity Report 109. USAID, Washington, D.C.

G

Girardet, H. (2004). *Cities People Planet: Liveable Cities for a Sustainable World.* Chichester, UK: John Wiley and Sons.

Glasze, G. Webster, C.J. & Frantz, K. (Eds.). (2005). *Private Cities: Local and Global Perspectives.* London: Routledge.

Gonzalez, D. (2005, 2 September). "From margins of society to center of the tragedy". *The New York Times.*

Government of Brazil (2005). *Millennium Development Goals: Brazilian Monitoring Report.* Unpublished document.

Government of Chile (2004, December). Report to UN-HABITAT by the Minsterio de Vivienda y Urbanismo on Chile's progress toward Millennium Development Goal 7, target 11. Unpublished document.

Government of Indonesia (2005, October). *Settlement, Drinking Water and Sanitation: Best Practices.* Jakarta: Ministry of Works.

Government of Kenya/UNCHS (2001). *Nairobi Situation Analysis Consultative Report: Collaborative Nairobi Slum Upgrading Initiative.* Nairobi.

Goytia, C. (2005, 10 May). "Titling views from Argentina". Message sent to mailing list of Geoffrey Payne and Associates, http://www.gpa.org.uk.

Greater New Orleans Community Data Center (2005). *Orleans Parish: Income and Poverty.* Retrieved 6 September 2005 from http://www.gnocdc.org.

H

Halfani, M. (2004). *Overview of Decentralization Policy Issues.* Presentation made on behalf of UN-HABITAT at the Inaugural Meeting of AGRED (Advisory group of Experts on Decentralization), 9-10 March, Gatineau, Canada.

Halifax Regional Municipality (2004, 12 October). "Entire metro transit bus fleet to adopt use of biodiesel fuel". [Media Release]. Retrieved 26 October 2005 from http://www.halifax.ca.

Hallman, K. (2004). *Socioeconomic Disadvantages and Unsafe Sex Behaviour Among Young Women and Men in South Africa.* Policy Research Paper No. 190. New York: The Population Council.

Hansen, S. & Bathia, R. (2004, March). *Water and Poverty in a Macro-Economic Context.* Paper commissioned by the Royal Norwegian Ministry of Environment.

Hardoy, J. E., Mitlin, D., & Satterthwaite, D. (2001). *Environmental Problems in an Urbanizing World.* London: Earthscan.

Hargeisa Municipality (Somalia). (2003). *Hargeisa Municipality Statistical Abstract.*

Hiasat, Y. (2005). "Jordan's commitment to countrywide slum upgrading". *Habitat Debate,* 11(3).

Hinrichsen, D. (2002). "Singapore – the global city". *People & the Planet.* Retrieved 12 November 2005 from http://www.peopleandplanet.net.

Housing Statistics Indonesia (Biro Pusat Statistik) (2000). "Statistics of permanent and non-permanent houses in Indonesia". Central Bureau of Statistics document.

Huchzermeyer, M. (2006). *Slum Upgrading Initiatives in Kenya Within the Wider Housing Market: A Housing Rights Concern.* Discussion Paper No. 1/2006, Kenya Housing Rights Project, COHRE Africa Programme Center on Housing Rights and Evictions, Geneva.

I

Inter-American Development Bank (2004, March). *The Challenge of an Urban Continent.* Washington, D.C.

International Energy Agency. (1996). *World Energy Outlook.* Paris: IEA/OECD.

International Labour Organization (2002). "Women and men in the informal economy: a statistical picture". Retrieved November 2005 from http://www.eldis.org.

International Labour Organization (2004). *Global Employment Trends for Youth.* Geneva: International Labour Organization.

IRIN News (2005, 6 October). "Zimbabwe: access to treatment a concern for displaced living with AIDS". United Nations Office for the Coordination of Humanitarian Affairs (OCHA). Retrieved 10 February 2006 from http://www.irinnews.org.

J

Jonsson, Å. & Satterthwaite, D. (2000). *Income-Based Poverty Lines: How well do the levels set internationally and within each country reflect (a) the cost of living in the larger/more prosperous/more expensive cities; and (b) the cost that the urban poor pay?* Paper prepared for the Panel on Urban Population Dynamics, Committee on Population, National Research Council/National Academy of Sciences. Washington, D.C.

K

Kantai, P. (2003). "Dandora burning". *EcoForum,* 26(3), 24-31.

Karasov, C. (2000). "On a different scale: putting China's environmental crisis in perspective". *Environmental Health Perspectives,* 108(10). Retrieved 10 September 2005 from http://ehp.niehs.nih.gov.

Kathuria, S. & Oberai, R. (2004, March). *How to Curb Crime and Its Impact on Private Sector Growth.* Moderated online discussion, archived on http://rru.worldbank.org/Discussions/Topics/Topic32.aspx.

Kessides, C. (1997). *World Bank Experience with the Provision of Infrastructure Services for the Urban Poor: Preliminary Identification and Review of Best Practices.* Washington, D.C.: World Bank.

Kirby, A. (2004, 13 December). "Pollution: a life and death issue". BBC News. Retrieved November 2005 from http://news.bbc.co.uk.

Kothari, M. (2005). Cited in *Habitat Debate,* 11(2).

L

Landman, K. (2003). *Alley-Gating and Neighbourhood Gating: Are They Two Sides of the Same Face?* Paper delivered at the conference, Gated Communities: Building Social Division or Safer Communities? 18-19 September, Glasgow.

Langley, E. (2002). *The Changing Visage of French Housing Policy and Finance: A Half-Century of Comprehensive, Complex and Compelling Home Building.* Joint Center for Housing Studies, Harvard University, Kennedy School of Government. Retrieved 15 September 2005 from http://www.jchs.harvard.edu.

Leaf, M. (2005). "Vancouver, Canada: multicultural collaboration and mainstreaming". In M. Balbo (Ed.), *International Migrants and the City.* Nairobi: UN-HABITAT.

Leavitt, M. (2000). *Melting Pot.* Greater New Orleans Community Data Center. Retrieved 6 September 2005 from http://www.gnocdc.org.

Lee, L. & Ghanime, L. (2004, September). *Millennium Development Goal 7 Summary Review, 67 Country MDG Reports.* Country Reporting on MDG7, Ensuring Environmental Sustainability (2nd ed.). UNDP.

Linchu, Z. & Zhi, Q. (2003). "The case of Shanghai, China". In *Understanding Slums: Case Studies for the Global Report on Human Settlements 2003: The Challenge of Slums.* Development Planning Unit, University College London/UN-HABITAT.

London Research Centre, Housing and Research Section (2005, September). *Indicator 15: Housing Overcrowding - Living at Over 1 Person Per Room.* Brent Council. Retrieved from http://www.brent.gov.uk.

Lugano, E. & Sayer, G. (2003, October). "Girls' education in Nairobi's informal settlements". *Links.* Retrieved 15 December, 2005, from www.oxfam.org.uk.

Luthra, D. (2005, 25 June). "Sri Lanka's slow tsunami response". *BBC News.*

M

Manase, G., Mulenga, M. & Fawcett, B. (2001). *Linking Urban Sanitation Agencies with Poor Community Needs.* Paper prepared for the 27th WEDC Conference, Lusaka, Zambia. Retrieved December 2005 from http://wedc.lboro.ac.uk.

Marcuse, P. (2001). "Dangers posed by 'really existing globalization'". *Habitat Debate,* 7(4).

Martinez-Martin, J. (2005). *Monitoring Intra-Urban Inequalities with GIS-Based Indicators: With a Case Study in Rosario, Argentina*. ITC Dissertation Series No. 127, Thesis, Utrecht University and ITC.

Martinot, E. (2005a). *Renewables 2005: Global Status Report*. Washington, D.C.: Worldwatch Institute.

Martinot, E. (2005b). "Solar (and sustainable) cities". Retrieved 26 October 2005 from http://www.martinot.info/solarcities.htm.

McMahon, J. M., Tortu, S., Pouget, E. R., Hamid, R., & Torres, L. (2004). *Increased Sexual Risk Behaviour and High HIV Seroincidence Among Drug-Using Low-Income Women with Primary Heterosexual Partners*. Paper presented at the 15th International AIDS Conference, 11-16 July, Bangkok.

Mehta, S. (2004). *Maximum City: Bombay Lost and Found*. London: Review.

Mendes, W. & Holden, M. (2005). *Vancouver: Promoting Multiculturalism and Social Inclusion Through Policy and Planning*. Paper commissioned by UN-HABITAT for this Report.

Milanovik, B. (2005, 21 November). "World apart". *Newsweek*.

Ministerio de Vivienda y Urbanismo (2004). *Metas Del Milenio*. Meta Nº 11, Informe del Gobierno de Chile, Diciembre.

Mitullah, W. (2003). "The case of Nairobi, Kenya". In *Understanding Slums: Case Studies for the Global Report on Human Settlements 2003: The Challenge of Slums*. Development Planning Unit, University College London/UN-HABITAT.

Moreau, R. & Mazumdar, S. (2005, 25 April). "Bombay dreams". *Newsweek*.

Moreno, E. L. (2005). "How far is the world from the slum target?" *Habitat Debate*, 11(3).

Moser, C. (2005). "City violence and the poor". In *In Focus: Poverty and the City*. Brasilia: UNDP/International Poverty Centre.

Moser, C. & Rodgers, D. (2005, March). *Change and Violence in Non-Conflict Situations*. Working Paper 245. London: Overseas Development Institute.

Mosley, W.H. & Chen, L. (1983). "Child survival: strategies for research". *Population and Development Review*, Supplement to Vol. 10.

Mougeot, L. J. (Ed.) (2005a). *AGROPOLIS: The Social, Political and Environmental Dimensions of Urban Agriculture*. London and Ottawa: Earthscan-IDRC.

Mougeot, L. J. (2005b). *Ecological Footprint and Urban Agriculture*. Paper commissioned by UN-HABITAT for this Report.

Mukherjee, N. (2001). *Achieving Sustained Sanitation for the Poor*. Water and Sanitation Programme for East Asia and the Pacific. Jakarta, Indonesia.

Musa, P. (2005, 17 April). "Sewage vegetables flood city markets". *Daily Nation*. [Nairobi, Kenya]

Myers, D. & Baer, W. (1996). "The changing problem of overcrowded housing". *Journal of the American Planning Association*, 62, 66-84.

N

Nalo, D. S. (2002). *Report of the Joint CBS/UN-HABITAT Fieldwork for the Delineation of the Urban Slums/Informal Settlements: A Case Study of Nairobi*. Unpublished technical report.

Narayan, D. (2000). *Voices of the Poor - Can Anyone Hear Us?* Washington, D.C., World Bank.

National Institute of Standards and Technology, Partnership for Advancing Technology in Housing (1999). *Improving Durability in Housing*. Background paper for the National Forum on Durability Research, Maryland, USA.

Netherlands Department for Housing (2005). *Dutch Urban Indicators Questionnaire of Amsterdam and Den Haag*. Unpublished document produced for UN-HABITAT.

Nicolae, V. (2005). "Prejudice, suspicion and discrimination – the plight of Europe's Roma communities". *Habitat Debate*, 11(2).

Nitti, R. & Sarkar, S. (2003, 7 January). "Reaching the poor through sustainable partnerships: the slum sanitation program in Mumbai, India." *Urban Notes: Thematic group on services to the urban poor*. Washington D.C.: World Bank. Retrieved October 2005 from http://www.worldbank.org/urban/upgrading/urban-notes.html.

Norris, M. & Shiels, P. (2004). *Regular National Report on Housing Developments in European Countries: Synthesis Report*. Department of the Environment, Heritage and Local Government, Ireland.

O

O'Higgins, N. (2002). *Government Policies that Promote Youth Employment*. Paper prepared for the World Youth Summit on Government Policy and Youth Employment, 7-11 September, Alexandria, Egypt.

Orc Macro (2004). *Demographic Health Surveys (1990-2003)*. Calverton, Maryland: Orc Macro.

P

Perlman, J.E. (2005). "The chronic poor in Rio de Janeiro: What has changed in 30 years?" In M. Keiner, M. Koll-Schretzenmayr & W.A. Schmid (Eds.), *Managing Urban Futures: Sustainability and Urban Growth in Developing Countries*. Aldershot and Burlington, VT: Ashgate.

Pettifor, A. E., Rees, H.V., Steffenson, A., Hlongwa-Madikizela, L., MacPhail, C., Vermaak, K., & Kleinshmidt, I. (2004). *HIV and Sexual Behaviour Among Young South Africans: A National Survey of 15-24 Year Olds*. Johannesburg: Reproductive Health Research Unit, University of the Witwatersrand.

Philips, T. (2005, 29 October). "Blood, sweat and fears in favelas of Rio". *The Guardian*.

Phombeah, G. (2005, 2 September). "Living amidst the rubbish of Kenya's slum". BBC News web feature. Retrieved 4 September 2005 from http://www.bbcnews.co.uk.

Pieterse, E. (2000). *Participatory Urban Governance: Practical Approaches, Regional Trends and UMP Experiences*. Urban Management Programme Discussion Paper 25. Nairobi: UNCHS (Habitat)/UNDP/World Bank.

Planning Commission of India (2002). *National Human Development Report 2001*. New Delhi.

Presidency of the Republic, Government of the Federative Republic of Brazil (2004). *Brazilian Monitoring Report on the Millennium Development Goals*. Institute for Applied Economic Research (IPEA) and National Institute of Geography and Statistics (IBGE). Unpublished document.

Q

Quan, J., Tan, S. F., & Toulmin, C. (Eds.) (2005, October). *Land in Africa: Market Asset or Secure Livelihood? Proceedings and Summary of Conclusions from the Land in Africa Conference held in London, November 8-9, 2004*. London: IIED. Retrieved 15 February 2006 from http://www.iied.org.

R

Rakodi, C., Gatabaki-Kamau, R., & Devas, N. (2000). "Poverty and political conflict in Mombasa". *Environment & Urbanization*, 12(1).

Rakodi, C. (2005). "The urban challenge in Africa". In M. Keiner, M. Koll-Schretzenmayr & W.A. Schmid (Eds.), *Managing Urban Futures: Sustainability and Urban Growth in Developing Countries*. Aldershot and Burlington, VT: Ashgate.

Ravallion, M. (2001, April). *On the Urbanization of Poverty*. World Bank Development Research Group, Policy Research Working Paper 2586.

Ray, K. (2003). "Water and sanitation in cities: translating global goals into local action". *Habitat Debate*, 9(3).

Risbud, N. (2003). "The case of Mumbai, India". In *Understanding Slums: Case Studies for the Global Report on Human Settlements 2003: The Challenge of Slums*. Development Planning Unit, University College London/UN-HABITAT.

Robertshaw, R., Louw, A., Shaw, M., Mashiyane, M., & Bretell, S. (2001, August). *Reducing Crime in Durban: A Victim Survey and Safer City Strategy*. ISS Monograph Series, No. 58.

Ross, N. A., Nobrega, K., & Dunn, J. (2001). "Income segregation, income inequality and mortality in North American metropolitan areas". *GeoJournal*, 53(2), 117-124.

S

Sachs, J. (2005). *The End of Poverty: Economic Possibilities for Our Time*. New York: Penguin.

Satterthwaite, D. (1999). *The Earthscan Reader in Sustainable Cities*. London: Earthscan.

Satterthwaite, D. (2004). *The Underestimation of Urban Poverty*. Poverty Reduction in Urban Areas Series, Working Paper 14. London: International Institute for Environment and Development.

Satterthwaite, D., McGranahan, G., & Mitlin, D. (2005, April). *Community-Driven Development for Water and Sanitation in Urban Areas*. WSSCC, Geneva. Retrieved October 2005 from http://www.wsscc.org.

Satterthwaite, D. & Tacoli, C. (2003). *The Urban Part of Rural Development: The Role of Small and Intermediate Urban Centres in Rural and Regional Development and Poverty Reduction*. International Institute for Environment and Development, Rural-Urban Briefing Papers, No. 8. Retrieved November 2005 from http://www.iied.org.

Scaramella, M. (2003). "The Case of Naples, Italy". In *Understanding Slums: Case Studies for the Global Report on Human Settlements 2003: The Challenge of Slums*. Development Planning Unit, University College London/UN-HABITAT.

Secretariat of the Third World Water Forum (2003). *Water Voice*. Retrieved September 2005 from http://www.worldwatercouncil.org.

Sharma, K. (2005a, 24 January). "Mumbai's tragedy". *The Hindu*.

Sharma, K. (2005b, 22 February). "Forget Shanghai, remember Mumbai". *The Hindu*.

Shisana, O. (2002). *Nelson Mandela/HSRC Study of HIV/AIDS: South African national HIV Prevalence, Behavioural Risks and Mass Media*. Household Survey 2002. Cape Town: Human Sciences Research Council.

Souza, C. (2002). "Is the participatory budgeting process in Brazil over-rated?" *Habitat Debate*, 8(1).

Statistics Canada (2001). Statistics retrieved 17 November 2005 from http://www.statcan.ca.

T

Taylor, P. (2000). "UNCHS (Habitat) – the global campaign for good urban governance". *Environment & Urbanization*, 12(1).

Taylor, P. J. (2005). "Leading world cities: empirical evaluations of urban nodes in multiple networks". *Urban Studies*, 42(9), 1593-1608.

Teather, D. (2005, 5 September). "Neighbouring states struggle to cope with influx of people". *Guardian*.

Tebbal, F. (2005). "Are countries working effectively on the Millennium targets?" *Habitat Debate*, 11(3).

Thompson, J., Porras, I. T., Wood, E., Tumwine, J. K., Mujwahuzi, M. R.,Katui-Katua, M., & Johnstone, N. (2000). "Waiting at the tap: changes in urban water use in East Africa over three decades". *Environment & Urbanization*, 12(2), 37-52.

TIME (2005, 27 June). "A new China rising".

U

UN-HABITAT (1995). *Crowding and Health in Low Income Settlements*. Nairobi: UN-HABITAT.

UN-HABITAT (2001). *The State of the World's Cities Report 2001*. Nairobi: UN-HABITAT.

UN-HABITAT (2002a). "Livable Region Strategic Plan (LRSP) for the Greater Vancouver Regional District". Best Practices Database. Retrieved 17 November 2005 from http://www.bestpractices.org.

UN-HABITAT (2002b). *Report of the Expert Group Meeting on "Urban Indicators: Slums, Secure Tenure and Global Sample of Cities".* Unpublished report.

UN-HABITAT (2002c, 20 March). *The Role of Cities in National and International Development.* HSP/WUF/1/DLG.II/Paper 2. Nairobi.

UN-HABITAT (2003a). *The Challenge of Slums: Global Report on Human Settlements.* London and Sterling, VA: Earthscan.

UN-HABITAT (2003b). *Guide to Monitoring Target 11: Improving the Lives of 100 Million Slum Dwellers.* Nairobi: UN-HABITAT.

UN-HABITAT (2003c). *Slums of the World.* Nairobi: UN-HABITAT.

UN-HABITAT (2003d). *Water and Sanitation in the World's Cities.* London and Sterling, VA: Earthscan.

UN-HABITAT (2004a). *Monitoring Urban Inequities Programme.* Unpublished concept paper by UN-HABITAT.

UN-HABITAT (2004b). *Report of the Urban Inequities Survey in Addis Ababa.* Unpublished document.

UN-HABITAT (2004c). *Urban Land for All.* Nairobi: UN-HABITAT.

UN-HABITAT (2004d, December), *Progress Report On Removing Discrimination Against Women in Respect of Property and Inheritance Rights.* Tools on Improving Women's Secure Tenure Series 1, Nr. 2. Nairobi: UN-HABITAT.

UN-HABITAT (2005a). *Good Policies and Enabling Legislation for Attaining the Millennium Development Goals.* Nairobi: UN-HABITAT.

UN-HABITAT (2005b). *Operational Activities Report 2005.* Nairobi: UN-HABITAT.

UN-HABITAT (2005c). *Urban Indicators Programme, Data Base III.* Data from 2003. Unpublished database, Monitoring Systems Branch.

UN-HABITAT/DFID (2002). *Sustainable Urbanization: Achieving Agenda 21.* Nairobi: UN-HABITAT/DFID.

UN Millennium Project (2005a). *A Home in the City: Task Force Report on Improving the Lives of Slum Dwellers.* London and Sterling, VA: Earthscan.

UN Millennium Project (2005b). *Investing in Development: A Practical Plan to Achieve the Millennium Development Goals.* Retrieved from http://www.unmillenniumproject.org.

UN Office of the High Commissioner for Human Rights & UN-HABITAT (2002, October). *Global Housing Rights Challenge: Monitoring and Evaluation of Progress.* Geneva.

UNAIDS (2004). *Report on the Global AIDS Epidemic.* Geneva.

UNAIDS/WHO (2004a). *Epidemiological Fact Sheet – 2004 Update: Kenya.*

UNAIDS/WHO (2004b, December). *AIDS Epidemic Update.* Geneva: UNAIDS.

UNDP (2000). *Azerbaijan Human Development Report 2000* Retrieved 16 February 2006 from http://www.un-az.org.

UNEP (2003). *GEO Year Book, Freshwater: Meeting Our Goals, Sustaining Our Future.* Retrieved from http://www.unep.org.

UNEP (2005). *Rapid Environmental Assessment of the Tsunami.* Retrieved from http://www.unep.org.

UNESCO (2003). *Facts and Figures - Water and Cities.* 2003 International Year of Fresh Water Web site: http://www.wateryear2003.org.

UNFPA (2003). *Achieving The Millennium Development Goals.* New York. Retrieved September 2005 from http://www.unfpa.org.

UNICEF (1997). *Progress of Nations: Sanitation League Table.* New York: UNICEF.

UNICEF (2003). *Progress Since the World Summit for Children.* New York: UNICEF.

UNICEF (2004). *The State of the World's Children 2005: Children Under Threat.* New York: UNICEF.

UNICEF (2005). *Progress for Children: A Report Card on Gender Parity and Primary Education. (No.2).* New York: UNICEF.

United Kingdom (2005). "The impact of overcrowding on health and education". Retrieved September 2005 from http://www.odpm.gov.uk.

United Kingdom Parliamentary Office of Science and Technology. (2002, December). *Access to Sanitation in Developing Countries.* POSTnote No. 190. Retrieved 20 December 2005 from http://www.parliament.uk/parliamentary_offices/post/environment.cfm#2002

United Nations (2002). Monterrey Consensus of International Conference on Financing for Development.

United Nations (2005a). *The Inequality Predicament: Report on the World Social Situation 2005.* New York: United Nations.

United Nations (2005b). *The Millennium Development Goals Report 2005.* New York: United Nations.

United Nations (2005c, July). *Report of the Fact-Finding Mission to Zimbabwe to Assess the Scope and Impact of Operation Murambatsyina by the UN Special Envoy on Human Settlements Issues in Zimbabwe, Mrs. Anna Kajumulo Tibaijuka.* Nairobi. Retrieved from http://www.unhabitat.org.

United Nations Department of Economic and Social Affairs, Population Division (2004). *World Urbanization Prospects: The 2003 Revision.* New York. Retrieved September 2005 from http://esa.un.org/unup.

United Nations Department of Economic and Social Affairs, Population Division (2005). *World Population Prospects: The 2004 Revision.* New York.

United Nations Department of Economic and Social Affairs, Statistics Division (2005). *Progress Towards the Millennium Development Goals, 1990-2005: Goal 7 – Ensure Environmental Sustainability.* Retrieved 31 August 2005 from http://unstats.un.org/unsd/mi/goals_2005/Goal_7_2005.pdf.

United Nations General Assembly Economic and Social Council (2004, 6 December). *World Youth Report 2005: Report of the Secretary-General.* A/60/61-E/2005/7. Retrieved 20 January 2006 from www.un.org/youth.

United Nations Office for the Coordination of Humanitarian Affairs (OCHA) (2005a). *Disaster Reduction and the Human Cost of Disaster: IRIN Web Special.* Retrieved June 2005 from http://www.irinnews.org.

United Nations Office for the Coordination of Humanitarian Affairs (OCHA) (2005b). *IRIN News.* Retrieved December 2005 from http://www.irinnews.org.

US Census Bureau (2000). "Population profile of the United States: 2000". [Media Release]. Retrieved December 2005 from http://www.census.gov/population/pop-profile/2000/chap07.pdf.

US Conference of Mayors (2004). *US Metro Economies: A Decade of Prosperity.* Retrieved December 2005 from http://www.usmayors.org/uscm.

US Department of Housing and Urban Development, Office of Policy Development and Research (2005). *Assessing Housing Durability: A Pilot Study.* Washington, D.C.

US Library of Congress, Federal Research Division (1998). *Sri Lanka: A Country Study.* Washington, D.C.

Vanderschueren, F. (2000, May). *The Prevention of Urban Crime.* Paper presented at the Africities 2000 Summit. Windhoek, Namibia.

Vanderschueren, F. & Vezina, C. (2003). *Reviewing Safer Cities.* Background Paper presented at the International Conference on Sustainable Safety: Municipalities at the Crossroads, 25-28 November 2003, Durban, South Africa.

Walsh, B., Lewis, K.R. & McQuaid, J. (2005, 1 September). "Rebuilding New Orleans: There may be a will, but is there a way?" *Newhouse News Service.*

Warah, R. (2002). "If you want to mobilize people, go to the public toilets". *Habitat Debate,* 8(2).

Watkins, K. (2000). *The Oxfam Education Report.* Oxford: Oxfam.

Watts, J. (2005, 31 October). "Satellite data reveals Beijing as pollution capital of the world". *The Guardian.*

WHO (2002). *World Report on Violence and Health. Red Posters from the Global Campaign for Violence Prevention.* Geneva: WHO.

WHO (2005). *The Protocol on Water and Health: Making a Difference.* WHO Regional Office for Europe.

WHO/UNICEF Joint Monitoring Programme for Water Supply and Sanitation (2000). *Global Water Supply and Sanitation Assessment 2000 Report.* Geneva: WHO.

WHO/UNICEF Joint Monitoring Programme for Water Supply and Sanitation (2004). *Meeting the MDG Drinking Water and Sanitation Target: A Mid-Term Assessment of Progress.* Geneva. Retrieved September 2005 from https://www.unicef.org.uk/publications.

Wilkinson, R. & Marmot, M. (Eds.) (2003). *Social Determinants of Health. The Solid Facts.* (2nd ed.) Geneva: World Health Organization.

World Bank (2000). *Cities in Transition.* Washington, D.C.

World Bank (2000/2001). *World Development Report 2000/2001: Attacking Poverty.* Washington, D.C.: World Bank.

World Bank (2003, June). *Land Policies for Growth and Poverty Reduction.* Washington, D.C.: World Bank.

World Bank (2005). *World Development Report 2006.* Washington, D.C.: Oxford University Press.

World Health Organization (1999). "Creating healthy cities in the 21st century". In: D. Satterthwaite (Ed.), *The Earthscan Reader in Sustainable Cities.* London: Earthscan.

World Health Organization (2002). *World Health Report 2002.* Geneva: WHO.

World Health Organization. (2004). *World Report on Road Traffic Injury Prevention.* Geneva: WHO.

Worldwatch Institute (2005). *State of the World 2005: Redefining Global Security.* New York and London: W.W. Norton & Company.

Worldwatch Institute (2006a, 24 January). "Rising toll from disasters underscores need for humanitarian, political action". [Media Release].

Worldwatch Institute (2006b). *State of the World 2006: Special Focus on India and China.* New York: W.W. Norton & Co.

YUVA & Montgomery Watson Consultants, India (2000-2001). *Final Report for Slum Sanitation Undertaken for Mumbai Sewerage Disposal Project (MSDP).* Mumbai.

Zulu, E. M., Dodoo, N. A, & Ezeh, A. C. (2002). "Sexual risk-taking in the slums of Nairobi, Kenya 1993-98". *Population Studies,* 56, 311-332.

Statistical Annex

Monitoring the Habitat Agenda and the Millennium Development Goals

General disclaimer

The designations employed and presentation of the data in the Statistical Annex do not imply the expression of any opinion whatsoever on the part of the Secretariat of the United Nations concerning the legal status of any country, city or area or of its authorities, or concerning the delimitation of its frontiers or boundaries.

Indicators for the Habitat Agenda and the Millennium Development Goals	Indicators	MDGs
1. Shelter		
Promote the right to adequate housing	Key indicator 1: durable structures	Goal 7, targets 11, 32
	Key indicator 2: overcrowding	Goal 7, targets 11, 32
	checklist 1: right to adequate housing	
	extensive indicator 1: housing price and rent-to-income	
Provide security of tenure	Key indicator 3: secure tenure	Goal 7, targets11, 32
	extensive indicator 2: authorized housing	
	extensive indicator 3: evictions	
Provide equal access to credit	checklist 2: housing finance	
Provide equal access to land	extensive indicator 4: land price-to-income	
Promote access to basic services	Key indicator 4: access to safe water	Goal 7, targets 10,30
	Key indicator 5: access to improved sanitation	Goal 7, targets 10,31
	Key indicator 6: connection to services	
2. Social development and eradication of poverty		
Provide equal opportunities for a safe and healthy life	Key indicator 7: under-five mortality	Goal 4, targets 5,13
	Key indicator 8: homicides	
	checklist 3: urban violence	
	extensive indicator 5: HIV prevalence	Goal 6, targets 7, 18
Promote social integration and support disadvantaged groups	Key indicator 9: poor households	Goal 1, targets 11, 1
Promote gender equality in human settlements development	Key indicator 10: literacy rates	Goal 3, targets 4, 10
	checklist 4: gender inclusion	
	extensive indicator 6: school enrolment	Goal 3, targets 4, 10
	extensive indicator 7: women councillors	Goal 3, targets 4, 12
3. Environmental Management		
Promote geographically-balanced settlement structures	Key indicator 11: urban population growth	
	Key indicator 12: planned settlements	
Manage supply and demand for water in an effective manner	Key indicator 13: price of water	
	extensive indicator 8: water consumption	
Reduce urban pollution	Key indicator 14: wastewater treated	
	Key indicator 15: solid waste disposal	
	extensive indicator 9: regular solid waste collection	
Prevent disasters and rebuild settlements	checklist 5: disaster prevention and mitigation instruments	
	extensive indicator 10: houses in hazardous locations	
Promote effective and environmentally sound transportation systems	Key indicator 16: travel time	
	extensive indicators 11: transport modes	
Support mechanisms to prepare and implement local environmental plans and local Agenda 21 initiatives	Checklist 6: local environmental plans	
4. Economic Development		
Strengthen small and micro-enterprises, particularly those developed by women	Key indicator 17: informal employment	
Encourage public-private sector partnership and stimulate productive employment opportunities	Key indicator 18: city product	
	Key indicator 19: unemployment	
5. Governance		
Promote decentralisation and strengthen local authorities	Key indicator 20: local government revenue	
	Checklist 7: decentralization	Goal 8, targets 16, 45
Encourage and support participation and civic engagement	Checklist 8: citizen participation	
	extensive indicator 12: voter participation	
	extensive indicator 13: civic associations	
Ensure transparent, accountable and efficient governance of towns, cities and metropolitan areas	Checklist 9: transparency and accountability	
Not Habitat Agenda but MDGs indicators	Child malnutrition	Goal 1, targets 2, 4
	Immunization against measles	Goal 4, targets 5,15
	Births attended by skilled health personnel	Goal 5, targets 6, 17
	Solid fuel	Goal 7, targets 9, 29
Not Habitat Agenda nor MDGs indicators	Prevalence of diarrhoea and prevalence of Acute respiratory infections (ARI)	

TABLE 1 POPULATION OF SLUM AREAS AT MID-YEAR, BY REGION AND COUNTRY; 1990, 2001 AND SLUM ANNUAL GROWTH RATE

	1990					2001					
	Total Population (thousands)	Urban Population (thousands)	Percentage Urban	Percentage Slum	Slum Population (thousands)	Total Population (thousands)	Urban Population (thousands)	Percentage Urban	Percentage Slum	Slum Population (thousands)	Slum Annual Growth Rate (%)
WORLD	**5,254,807**	**2,285,693**	**43.5**	**31.3**	**714,972**	**6,134,124**	**2,923,184**	**47.7**	**31.2**	**912,918**	2.22
Developed regions	**933,494**	**694,260**	**74.4**	**6.0**	**41,750**	**985,592**	**753,909**	**76.5**	**6.0**	**45,191**	0.72
EURASIA (Countries in CIS)	**281,610**	**184,261**	**65.4**	**10.3**	**18,929**	**282,639**	**181,182**	**64.1**	**10.3**	**18,714**	-0.10
European countries in CIS	**214,807**	**152,222**	**70.9**	**6.0**	**9,208**	**208,208**	**147,673**	**70.9**	**6.0**	**8,878**	-0.33
Asian countries in CIS	**66,803**	**32,039**	**48.0**	**30.3**	**9,721**	**74,431**	**33,509**	**45.0**	**29.4**	**9,836**	0.11
Developing regions	**4,039,703**	**1,407,172**	**34.8**	**46.5**	**654,294**	**4,865,893**	**1,988,093**	**40.9**	**42.7**	**849,013**	2.37
Northern Africa	**118,347**	**57,602**	**48.7**	**37.7**	**21,719**	**145,581**	**75,693**	**52.0**	**28.2**	**21,355**	-0.15
Sub-Saharan Africa	**501,133**	**139,644**	**27.9**	**72.3**	**100,973**	**667,022**	**231,052**	**34.6**	**71.9**	**166,208**	4.53
Latin America and the Caribbean	**440,419**	**312,995**	**71.1**	**35.4**	**110,837**	**526,594**	**399,322**	**75.8**	**31.9**	**127,566**	1.28
Eastern Asia	**1,226,423**	**367,210**	**29.9**	**41.1**	**150,761**	**1,364,438**	**533,182**	**39.1**	**36.4**	**193,824**	2.28
Eastern Asia excluding China (optional)	**71,118**	**50,641**	**71.2**	**25.3**	**12,831**	**79,466**	**61,255**	**77.1**	**25.4**	**15,568**	1.76
Southern Asia	**1,173,908**	**311,867**	**26.6**	**63.7**	**198,663**	**1,449,417**	**428,677**	**29.6**	**59.0**	**253,122**	2.20
South-Eastern Asia	**440,461**	**133,195**	**30.2**	**36.8**	**48,986**	**529,764**	**202,854**	**38.3**	**28.0**	**56,781**	1.34
Western Asia	**132,946**	**83,229**	**62.6**	**26.4**	**22,006**	**175,322**	**115,241**	**65.7**	**25.7**	**29,658**	2.71
Oceania	**6,066**	**1,430**	**23.6**	**24.5**	**350**	**7,755**	**2,072**	**26.7**	**24.1**	**499**	3.24
Optional grouping											
Landlocked Developing Countries (LLDCs)	**297,396**	**96,106**	**32.3**	**48.4**	**46,509**	**275,262**	**83,708**	**30.4**	**56.5**	**47,303**	0.15
Small Island Developing States (SIDS)	**44,908**	**23,852**	**53.1**	**24.0**	**5,735**	**52,644**	**30,083**	**57.1**	**24.4**	**7,327**	2.23
Least Developed Countries (LDCs)	**515,348**	**107,341**	**20.8**	**76.3**	**81,925**	**685,365**	**179,295**	**26.2**	**78.2**	**140,121**	4.88
List of countries											
EURASIA (Countries in CIS)	281,610	184,261	65.4	10.3	18,929	282,639	181,182	64.1	10.3	18,714	-0.10
Developing regions											
Northern Africa	118,347	57,602	48.7	37.7	21,719	145,581	75,693	52.0	28.2	21,355	-0.15
Algeria	24,855	12,776	51.4	11.8	1,508	30,841	17,801	57.7	11.8	2,101	3.02
Egypt	56,223	24,499	43.6	57.5	14,087	69,080	29,475	42.7	39.9	11,762	-1.64
Libyan Arab Jamahiriya	4,311	3,528	81.8	35.2	1,242	5,408	4,757	88.0	35.2	1,674	2.72
Morocco	24,624	11,917	48.4	37.4	4,457	30,430	17,082	56.1	32.7	5,579	2.04
Tunisia	8,156	4,726	57.9	9.0	425	9,562	6,329	66.2	3.7	234	-5.43
Western Sahara	178	156	87.6	-	-	260	249	95.7	2.0	5	
Sub-Saharan Africa	501,133	139,644	27.9	72.3	100,973	667,022	231,052	34.6	71.9	166,208	4.53
Angola	9,570	2,639	27.6	83.1	2,193	13,527	4,715	34.9	83.1	3,918	5.28
Benin	4,655	1,605	34.5	80.3	1,288	6,446	2,774	43.0	83.6	2,318	5.34
Botswana	1,240	525	42.3	59.2	311	1,554	768	49.4	60.7	466	3.69
Burkina Faso	9,008	1,221	13.6	80.9	987	11,856	1,999	16.9	76.5	1,528	3.97
Burundi	5,636	353	6.3	83.3	294	6,502	603	9.3	65.3	394	2.66
Cameroon	11,614	4,679	40.3	62.1	2,906	15,203	7,558	49.7	67.0	5,064	5.05
Cape Verde	341	151	44.3	70.3	106	437	277	63.5	69.6	193	5.42
Central African Rep	2,945	1,104	37.5	94.0	1,038	3,782	1,575	41.7	92.4	1,455	3.07
Chad	5,829	1,227	21.0	99.3	1,218	8,135	1,964	24.1	99.1	1,947	4.26
Comoros	527	147	27.9	61.7	91	727	246	33.8	61.2	151	4.61
Congo	2,230	1,243	55.7	84.5	1,050	3,110	2,056	66.1	90.1	1,852	5.15
Côte d'Ivoire	12,582	5,014	39.9	50.5	2,532	16,349	7,197	44.0	67.9	4,884	5.97
Dem Rep of the Congo	36,999	10,340	27.9	51.9	5,366	52,522	16,120	30.7	49.5	7,985	3.61
Djibouti	504	408	81.0	-	-	644	542	84.2			
Equatorial Guinea	352	126	35.8	89.1	112	470	232	49.3	86.5	201	5.28
Eritrea	3,103	490	15.8	69.9	342	3,816	730	19.1	69.9	510	3.62
Ethiopia	47,509	6,044	12.7	99.0	5,984	64,459	10,222	15.9	99.4	10,159	4.81
Gabon	935	637	68.1	56.1	357	1,262	1,038	82.3	66.2	688	5.95
Gambia	928	231	24.9	67.0	155	1,337	418	31.3	67.0	280	5.39
Ghana	15,138	5,078	33.5	80.4	4,083	19,734	7,177	36.4	69.6	4,993	1.83
Guinea	6,139	1,439	23.4	79.6	1,145	8,274	2,312	27.9	72.3	1,672	3.44
Guinea-Bissau	946	225	23.8	93.4	210	1,227	397	32.3	93.4	371	5.17
Kenya	23,574	5,660	24.0	70.4	3,985	31,293	10,751	34.4	70.7	7,605	5.88
Lesotho	1,682	338	20.1	49.8	168	2,057	592	28.8	57.0	337	6.32
Liberia	2,144	900	42.0	70.2	632	3,108	1,414	45.5	55.7	788	2.00
Madagascar	11,956	2,818	23.6	90.9	2,562	16,437	4,952	30.1	92.9	4,603	5.33
Malawi	9,434	1,092	11.6	94.6	1,033	11,572	1,745	15.1	91.1	1,590	3.92
Mali	8,778	2,091	23.8	94.1	1,968	11,677	3,606	30.9	93.2	3,361	4.87
Mauritania	1,992	877	44.0	94.3	827	2,747	1,624	59.1	94.3	1,531	5.60
Mauritius	1,057	428	40.5	-		1,171	486	41.6	-	-	
Mozambique	13,645	2,880	21.1	94.5	2,722	18,644	6,208	33.3	94.1	5,841	6.94
Namibia	1,375	366	26.6	42.3	155	1,788	561	31.4	37.9	213	2.88
Niger	7,707	1,241	16.1	96.0	1,191	11,227	2,366	21.1	96.2	2,277	5.89
Nigeria	85,953	30,120	35.0	80.0	24,096	116,929	52,539	44.9	79.2	41,595	4.96
Réunion	604	386	63.9	-	-	732	528	72.1			
Rwanda	6,766	360	5.3	82.2	296	7,949	497	6.3	87.9	437	3.55
Saint Helena	6	3	50.0	-	-	6	5	71.9	2.0	0	
Sao Tome & Principe	115	45	39.1	-	-	140	67	47.7	2.0	1	
Senegal	7,327	2,933	40.0	77.6	2,276	9,662	4,653	48.2	76.4	3,555	4.05
Seychelles	70	37	52.9	-	-	81	53	64.6	2.0	1	
Sierra Leone	4,061	1,218	30.0	90.9	1,107	4,587	1,714	37.3	95.8	1,642	3.58
Somalia	7,163	1,734	24.2	96.3	1,670	9,157	2,557	27.9	97.1	2,482	3.60
South Africa	36,376	17,763	48.8	46.2	8,207	43,792	25,260	57.7	33.2	8,376	0.19
Sudan	24,818	6,606	26.6	86.4	5,708	31,809	11,790	37.1	85.7	10,107	5.19
Swaziland	769	183	23.8	-	-	938	250	26.7			
Togo	3,453	984	28.5	80.9	796	4,657	1,579	33.9	80.6	1,273	4.27
Uganda	17,245	1,925	11.2	93.8	1,806	24,023	3,486	14.5	93.0	3,241	5.32
U. Rep of Tanzania	26,043	5,652	21.7	99.1	5,601	35,965	11,982	33.3	92.1	11,031	6.16
Zambia	8,049	3,172	39.4	72.0	2,284	10,649	4,237	39.8	74.0	3,136	2.88
Zimbabwe	10,241	2,906	28.4	4.0	116	12,852	4,630	36.0	3.4	157	2.76
Latin America and the Caribbean	**440,419**	**312,995**	**71.1**	**35.4**	**110,837**	**526,594**	**399,322**	**75.8**	**31.9**	**127,566**	**1.28**
Anguilla	8	8	100.0	40.6	3	12	12	100.0	40.6	5	3.69
Antigua and Barbuda	63	22	34.9	6.9	2	65	24	37.1	6.9	2	0.79
Argentina	32,527	28,141	86.5	30.5	8,597	37,488	33,119	88.3	33.1	10,964	2.21
Aruba	66	33	50.0	2.0	1	104	53	51.0	2.0	1	4.31
Bahamas	255	213	83.5	2.0	4	308	274	88.9	2.0	5	2.29

	1990					2001					
	Total Population (thousands)	Urban Population (thousands)	Percentage Urban	Percentage Slum	Slum Population (thousands)	Total Population (thousands)	Urban Population (thousands)	Percentage Urban	Percentage Slum	Slum Population (thousands)	Slum Annual Growth Rate (%)
Barbados	257	115	44.7	1.0	1	268	136	50.5	1.0	1	1.52
Belize	186	89	47.8	54.2	48	231	111	48.1	62.0	69	3.23
Bolivia	6,573	3,653	55.6	70.0	2,555	8,516	5,358	62.9	61.3	3,284	2.28
Brazil	147,957	110,610	74.8	45.0	49,806	172,559	141,041	81.7	36.6	51,676	0.34
British Virgin Islands	17	9	52.9	3.0	0	24	15	62.0	3.0	0	4.64
Cayman Islands	26	26	100.0	2.0	1	40	40	100.0	2.0	1	3.92
Chile	13,100	10,908	83.3	4.0	432	15,402	13,254	86.1	8.6	1,143	8.85
Colombia	34,970	24,029	68.7	26.0	6,239	42,803	32,319	75.5	21.8	7,057	1.12
Costa Rica	3,049	1,637	53.7	11.9	195	4,112	2,448	59.5	12.8	313	4.31
Cuba	10,629	7,828	73.6	2.0	156	11,237	8,482	75.5	2.0	169	0.73
Dominica	71	48	67.6	16.6	8	71	50	71.4	14.0	7	-1.17
Dominican Republic	7,061	4,126	58.4	56.4	2,327	8,507	5,615	66.0	37.6	2,111	-0.88
Ecuador	10,264	5,654	55.1	28.1	1,588	12,880	8,171	63.4	25.6	2,095	2.52
El Salvador	5,112	2,517	49.2	44.7	1,126	6,400	3,935	61.5	35.2	1,386	1.89
Falkland Is (Malvinas)	2	2	100.0	2.0	0	2	2	81.3	2.0	0	0.00
French Guiana	116	87	75.0	12.9	11	170	128	75.2	12.9	16	3.51
Greenland	56	44	78.6	18.5	8	56	46	82.3	18.5	9	0.40
Grenada	91	31	34.1	6.9	2	94	36	38.4	6.9	2	1.36
Guadeloupe	391	385	98.5	6.9	27	431	430	99.6	6.9	30	1.00
Guatemala	8,749	3,333	38.1	65.8	2,192	11,687	4,668	39.9	61.8	2,884	2.49
Guyana	731	243	33.2	4.9	12	763	280	36.7	4.9	14	1.29
Haiti	6,907	2,035	29.5	84.9	1,728	8,270	3,004	36.3	85.7	2,574	3.63
Honduras	4,870	2,036	41.8	24.0	488	6,575	3,531	53.7	18.1	638	2.43
Jamaica	2,369	1,219	51.5	29.2	356	2,598	1,470	56.6	35.7	525	3.53
Martinique	360	326	90.6	2.0	6	386	367	95.2	2.0	7	1.08
Mexico	83,223	60,303	72.5	23.1	13,923	100,368	74,846	74.6	19.6	14,692	0.49
Montserrat	11	1	9.1	10.7	0	3	-	13.1	8.8	-	
Netherlands Antilles	188	128	68.1	1.0	1	217	151	69.3	1.0	2	1.50
Nicaragua	3,824	2,029	53.1	80.7	1,638	5,208	2,943	56.5	80.9	2,382	3.41
Panama	2,398	1,288	53.7	30.8	397	2,899	1,639	56.5	30.8	505	2.19
Paraguay	4,219	2,054	48.7	36.8	756	5,636	3,194	56.7	25.0	797	0.48
Peru	21,569	14,862	68.9	60.4	8,979	26,093	19,084	73.1	68.1	12,993	3.36
Puerto Rico	3,528	2,516	71.3	2.0	50	3,952	2,987	75.6	2.0	59	1.56
Saint Kitts and Nevis	42	14	33.3	5.0	1	38	13	34.2	5.0	1	-0.67
Saint Lucia	131	49	37.4	11.9	6	149	57	38.0	11.9	7	1.37
St Vincent & the Grenadines	106	43	40.6	5.0	2	114	64	56.0	5.0	3	3.62
Saint-Pierre-et-Miquelon	6	6	100.0	8.7	1	7	6	92.2	8.7	1	
Suriname	402	263	65.4	6.9	18	419	313	74.8	6.9	22	1.58
Trinidad and Tobago	1,215	840	69.1	34.7	292	1,300	969	74.5	32.0	310	0.55
Turks and Caicos Islands	12	5	41.7	2.0	0	17	8	45.6	2.0	0	4.27
Uruguay	3,106	2,763	89.0	6.9	191	3,361	3,097	92.1	2.0	62	-10.27
US Virgin Islands	104	46	44.2	2.0	1	122	57	46.7	6.9	4	13.25
Venezuela	19,502	16,378	84.0	40.7	6,664	24,632	21,475	87.2	40.7	8,738	2.46
Eastern Asia	**1,226,423**	**367,210**	**29.9**	**41.1**	**150,761**	**1,364,438**	**533,182**	**39.1**	**36.4**	**193,824**	**2.28**
China	1,155,305	316,569	27.4	43.6	137,929	1,284,972	471,927	36.7	37.8	178,256	2.33
Hong Kong SAR of China	5,705	5,701	99.9	2.0	113	6,961	6,961	100.0	2.0	139	1.82
Macao SAR of China	372	367	98.7	2.0	7	449	444	98.9	2.0	9	1.73
Korea, Dem People's Rep of	19,956	11,651	58.4	1.0	117	22,428	13,571	60.5	0.7	95	-1.86
Korea, Rep of	42,869	31,658	73.8	37.0	11,728	47,069	38,830	82.5	37.0	14,385	25.55
Mongolia	2,216	1,264	57.0	68.5	866	2,559	1,449	56.6	64.9	940	0.75
Southern Asia	**1,173,908**	**311,867**	**26.6**	**63.7**	**198,663**	**1,449,417**	**428,677**	**29.6**	**59.0**	**253,122**	**2.20**
Afghanistan	13,675	2,495	18.2	98.5	2,458	22,474	5,019	22.3	98.5	4,945	6.35
Bangladesh	110,025	21,750	19.8	87.3	18,988	140,369	35,896	25.6	84.7	30,403	4.28
Bhutan	1,696	87	5.1	70.0	61	2,141	158	7.4	44.1	70	1.22
India	844,886	215,747	25.5	60.8	131,174	1,025,096	285,608	27.9	55.5	158,418	1.72
Iran (Islamic Republic of)	58,435	32,917	56.3	51.9	17,094	71,369	46,204	64.7	44.2	20,406	1.61
Maldives	216	56	25.9	0.0	-	300	84	28.0	0.0	-	
Nepal	18,142	1,624	9.0	96.9	1,574	23,593	2,874	12.2	92.4	2,656	4.76
Pakistan	109,811	33,565	30.6	78.7	26,416	144,971	48,425	33.4	73.6	35,627	2.72
Sri Lanka	17,022	3,626	21.3	24.8	899	19,104	4,409	23.1	13.6	597	-3.72
South-Eastern Asia	**440,461**	**133,195**	**30.2**	**36.8**	**48,986**	**529,764**	**202,854**	**38.3**	**28.0**	**56,781**	**1.34**
Brunei Darussalam	257	169	65.8	2.0	3	335	244	72.8	2.0	5	3.34
Cambodia	9,630	1,213	12.6	71.7	870	13,441	2,348	17.5	72.2	1,696	6.07
Indonesia	182,474	55,819	30.6	32.2	17,964	214,840	90,356	42.1	23.1	20,877	1.37
Lao People's Dem Republic	4,132	638	15.4	66.1	422	5,403	1,066	19.7	66.1	705	4.67
Malaysia	17,845	8,891	49.8	2.0	177	22,633	13,154	58.1	2.0	262	3.56
Myanmar	40,517	9,984	24.6	31.1	3,105	48,364	13,606	28.1	26.4	3,596	1.34
Philippines	61,040	29,774	48.8	54.9	16,346	77,131	45,812	59.4	44.1	20,183	1.92
Singapore	3,016	3,016	100.0	0.0	-	4,108	4,108	100.0	0.0	-	
Thailand	54,736	10,244	18.7	19.5	1,998	63,584	12,709	20.0	2.0	253	-18.79
Timor-Leste	740	58	7.8	2.0	1	750	56	7.5	12.0	7	16.00
Viet Nam	66,074	13,389	20.3	60.5	8,100	79,175	19,395	24.5	47.4	9,197	1.15
Western Asia	**132,946**	**83,229**	**62.6**	**26.4**	**22,006**	**175,322**	**115,241**	**65.7**	**25.7**	**29,658**	**2.71**
Bahrain	490	429	87.6	0.0	-	652	603	92.5	2.0	12	
Cyprus	681	442	64.9	0.0	-	790	555	70.2	0.0	-	
Iraq	17,271	12,027	69.6	56.7	6,825	23,584	15,907	67.4	56.7	9,026	2.54
Israel	4,514	4,074	90.3	2.0	81	6,172	5,666	91.8	2.0	113	3.00
Jordan	3,254	2,350	72.2	16.5	388	5,051	3,979	78.7	15.7	623	4.32
Kuwait	2,143	2,034	94.9	3.0	60	1,971	1,894	96.1	3.0	56	-0.65
Lebanon	2,713	2,284	84.2	50.0	1,142	3,556	3,203	90.1	50.0	1,602	3.07
Occupied Palestinian Territory	2,154	1,379	64.0	-	-	3,311	2,222	67.1	60.0	1,333	
Oman	1,785	1,109	62.1	60.5	671	2,622	2,006	76.5	60.5	1,214	5.39
Qatar	453	407	89.8	2.0	8	575	534	92.9	2.0	11	2.47
Saudi Arabia	15,400	12,046	78.2	19.8	2,385	21,028	18,229	86.7	19.8	3,609	3.77
Syrian Arab Republic	12,386	6,061	48.9	10.4	629	16,610	8,596	51.8	10.4	892	3.18
Turkey	56,098	34,324	61.2	23.3	7,997	67,632	44,755	66.2	17.9	8,011	0.02
United Arab Emirates	2,014	1,615	80.2	2.0	32	2,654	2,314	87.2	2.0	46	3.27
Yemen	11,590	2,648	22.8	67.5	1,787	19,114	4,778	25.0	65.1	3,110	5.03

TABLE 2 SLUM POPULATION PROJECTIONS, 1990-2020

	SLUM POPULATION (THOUSAND)						SLUM PROJECTION TAGET 11 (THOUSAND)						SCENARIOS 2020		
	1990	2001	2005	2010	2015	2020	1990	2001	2005	2010	2015	2020	No Change	Moderate Change 100 mill.	Reduce % by half 1990-2020
WORLD	**714,972**	**912,918**	**997,767**	**1,115,002**	**1,246,012**	**1,392,416**	**714,972**	**912,918**	**976,858**	**1,070,494**	**1,175,132**	**1,292,065**	**1,392,416**	**1,292,065**	**705,745**
Developed regions	**41,750**	**45,191**	**46,511**	**48,216**	**49,983**	**51,815**	**41,750**	**45,191**	**45,507**	**46,167**	**46,851**	**47,560**	**51,815**	**47,560**	**26,137**
EURASIA (Countries in CIS)	**18,929**	**18,714**	**18,637**	**18,541**	**18,445**	**18,350**	**18,929**	**18,714**	**18,228**	**17,725**	**17,225**	**16,727**	**18,350**	**16,727**	**9,039**
European countries in CIS	**9,208**	**8,878**	**8,761**	**8,617**	**8,475**	**8,336**	**9,208**	**8,878**	**8,568**	**8,234**	**7,906**	**7,583**	**8,336**	**7,583**	
Asian countries in CIS	**9,721**	**9,836**	**9,879**	**9,932**	**9,986**	**10,040**	**9,721**	**9,836**	**9,663**	**9,499**	**9,334**	**9,168**	**10,040**	**9,168**	
Developing regions	**654,294**	**849,013**	**933,376**	**1,050,714**	**1,182,803**	**1,331,498**	**654,294**	**849,013**	**913,874**	**1,009,026**	**1,116,140**	**1,236,719**	**1,331,498**	**1,236,719**	**670,570**
Northern Africa	**21,719**	**21,355**	**21,224**	**21,062**	**20,901**	**20,741**	**21,719**	**21,355**	**20,758**	**20,133**	**19,513**	**18,898**	**20,741**	**18,898**	**17,286**
Sub-Saharan Africa	**100,973**	**166,208**	**199,231**	**249,886**	**313,419**	**393,105**	**100,973**	**166,208**	**195,245**	**240,808**	**297,955**	**369,631**	**393,105**	**369,631**	**150,654**
Latin America and the Caribbean	**110,837**	**127,566**	**134,257**	**143,116**	**152,559**	**162,626**	**110,837**	**127,566**	**131,390**	**137,174**	**143,340**	**149,913**	**162,626**	**149,913**	**81,385**
Eastern Asia	**150,761**	**193,824**	**212,368**	**238,061**	**266,863**	**299,150**	**150,761**	**193,824**	**207,923**	**228,583**	**251,742**	**277,704**	**299,150**	**277,704**	**157,527**
Eastern Asia excluding China (optional)	**12,831**	**15,568**	**16,702**	**18,236**	**19,911**	**21,739**	**12,831**	**15,568**	**16,348**	**17,494**	**18,744**	**20,109**	**21,739**	**20,109**	
Southern Asia	**198,663**	**253,122**	**276,432**	**308,611**	**344,537**	**384,644**	**198,663**	**253,122**	**270,637**	**296,283**	**324,914**	**356,877**	**384,644**	**356,877**	**178,762**
South-Eastern Asia	**48,986**	**56,781**	**59,913**	**64,073**	**68,521**	**73,279**	**48,986**	**56,781**	**58,636**	**61,420**	**64,398**	**67,583**	**73,279**	**67,583**	**58,302**
Western Asia	**22,006**	**29,658**	**33,057**	**37,860**	**43,360**	**49,659**	**22,006**	**29,658**	**32,371**	**36,379**	**40,968**	**46,224**	**49,659**	**46,224**	**26,290**
Oceania	**350**	**499**	**568**	**668**	**786**	**924**	**350**	**499**	**557**	**643**	**744**	**863**	**924**	**863**	**363**
Developing regions															
Northern Africa	**21,719**	**21,355**	**21,224**	**21,062**	**20,901**	**20,741**	**21,719**	**21,355**	**20,758**	**20,133**	**19,513**	**18,898**	**20,741**	**18,898**	**10,513**
Algeria	1,508	2,101	2,370	2,755	3,204	3,725	1,508	2,101	2,321	2,649	3,030	3,474	3,725	3,474	1,888
Egypt	14,087	11,762	11,015	10,148	9,349	8,613	14,087	11,762	10,766	9,671	8,662	7,733	8,613	7,733	4,365
Libyan Arab Jamahiriya	1,242	1,674	1,867	2,138	2,450	2,806	1,242	1,674	1,828	2,055	2,314	2,612	2,806	2,612	1,422
Morocco	4,457	5,579	6,054	6,705	7,425	8,223	4,457	5,579	5,927	6,435	6,998	7,621	8,223	7,621	4,168
Tunisia	425	234	188	144	110	84	425	234	184	136	99	71	84	71	42
Western Sahara	-	5					-	5							
Sub-Saharan Africa	**100,973**	**166,208**	**199,231**	**249,886**	**313,419**	**393,105**	**100,973**	**166,208**	**195,245**	**240,808**	**297,955**	**369,631**	**393,105**	**369,631**	**199,245**
Angola	2,193	3,918	4,839	6,300	8,201	10,677	2,193	3,918	4,743	6,077	7,814	10,075	10,677	10,075	5,412
Benin	1,288	2,318	2,870	3,749	4,896	6,394	1,288	2,318	2,814	3,617	4,666	6,035	6,394	6,035	3,241
Botswana	311	466	540	650	781	939	311	466	529	625	740	879	939	879	476
Burkina Faso	987	1,528	1,791	2,185	2,665	3,250	987	1,528	1,755	2,104	2,529	3,047	3,250	3,047	1,647
Burundi	294	394	438	501	572	653	294	394	429	481	540	608	653	608	331
Cameroon	2,906	5,064	6,197	7,977	10,268	13,217	2,906	5,064	6,074	7,693	9,777	12,459	13,217	12,459	6,699
Cape Verde	106	193	240	314	412	540	106	193	235	303	393	510	540	510	274
Central African Rep	1,038	1,455	1,646	1,919	2,238	2,610	1,038	1,455	1,612	1,845	2,117	2,435	2,610	2,435	1,323
Chad	1,218	1,947	2,308	2,856	3,534	4,373	1,218	1,947	2,262	2,751	3,357	4,106	4,373	4,106	2,216
Comoros	91	151	181	228	287	361	91	151	177	220	273	340	361	340	183
Congo	1,050	1,852	2,276	2,945	3,810	4,930	1,050	1,852	2,231	2,840	3,629	4,650	4,930	4,650	2,499
Côte d'Ivoire	2,532	4,884	6,203	8,361	11,271	15,194	2,532	4,884	6,082	8,074	10,760	14,381	15,194	14,381	7,701
Dem Rep of the Congo	5,366	7,985	9,227	11,054	13,243	15,865	5,366	7,985	9,039	10,637	12,552	14,846	15,865	14,846	8,041
Equatorial Guinea	112	201	248	323	420	547	112	201	243	311	400	516	547	516	277
Eritrea	342	510	590	707	847	1,016	342	510	578	680	803	950	1,016	950	515
Ethiopia	5,984	10,159	12,315	15,665	19,926	25,347	5,984	10,159	12,070	15,102	18,960	23,866	25,347	23,866	12,847
Gabon	357	688	872	1,174	1,581	2,129	357	688	855	1,134	1,509	2,015	2,129	2,015	1,079
Gambia	155	280	348	455	596	781	155	280	341	439	568	737	781	737	396
Ghana	4,083	4,993	5,372	5,886	6,450	7,067	4,083	4,993	5,258	5,647	6,073	6,540	7,067	6,540	3,582
Guinea	1,145	1,672	1,918	2,278	2,705	3,213	1,145	1,672	1,879	2,192	2,563	3,003	3,213	3,003	1,628
Guinea-Bissau	210	371	456	591	765	990	210	371	447	570	728	934	990	934	502
Kenya	3,985	7,605	9,620	12,905	17,311	23,223	3,985	7,605	9,432	12,460	16,522	21,972	23,223	21,972	11,771
Lesotho	168	337	434	596	817	1,121	168	337	426	576	781	1,062	1,121	1,062	568
Liberia	632	788	853	943	1,043	1,153	632	788	835	905	983	1,068	1,153	1,068	584
Madagascar	2,562	4,603	5,696	7,434	9,703	12,664	2,562	4,603	5,583	7,172	9,246	11,953	12,664	11,953	6,419
Malawi	1,033	1,590	1,860	2,262	2,752	3,348	1,033	1,590	1,822	2,178	2,611	3,138	3,348	3,138	1,697
Mali	1,968	3,361	4,083	5,208	6,643	8,474	1,968	3,361	4,002	5,022	6,322	7,981	8,474	7,981	4,295
Mauritania	827	1,531	1,915	2,534	3,353	4,437	827	1,531	1,878	2,446	3,198	4,193	4,437	4,193	2,249
Mozambique	2,722	5,841	7,710	10,909	15,437	21,842	2,722	5,841	7,563	10,549	14,775	20,753	21,842	20,753	11,071
Namibia	155	213	239	276	318	368	155	213	234	265	301	343	368	343	186
Niger	1,191	2,277	2,882	3,869	5,194	6,972	1,191	2,277	2,826	3,736	4,957	6,597	6,972	6,597	3,534
Nigeria	24,096	41,595	46,272	55,732	66,026	76,749	24,096	41,595	48,507	57,422	67,037	76,943	76,749	76,943	38,900
Rwanda	296	437	504	601	718	857	296	437	493	579	681	802	857	802	435
Sao Tome & Principe	-	1					-	1							
Senegal	2,276	3,555	4,181	5,120	6,270	7,679	2,276	3,555	4,096	4,930	5,952	7,203	7,679	7,203	3,892
Seychelles	-	1					-	1							
Sierra Leone	1,107	1,642	1,895	2,266	2,711	3,243	1,107	1,642	1,856	2,181	2,569	3,034	3,243	3,034	1,644
Somalia	1,670	2,482	2,867	3,433	4,111	4,923	1,670	2,482	2,809	3,304	3,896	4,606	4,923	4,606	2,495
South Africa	8,207	8,376	8,439	8,517	8,597	8,677	8,207	8,376	8,254	8,147	8,039	7,930	8,677	7,930	4,398
Sudan	5,708	10,107	12,441	16,131	20,915	27,118	5,708	10,107	12,195	15,560	19,923	25,580	27,118	25,580	13,745
Togo	796	1,273	1,510	1,870	2,315	2,866	796	1,273	1,480	1,801	2,199	2,691	2,866	2,691	1,452
Uganda	1,806	3,241	4,010	5,231	6,825	8,904	1,806	3,241	3,931	5,047	6,503	8,403	8,904	8,403	4,513
U. Rep of Tanzania	5,601	11,031	14,113	19,205	26,133	35,561	5,601	11,031	13,840	18,551	24,962	33,685	35,561	33,685	18,024
Zambia	2,284	3,136	3,519	4,065	4,695	5,423	2,284	3,136	3,446	3,907	4,439	5,053	5,423	5,053	2,749
Zimbabwe	116	157	176	202	232	266	116	157	172	194	219	247	266	247	135
Latin America and the Caribbean	**110,837**	**127,566**	**134,257**	**143,116**	**152,559**	**162,626**	**110,837**	**127,566**	**131,390**	**137,174**	**143,340**	**149,913**	**162,626**	**149,913**	**82,427**
Anguilla	3	5	6	7	8	10	3	5	6	7	8	9	10	9	5
Antigua and Barbuda	2	2	2	2	2	2	2	2	2	2	2	2	2	2	1
Argentina	8,597	10,964	11,978	13,379	14,943	16,690	8,597	10,964	11,727	12,844	14,092	15,486	16,690	15,486	8,459
Aruba	1	1	1	2	2	2	1	1	1	1	2	2	2	2	1
Bahamas	4	5	6	7	8	8	4	5	6	6	7	8	8	8	4
Barbados	1	1	1	2	2	2	1	1	1	1	2	2	2	2	1
Belize	48	69	78	92	108	127	48	69	77	88	102	119	127	119	64
Bolivia	2,555	3,284	3,597	4,032	4,519	5,064	2,555	3,284	3,522	3,871	4,263	4,701	5,064	4,701	2,567
Brazil	49,806	51,676	52,374	53,259	54,159	55,074	49,806	51,676	51,234	50,958	50,677	50,392	55,074	50,392	27,914

	SLUM POPULATION (THOUSAND)						SLUM PROJECTION TAGET 11 (THOUSAND)						SCENARIOS		
	1990	2001	2005	2010	2015	2020	1990	2001	2005	2010	2015	2020	No Change	Moderate Change 100 mill.	Reduce % by half 1990-2020
British Virgin Islands	0	0	1	1	1	1	0	0	1	1	1	1	1	1	1
Cayman Islands	1	1	1	1	1	2	1	1	1	1	1	2	2	2	1
Chile	432	1,143	1,628	2,534	3,943	6,136	432	1,143	1,598	2,456	3,791	5,868	6,136	5,868	3,110
Colombia	6,239	7,057	7,381	7,806	8,256	8,732	6,239	7,057	7,223	7,480	7,752	8,039	8,732	8,039	4,426
Costa Rica	195	313	372	461	572	710	195	313	364	444	544	667	710	667	360
Cuba	156	169	174	180	187	194	156	169	170	173	175	178	194	178	98
Dominica	8	7	7	6	6	6	8	7	7	6	6	5	6	5	3
Dominican Republic	2,327	2,111	2,038	1,950	1,865	1,785	2,327	2,111	1,992	1,861	1,735	1,615	1,785	1,615	905
Ecuador	1,588	2,095	2,317	2,629	2,982	3,382	1,588	2,095	2,269	2,525	2,815	3,144	3,382	3,144	1,714
El Salvador	1,126	1,386	1,495	1,644	1,807	1,986	1,126	1,386	1,464	1,577	1,702	1,839	1,986	1,839	1,006
French Guiana	11	16	19	23	27	32	11	16	19	22	26	30	32	30	16
Greenland	8	9	9	9	9	9	8	9	8	8	8	8	9	8	5
Grenada	2	2	3	3	3	3	2	2	3	3	3	3	3	3	2
Guadeloupe	27	30	31	33	34	36	27	30	30	31	32	33	36	33	18
Guatemala	2,192	2,884	3,186	3,609	4,089	4,632	2,192	2,884	3,120	3,467	3,860	4,305	4,632	4,305	2,348
Guyana	12	14	15	16	17	18	12	14	14	15	16	16	18	16	9
Haiti	1,728	2,574	2,976	3,568	4,277	5,128	1,728	2,574	2,916	3,434	4,054	4,799	5,128	4,799	2,599
Honduras	488	638	703	793	896	1,012	488	638	688	762	846	940	1,012	940	513
Jamaica	356	525	604	721	860	1,026	356	525	592	693	815	959	1,026	959	520
Martinique	6	7	8	8	8	9	6	7	7	8	8	8	9	8	5
Mexico	13,923	14,692	14,983	15,353	15,733	16,123	13,923	14,692	14,657	14,694	14,732	14,771	16,123	14,771	8,172
Netherlands Antilles	1	2	2	2	2	2	1	2	2	2	2	2	2	2	1
Nicaragua	1,638	2,382	2,730	3,237	3,837	4,550	1,638	2,382	2,674	3,114	3,635	4,253	4,550	4,253	2,306
Panama	397	505	552	615	687	766	397	505	540	591	648	711	766	711	388
Paraguay	756	797	812	832	852	873	756	797	795	796	798	800	873	800	443
Peru	8,979	12,993	14,862	17,581	20,796	24,601	8,979	12,993	14,558	16,911	19,695	22,988	24,601	22,988	12,469
Puerto Rico	50	59	63	68	74	80	50	59	62	66	70	74	80	74	41
Saint Kitts and Nevis	1	1	1	1	1	1	1	1	1	1	1	1	1	1	0
Saint Lucia	6	7	7	8	8	9	6	7	7	7	8	8	9	8	4
St Vincent & the Grenadines	2	3	4	4	5	6	2	3	4	4	5	6	6	6	3
Saint-Pierre-et-Miquelon	1	1					1	1							
Suriname	18	22	23	25	27	29	18	22	23	24	26	27	29	27	15
Trinidad and Tobago	292	310	317	326	335	344	292	310	310	312	314	315	344	315	174
Uruguay	191	62	41	24	15	9	191	62	40	23	13	6	9	6	4
US Virgin Islands	1	4	7	13	25	49	1	4	7	13	24	47	49	47	25
Venezuela	6,664	8,738	9,642	10,906	12,336	13,952	6,664	8,738	9,441	10,475	11,645	12,967	13,952	12,967	7,072
Eastern Asia	**150,761**	**193,824**	**212,368**	**238,061**	**266,863**	**299,150**	**150,761**	**193,824**	**207,923**	**228,583**	**251,742**	**277,704**	**299,150**	**277,704**	**151,624**
China	137,929	178,256	195,682	219,878	247,066	277,616	137,929	178,256	191,590	211,141	233,109	257,793	277,616	257,793	140,709
Hong Kong SAR of China	113	139	149	163	179	196	113	139	146	156	168	181	196	181	99
Macao SAR of China	7	9	9	10	11	12	7	9	9	10	11	11	12	11	6
Korea, Dem People's Rep of	117	95	88	80	73	67	117	95	86	77	68	60	67	60	34
Korea, Rep of	11,728	14,385	15,494	17,002	18,655	20,470	11,728	14,385	15,167	16,313	17,569	18,948	20,470	18,948	10,779
Mongolia	866	940	969	1,006	1,044	1,084	866	940	948	963	979	995	1,084	995	571
Southern Asia	**198,663**	**253,122**	**276,432**	**308,611**	**344,537**	**384,644**	**198,663**	**253,122**	**270,637**	**296,283**	**324,914**	**356,877**	**384,644**	**356,877**	**194,957**
Afghanistan	2,458	4,945	6,375	8,760	12,036	16,536	2,458	4,945	6,252	8,464	11,502	15,676	16,536	15,676	8,381
Bangladesh	18,988	30,403	36,079	44,687	55,348	68,553	18,988	30,403	35,353	43,047	52,576	64,378	68,553	64,378	34,746
Bhutan	61	70	73	78	83	88	61	70	72	75	78	81	88	81	45
India	131,174	158,418	169,671	184,868	201,425	219,466	131,174	158,418	166,079	177,332	189,592	202,950	219,466	202,950	111,236
Iran (Islamic Republic of)	17,094	20,406	21,763	23,587	25,564	27,707	17,094	20,406	21,301	22,621	24,052	25,603	27,707	25,603	14,043
Nepal	1,574	2,656	3,213	4,077	5,172	6,562	1,574	2,656	3,149	3,930	4,920	6,177	6,562	6,177	3,326
Pakistan	26,416	35,627	39,722	45,507	52,136	59,730	26,416	35,627	38,897	43,728	49,262	55,602	59,730	55,602	30,274
Sri Lanka	899	597	515	428	355	295	899	597	503	406	325	258	295	258	149
South-Eastern Asia	**48,986**	**56,781**	**59,913**	**64,073**	**68,521**	**73,279**	**48,986**	**56,781**	**58,636**	**61,420**	**64,398**	**67,583**	**73,279**	**67,583**	**37,141**
Brunei Darussalam	3	5	6	7	8	9	3	5	5	6	7	9	9	9	5
Cambodia	870	1,696	2,162	2,929	3,968	5,375	870	1,696	2,120	2,829	3,789	5,089	5,375	5,089	2,724
Indonesia	17,964	20,877	22,049	23,608	25,277	27,064	17,964	20,877	21,579	22,632	23,759	24,965	27,064	24,965	13,718
Lao People's Dem Republic	422	705	850	1,073	1,355	1,711	422	705	833	1,034	1,289	1,610	1,711	1,610	867
Malaysia	177	262	302	361	431	515	177	262	296	347	408	482	515	482	261
Myanmar	3,105	3,596	3,794	4,056	4,336	4,635	3,105	3,596	3,713	3,888	4,075	4,275	4,635	4,275	2,349
Philippines	16,346	20,183	21,792	23,984	26,397	29,053	16,346	20,183	21,333	23,015	24,866	26,904	29,053	26,904	14,725
Thailand	1,998	253	119	47	18	7	1,998	253	115	42	13	2	7	2	4
Timor-Leste	1	7	13	28	63	140	1	7	13	28	61	136	140	136	71
Viet Nam	8,100	9,197	9,632	10,204	10,811	11,453	8,100	9,197	9,426	9,779	10,152	10,548	11,453	10,548	5,805
Western Asia	**22,006**	**29,658**	**33,057**	**37,860**	**43,360**	**49,659**	**22,006**	**29,658**	**32,371**	**36,379**	**40,968**	**46,224**	**49,659**	**46,224**	**25,169**
Bahrain	-	12					-	12							
Iraq	6,825	9,026	9,992	11,346	12,884	14,630	6,825	9,026	9,784	10,899	12,166	13,604	14,630	13,604	7,415
Israel	81	113	127	148	172	199	81	113	124	142	162	186	199	186	101
Jordan	388	623	741	920	1,141	1,416	388	623	726	886	1,084	1,330	1,416	1,330	718
Kuwait	60	56	55	53	51	50	60	56	54	51	48	45	50	45	25
Lebanon	1,142	1,602	1,811	2,112	2,463	2,872	1,142	1,602	1,774	2,031	2,330	2,679	2,872	2,679	1,456
Occupied Palestinian Territory	-	1,333					-	1,333							
Oman	671	1,214	1,506	1,972	2,581	3,379	671	1,214	1,476	1,902	2,460	3,190	3,379	3,190	1,713
Qatar	8	11	12	13	15	17	8	11	11	13	14	16	17	16	9
Saudi Arabia	2,385	3,609	4,196	5,066	6,115	7,382	2,385	3,609	4,111	4,876	5,799	6,914	7,382	6,914	3,742
Syrian Arab Republic	629	892	1,012	1,187	1,391	1,630	629	892	992	1,141	1,316	1,522	1,630	1,522	826
Turkey	7,997	8,011	8,016	8,022	8,029	8,035	7,997	8,011	7,841	7,671	7,501	7,332	8,035	7,332	4,072
United Arab Emirates	32	46	52	62	73	86	32	46	51	59	69	80	86	80	43
Yemen	1,787	3,110	3,803	4,892	6,292	8,092	1,787	3,110	3,728	4,717	5,990	7,628	8,092	7,628	4,102

TABLE 3 POPULATION OF SLUM AREAS AT MID-YEAR BY SHELTER DEPRIVATION, BY REGION AND COUNTRY 1990 AND 2001

	POPULATION OF SLUM AREAS AT MID-YEAR, 1990						POPULATION OF SLUM AREAS AT MID-YEAR, 2001					
			Percentage of slum by number of shelter deprivation						Percentage of slum by number of shelter deprivation			
	Percentage Urban	Percentage Slum	One Shelter Deprivation	Two Shelter Deprivation	Three Shelter Deprivation	Four Shelter Deprivation	Percentage Urban	Percentage Slum	One Shelter Deprivation	Two Shelter Deprivation	Three Shelter Deprivation	Four Shelter Deprivation
Northern Africa	48.7	37.7	30.1	4.9	2.2	0.5	52.0	28.2	25.0	3.0	0.2	-
Sub-Saharan Africa	27.9	72.3	32.1	27.0	10.8	2.4	34.6	71.9	35.5	24.0	10.6	1.9
Latin America and the Caribbean	71.1	35.4	23.3	8.9	2.5	0.7	75.8	31.9	21.1	8.0	2.4	0.4
Eastern Asia	29.9	41.1					39.1	36.4				
Southern Asia	26.6	63.7	38.8	20.7	4.1	0.1	29.6	59.0	38.8	17.5	2.7	-
South-Eastern Asia	30.2	36.8	17.8	12.0	5.4	1.6	38.3	28.0	20.7	5.7	1.4	0.2
Western Asia	62.6	26.4	19.8	4.8	1.6	0.3	65.7	25.7	19.7	4.0	1.6	0.3
Oceania	23.6	24.5					26.7	24.1				
Northern Africa												
Egypt	43.6	57.5	45.6	7.0	4.0	0.9	42.7	39.9	35.4	4.5	0.1	-
Morocco	48.4	37.4	30.3	5.8	1.0	0.3	56.1	32.7	29.0	3.0	0.5	0.1
Sub-Saharan Africa												
Benin	34.5	80.3	36.3	24.1	15.9	3.9	43.0	83.6	46.6	24.1	12.9	-
Burkina Faso	13.6	80.9	37.6	28.6	13.7	1.1	16.9	76.5	51.1	22.9	2.1	0.4
Cameroon	40.3	62.1	30.0	22.3	8.8	1.0	49.7	67.0	37.7	16.7	11.2	1.5
Central African Rep	37.5	94.0	16.2	36.6	37.3	3.9	41.7	92.4	15.9	36.0	36.7	3.8
Chad	21.0	99.3	10.3	25.6	45.9	17.5	24.1	99.1	10.3	25.5	45.9	17.5
Comoros	27.9	61.7	29.3	21.8	9.3	1.3	33.8	61.2	29.1	21.7	9.2	1.3
Côte d'Ivoire	39.9	50.5	32.1	13.3	4.4	0.7	44.0	67.9	39.4	25.0	3.8	-
Ethiopia	12.7	99.0	20.9	42.6	29.8	5.7	15.9	99.4	21.0	42.8	29.9	5.7
Gabon	68.1	56.1	39.2	13.2	3.5	0.2	82.3	66.2	46.3	15.6	4.1	0.3
Ghana	33.5	80.4	46.7	27.9	5.6	0.2	36.4	69.6	43.5	21.2	4.8	0.1
Guinea	23.4	79.6	34.9	34.1	9.4	1.2	27.9	72.3	31.7	31.0	8.5	1.1
Kenya	24.0	70.4	38.2	24.3	6.2	1.6	34.4	70.7	41.6	20.1	9.0	
Madagascar	23.6	90.9	20.8	26.4	27.0	16.7	30.1	92.9	21.2	27.0	27.6	17.1
Malawi	11.6	94.6	57.1	33.0	4.5		15.1	91.1	55.0	31.8	4.4	-
Mali	23.8	94.1	37.7	56.4			30.9	93.2	40.8	34.4	18.0	-
Mozambique	21.1	94.5	36.2	36.5	16.4	5.4	33.3	94.1	36.2	30.8	22.9	4.2
Namibia	26.6	42.3	9.3	18.1	13.5	1.5	31.4	37.9	8.3	16.2	12.1	1.3
Niger	16.1	96.0	30.8	43.0	19.2	3.1	21.1	96.2	30.8	43.1	19.2	3.1
Nigeria	35.0	80.0	31.4	35.1	11.3	2.3	44.9	79.2	36.8	30.7	9.9	1.7
Rwanda	5.3	82.2	29.0	26.9	21.9	4.5	6.3	87.9	45.7	30.4	12.3	-
Senegal	40.0	77.6	60.7	14.7	2.2		48.2	76.4	63.7	11.2	1.5	-
South Africa	48.8	46.2	34.1	10.3	1.6	0.1	57.7	33.2	24.5	7.4	1.2	0.1
Togo	28.5	80.9	55.7	21.4	3.3	0.6	33.9	80.6	55.5	21.3	3.3	0.5
U. Rep of Tanzania	21.7	99.1	28.5	44.4	22.4	3.8	33.3	92.1	54.0	24.6	12.1	1.3
Zambia	39.4	72.0	32.1	22.8	14.0	3.2	39.8	74.0	46.5	20.8	6.4	0.2
Zimbabwe	28.4	4.0	3.3	0.5	0.1	0.1	36.0	3.4	3.2	0.2	-	-
Latin America and the Caribbean												
Bolivia	55.6	70.0	45.0	24.9			62.9	61.3	26.3	20.4	11.3	3.4
Brazil	74.8	45.0	32.0	11.0	1.9	0.2	81.7	36.6	26.0	8.9	1.5	0.1
Colombia	68.7	26.0	19.9	4.4	1.3	0.4	75.5	21.8	18.2	3.0	0.5	0.1
Dominican Republic	58.4	56.4	42.4	12.2	1.7	0.1	66.0	37.6	28.2	8.2	1.1	0.1
Guatemala	38.1	65.8	33.1	20.2	11.1	1.3	39.9	61.8	31.1	19.0	10.4	1.2
Haiti	29.5	84.9	67.0	15.5	2.4		36.3	85.7	67.6	15.7	2.4	-
Nicaragua	53.1	80.7	23.8	23.7	21.1	12.1	56.5	80.9	33.7	29.1	17.9	0.3
Peru	68.9	60.4	20.9	19.3	14.1	6.1	73.1	68.1	32.6	20.3	11.9	3.3
Asia												
Bangladesh	19.8	87.3	36.7	32.4	18.0	0.3	25.6	84.7	28.1	37.5	18.9	0.2
India	25.5	60.8	37.8	20.1	2.9		27.9	55.5	39.5	15.2	0.8	-
Indonesia	30.6	32.2	18.8	11.2	2.1		42.1	23.1	18.0	4.2	0.9	-
Nepal	9.0	96.9	19.8	35.5	34.6	7.1	12.2	92.4	40.2	37.4	14.8	-
Pakistan	30.6	78.7	56.0	20.6	2.1		33.4	73.6	52.4	19.2	2.0	-
Philippines	48.8	54.9	16.7	18.3	14.4	5.5	59.4	44.1	32.9	8.9	2.0	0.2
Turkey	61.2	23.3	18.7	3.7	0.8	0.1	66.2	17.9	15.5	2.0	0.3	0.1
Viet Nam	20.3	60.5	37.9	15.7	5.6	1.3	24.5	47.4	29.7	12.3	4.4	1.0
Yemen	22.8	67.5	34.3	18.6	12.2	2.4	25.0	65.1	33.1	18.0	11.7	2.3

TABLE 4 PROPORTION OF URBAN HOUSEHOLDS WITH FINISHED MAIN FLOOR MATERIALS, SUFFICIENT LIVING AREA, SUSTAINABLE ACCESS TO SAFE WATER AND IMPROVED SANITATION

		Finished main floor materials					Access to sufficient living area					Access to safe water source					Access to improved sanitation				
Country	City	1990	1993	1998	2000	2003	1990	1993	1998	2000	2003	1990	1993	1998	2000	2003	1990	1993	1998	2000	2003
Northern Africa		97.5	97.7	98.0	98.1	98.3	71.0	78.2	85.8	90.6	90.5	96.0	95.8	95.3	95.2	94.9	84.0	85.3	87.3	88.2	89.4
Sub-Saharan Africa		82.2	83.9	86.1	87.0	89.1	72.9	73.6	73.6	73.7	73.1	82.0	82.0	82.0	82.0	82.0	54.0	54.3	54.7	54.8	55.1
Latin America and the Caribbean						98.2					88.2	93.0	93.5	94.3	94.7	95.2	82.0	82.5	83.3	83.7	84.2
Eastern Asia						98.4					91.5	99.0	97.5	95.0	94.0	92.5	64.0	65.3	67.3	68.2	69.4
Southern Asia						84.8					65.0	90.0	91.0	92.7	93.3	94.3	54.0	57.0	62.0	64.0	67.0
South-Eastern Asia		91.9	92.3	93.0	93.3	93.6	66.0	67.9	71.0	73.2	73.1	91.0	91.0	91.0	91.0	91.0	67.0	70.0	75.0	77.0	80.0
Western Asia		96.7	96.8	96.7	96.8	96.4	91.3	92.2	93.5	94.2	91.1	94.0	94.3	94.7	94.8	95.1	96.0	95.8	95.3	95.2	94.9
Northern Africa																					
Algeria	Oran											98.0	98.0	98.0	98.0	98.0					
Algeria	Constantine											98.0	98.0	98.0	98.0	98.0					
Algeria	Blida											98.0	98.0	98.0	98.0	98.0					
Algeria	Sétif											98.0	98.0	98.0	98.0	92.0					
Algeria	Tébessa											98.0	98.0	98.0	98.0	92.0					
Algeria	Wargla											98.0	98.0	98.0	98.0	92.0					
Algeria	Midyah											98.0	98.0	98.0	98.0	92.0					
Egypt	Cairo	98.9	98.8	98.6	98.5	98.4	70.9	78.1	89.9	94.7	94.7	99.0	99.2	99.6	99.7	99.9	74.2	77.3	82.5	84.5	87.6
Egypt	Alexandria	97.7	97.7	97.9	97.9	98.0	70.4	77.7	89.9	94.7	94.7	99.6	99.7	99.7	99.7	99.8	79.5	82.6	87.7	89.7	92.8
Egypt	Port Said	97.2	97.6	98.4	98.6	99.1	75.5	81.8	92.4	96.7	96.7	97.7	97.5	97.1	97.0	96.7	89.1	90.7	93.3	94.4	96.0
Egypt	Suez	99.4	99.0	98.2	97.9	97.4	72.9	80.2	92.5	97.4	97.4	99.4	99.5	99.5	99.5	99.6	81.4	83.4	86.6	87.9	89.8
Egypt	Assyut	75.1	77.9	82.5	84.4	87.1			87.7	94.9	94.9	92.9	94.9	98.3	99.6	99.6	61.3	61.6	62.2	62.4	62.7
Egypt	Aswan	70.7	74.1	79.8	82.1	85.5			85.8	93.6	93.6	97.2	97.8	98.8	99.1	99.7	62.3	63.8	66.2	67.1	68.6
Egypt	Beni Suef	52.2	64.3	84.5	92.6	92.6			85.6	92.3	92.3	87.3	90.6	95.9	98.1	98.1	47.7	55.0	67.1	72.0	79.3
Morocco	Casablanca	98.7	99.0	99.4	99.6	99.9			69.2	74.7	82.8	100.0	100.0	99.9	99.8	99.7	71.5	77.4	87.1	91.1	96.9
Morocco	Rabat	99.1	99.1	99.1	98.7	98.1			79.3	82.4	87.0	100.0	99.9	99.7	99.7	99.6	86.3	89.1	93.9	95.9	98.8
Morocco	Fes					99.6					74.4					99.5					99.4
Morocco	Marrakech					99.5					80.4					98.1					99.7
Morocco	Tangier					99.8					85.0					94.7					98.4
Morocco	Meknès					95.4					74.8					98.8					97.0
Sub-Saharan Africa																					
Angola	Luanda					51.6					62.9					51.9					59.5
Benin	Djougou	61.6	66.4	74.5	77.7	82.5	81.0	81.0	81.0	81.0	81.0	85.0	84.2	82.7	82.2	81.3	47.5	43.0	35.6	32.6	28.1
Benin	Porto-Novo	70.4	74.9	82.5	85.5	90.0	80.2	80.2	80.2	80.2	80.2				69.6	78.5	54.3	46.1	32.5	27.0	18.8
Burkina Faso	Ouagadougou	90.2	91.7	94.2	95.2	96.7	85.5	85.4	85.2	85.2	85.0	77.5	81.4	87.8	90.4	94.3	50.9	50.6	50.1	49.9	49.6
Cameroon	Yaounde	92.1	92.2	92.6	92.7	92.9	84.8	86.8	90.2	91.5	93.5	83.3	83.7	84.5	84.8	85.2	11.8	37.8	81.2	98.6	98.6
Côte d'Ivoire	Abidjan	99.6	99.3	98.9	98.8	98.5	65.2	68.4	73.9	76.0	79.3	98.9	99.2	99.6	99.8	99.8	76.3	76.9	78.0	78.5	79.2
Dem. Rep. of the Congo	Kinshasa					86.9					46.5					85.5					78.2
Dem. Rep. of the Congo	Butembo					21.6					55.7					70.1					82.5
Ethiopia	Addis Ababa					66.7					64.5					98.4					48.1
Ethiopia	Nazret					22.7					61.2					85.7					34.9
Gambia	Banjul																				96.0
Ghana	Accra	99.9	99.7	99.3	99.2	99.0	76.9	77.9	79.4	80.0	80.9	99.5	99.5	98.1	97.5	96.7	58.8	62.7	69.2	71.8	75.7
Guinea	Conakry					98.7					72.4					93.7					42.5
Lesotho	Maseru					85.4					90.2					90.3					45.2
Mali	Bamako	100.0	98.0	83.1	77.1	68.1	77.1	77.1	77.1	77.1	77.1	87.4	88.0	88.9	89.2	89.8	38.9	41.3	45.2	46.7	49.1
Mozambique	Maputo					83.2					81.7					96.7					46.3
Nigeria	Lagos	88.2	90.9	95.3	97.1	99.8	60.9	60.9	61.1	61.1	61.2	94.9	94.9	94.9	92.3	88.5	100.0	95.8	85.3	81.1	74.8
Nigeria	Ibadan					96.9	95.1	95.1	82.5	77.5	69.9	90.6	90.6	90.6	90.6	62.6	26.8	26.8	26.8	26.8	67.3
Nigeria	Ogbomosho	99.1	98.0	96.2	95.5	94.4	66.7	70.9	77.9	80.7	84.9	92.0	87.6	80.3	77.4	73.0	45.3	43.5	40.5	39.3	37.5
Nigeria	Zaria	99.1	99.0	98.9	98.8	98.7	46.7	51.7	60.0	63.3	68.3	96.2	97.1	98.5	99.1	99.1	52.9	54.1	56.0	56.8	57.9
Nigeria	Akure	94.4	95.7	97.8	98.7	98.7	100.0	100.0	89.1	83.5	75.1			72.4	77.2	84.5	92.5	83.0	67.0	60.6	51.1
Rwanda	Kigali	62.7	65.5	70.0	71.8	74.6	86.7	86.7	86.7	86.7	86.7	72.7	75.4	80.0	81.8	84.5	47.8	55.1	67.2	72.1	79.4
Senegal	Dakar	97.8	98.7	98.7	98.7	98.7	69.2	70.0	71.4	71.9	72.8	97.1	96.0	94.3	93.6	92.6	59.0	63.9	72.2	75.6	80.5
South Africa	Johannesburg					99.0					90.9					98.3					90.5
South Africa	Cape Town					97.2					85.7					98.8					94.7
South Africa	Durban										85.1					72.3					
South Africa	Pretoria					99.0					90.9					98.3					90.5
South Africa	Port Elizabeth					57.8					79.9					61.4					
South Africa	West Rand										81.9					98.4					83.4
Uganda	Kampala	69.6	74.1	81.7	84.7	89.3	62.7	62.7	62.7	62.7	62.7			75.6	87.3	87.3	55.0	56.6	59.2	60.2	61.8
United Republic of Tanzania	Dar es Salaam	79.2	82.2	87.1	89.1	92.1	76.9	79.8	84.6	86.5	89.4	93.9	91.0	86.2	84.3	81.4	51.8	52.1	52.5	52.7	53.0
United Republic of Tanzania	Arusha	57.0	56.3	55.0	54.5	53.7	65.7	70.5	78.4	81.6	86.4	92.8	94.6	97.5	98.7	98.7	51.9	51.7	51.4	51.3	51.0
Zambia	Ndola	96.9	96.9	89.5	86.5	82.1	76.2	76.2	76.2	76.2	76.2	98.5	95.9	91.5	89.7	87.1	84.5	84.2	83.8	83.6	83.3
Zambia	Chingola	96.5	96.5	96.5	94.6	91.6	79.7	79.7	79.7	79.7	79.7	95.2	93.7	91.2	90.2	88.6	84.6	86.4	89.5	90.7	92.6
Zimbabwe	Harare	93.8	94.6	96.1	96.7	97.6	85.0	85.0	85.0	85.0	85.0	97.7	98.0	98.6	98.9	99.2	98.2	97.8	97.3	97.0	96.7

TABLE 4 PROPORTION OF URBAN HOUSEHOLDS WITH FINISHED MAIN FLOOR MATERIALS, SUFFICIENT LIVING AREA, SUSTAINABLE ACCESS TO SAFE WATER AND IMPROVED SANITATION

		Finished main floor materials					Access to sufficient living area					Access to safe water source					Access to improved sanitation				
Country	City	1990	1993	1998	2000	2003	1990	1993	1998	2000	2003	1990	1993	1998	2000	2003	1990	1993	1998	2000	2003
Latin America and the Caribbean																					
Brazil	São Paulo					99.6					83.3	97.3	97.5	97.7	97.8	98.0	83.7	84.8	86.7	87.4	88.5
Brazil	Rio de Janeiro					99.6					89.7	95.6	96.0	96.6	96.9	97.3	67.5	70.2	74.6	76.3	79.0
Brazil	Belo Horizonte					97.6					91.2	93.6	94.7	96.6	97.3	98.5	83.3	85.7	89.8	91.4	93.9
Brazil	Fortaleza					95.3					90.5	68.9	74.6	84.2	88.0	93.7			36.8	43.8	54.2
Brazil	Curitiba					96.8					96.1	94.9	95.9	97.6	98.2	99.2	55.4	61.6	71.8	75.9	82.0
Brazil	Brasilia					99.6					88.7	85.1	87.7	92.0	93.8	96.4	72.1	75.3	80.5	82.6	85.7
Brazil	Goiânia					99.1					92.0	85.7	88.7	93.8	95.8	98.8	74.1	73.8	73.3	73.0	72.7
Brazil	São José dos Campos											97.3	97.4	97.5	97.5	97.6	85.1	86.4	88.6	89.5	90.8
Brazil	Nova Iguaçu															91.1					50.6
Brazil	Ribeirão Preto											97.7	97.8	97.8	97.9	97.9	92.9	93.5	94.5	94.9	95.5
Brazil	Vitoria					97.1					87.9					96.7					89.3
Brazil	Guarujá											91.7	92.8	94.8	95.5	96.7	72.5	71.9	70.9	70.6	70.0
Brazil	Rondonópolis											78.6	81.1	85.3	86.9	89.4	16.6	20.1	25.8	28.1	31.6
Chile	Santiago																86.6	87.9	90.0	90.9	92.2
Chile	Chillan																81.6	85.3	91.5	94.0	97.7
Colombia	Bogotá					95.5					90.8					100.0					100.0
Colombia	Medellín					99.9					93.8					100.0					99.7
Colombia	Neiva					96.9					91.2					100.0					99.5
Colombia	Valledupar					99.6					82.0					99.6					99.8
Ecuador	Guayaquil											95.9	96.2	96.6	96.8	97.1	51.6	51.0	50.1	49.7	49.1
Guatemala	Guatemala City					80.0					71.6	82.6	83.8	85.7	86.5	87.7	32.7	44.7	64.6	72.6	84.6
Mexico	Mexico											91.2	92.5	94.7	95.6	97.0	80.7	83.2	87.4	89.1	91.6
Mexico	Guadalajara											96.7	97.2	98.1	98.5	99.0	96.4	97.1	98.2	98.7	99.3
Mexico	Tijuana											67.8	74.8	86.5	91.2	98.2	57.0	63.2	73.5	77.6	83.8
Mexico	León											88.3	89.4	91.3	92.0	93.1	83.4	84.6	86.4	87.2	88.3
Mexico	Culiacán											91.8	93.3	95.8	96.8	98.2	70.6	75.3	83.1	86.3	91.0
Mexico	Hermosillo											94.2	94.2	94.2	94.3	94.3	73.3	77.2	83.7	86.3	90.2
Mexico	Villahermosa											88.5	90.2	93.0	94.1	95.8	81.7	82.4	83.5	83.9	84.6
Uruguay	Montevideo											90.1	91.6	94.2	95.2	96.7	94.5				
Venezuela	Caracas											85.4	87.3	90.5	91.8	93.7	88.0	95.2	96.4	96.9	97.6
Venezuela	Maracaibo											82.0	84.6	88.8	90.5	93.1	88.0	88.0	88.1	88.1	88.1
Venezuela	Valencia																	90.3	94.3	95.8	98.2
Eastern Asia																					
China	Shanghai										92.3										
China	Beijing										92.3										
China	Guangzhou										92.2										
China	Harbin										92.2										
China	Zhengzhou										92.3										
China	Lanzhou										92.4										
China	Xuzhou										92.3										
China	Yulin										92.3										
China	Yiyang										92.3										
China	Yueyang										92.3										
China	Datong										92.3										
China	Leshan										92.3										
China	Yongzhou										92.3										
China	Chifeng										92.2										
China	Huaibei										92.5										
China	Hegang										92.2										
China	Dandong										92.2										
China	Dezhou										92.4										
China	Anqing										92.3										
China	Shaoguan										92.3										
China	Changzhi										92.4										
Mongolia	Ulan Bator					98.4					44.1					97.0					75.3
South-Central Asia																					
Bangladesh	Dhaka					71.0					60.2					99.5					90.4
Bangladesh	Rajshahi					42.9					55.5					99.1					73.8
India	Mumbai										59.0	96.1	97.5	99.7	99.7	99.7	77.4	85.0	97.8	97.8	97.8
India	Kolkota										73.0	97.2	97.8	98.6	99.0	99.5	86.5	88.3	91.3	92.5	94.2
India	Delhi										73.3	99.6	99.5	99.2	99.1	99.0	81.1	85.8	93.7	96.8	99.0
India	Hyderabad										78.9	97.4	98.4	98.4	98.4	98.4	77.3	80.3	85.1	87.1	90.0
India	Pune (Poona)										68.9	99.8	99.2	98.4	98.0	97.5	79.0	77.7	75.4	74.4	73.1
India	Kanpur										64.8	91.7	94.8	94.8	94.8	94.8	61.7	69.1	81.4	86.3	93.7
India	Jaipur										78.5	99.4	99.6	99.6	99.6	99.6	96.6	95.3	93.0	92.1	90.7
India	Coimbatore										78.6	90.4	92.9	97.0	98.7	98.7	81.0	84.1	89.4	91.5	94.7
India	Kochi (Cochin)										93.5	90.3	92.5	96.1	97.5	99.7	91.2	93.4	97.1	98.5	98.5
India	Vijayawada										80.5	95.3	96.2	97.8	98.4	99.3	70.5	71.3	72.7	73.2	74.0

Country	City	Finished main floor materials					Access to sufficient living area					Access to safe water source					Access to improved sanitation				
		1990	1993	1998	2000	2003	1990	1993	1998	2000	2003	1990	1993	1998	2000	2003	1990	1993	1998	2000	2003
India	Amritsar										71.1						95.8	95.2	94.3	94.0	93.4
India	Srinagar										77.1	96.4	97.0	98.1	98.5	99.2	68.9	74.5	83.9	87.6	93.2
India	Jodhpur										77.9	99.2	99.0	98.6	98.5	98.2	77.4	82.5	91.0	94.4	99.5
India	Akola										65.8	88.0	89.9	93.1	94.4	96.3	53.5	60.1	71.0	75.4	82.0
India	Rajahmundry										80.2	83.6	88.4	96.4	99.6	99.6	45.7	49.3	55.2	57.6	61.2
India	Yamunanagar										74.2	96.9	98.1	98.1	98.1	98.1	70.5	74.9	82.4	85.3	89.8
India	Kharagpur										68.2	89.5	92.2	96.7	98.5	98.5	64.2	74.8	92.4	99.4	99.4
India	Hisar										69.5	91.1	94.3	99.7	99.7	99.7			78.2	85.8	97.1
India	Jalna										94.8					99.6					95.4
India	Karnal										82.9					99.8					87.4
India	Agartala										72.0	79.2	86.9	99.6	99.6	99.6	78.4	83.8	92.8	96.4	96.4
India	Gadag-Betigeri										74.9	97.0	98.0	99.8	99.8	99.8	73.7	76.6	81.4	83.4	86.3
India	Krishnanagar										73.3	94.0	95.1	97.0	97.8	98.9	70.7	74.6	81.1	83.7	87.6
Kazakhstan	Shimkent					37.4										82.2					80.2
Kazakhstan	Zhezkazgan					43.0										100.0					99.8
Pakistan	Karachi					99.6					42.3					96.6					90.0
Pakistan	Faisalabad					98.6					39.5					98.1					87.2
Pakistan	Islamabad					98.9					49.1					94.1					70.3
Tajikistan	Dushanbe					94.2					90.4					99.7					89.1
Uzbekistan	Tashkent					99.7					97.1					100.0					90.7
South-Eastern Asia																					
Cambodia	Phnom Penh					96.9										81.2					95.4
Cambodia	Siem Reab					96.9										57.9					45.8
Indonesia	Jakarta	97.8	98.3	99.0	99.4	99.5						99.0	99.0	99.0	99.0	99.0	94.9	95.8	97.2	97.8	96.7
Indonesia	Bandung	99.0	99.0	99.0	98.6	98.0						99.9	99.9	94.7	92.7	89.6	99.9	96.6	91.2	89.0	85.7
Indonesia	Surabaja	98.8	98.8	96.8	95.4	93.2						97.8	97.8	92.9	90.9	87.9	89.0	89.0	89.0	81.2	69.6
Indonesia	Medan	99.8	99.8	99.8	99.8	99.8						95.3	96.2	97.8	98.4	99.4	98.8	98.0	95.4	92.4	88.0
Indonesia	Palembang	95.7	96.5	97.8	98.4	99.1						90.7	92.3	95.1	96.2	97.8	97.8	97.8	95.3	93.2	90.2
Indonesia	Ujung Pandang	94.5	95.9	98.1	99.0	99.8						99.4	99.4	99.4	99.4	99.4	76.9	83.2	93.6	97.8	99.3
Indonesia	Bogor	95.4	95.9	96.7	97.0	97.5						93.9	94.0	94.3	94.5	94.7	62.8	71.2	85.2	90.8	99.1
Indonesia	Surakarta	88.7	89.1	89.7	90.0	90.4						98.1	99.8	97.5	96.6	95.3	96.0	96.0	88.0	81.4	71.3
Indonesia	Pekan Baru	99.9	99.9	99.9	99.8	99.7						97.2	97.2	97.1	96.2	94.9	99.5	99.5	95.1	92.2	87.7
Indonesia	Denpasar	97.4	97.7	98.1	98.3	98.5						98.3	98.3	96.9	95.4	93.2	98.3	98.3	97.4	96.2	94.6
Indonesia	Jambi	99.9	99.3	98.3	98.0	97.4						97.8	98.3	99.2	99.5	99.5	97.8	97.8	93.6	90.3	85.3
Indonesia	Purwokerto	75.7	77.2	79.6	80.5	82.0						88.9	88.9	86.1	82.5	77.0	53.9	58.7	66.8	70.1	75.0
Indonesia	Kediri	79.4	82.9	88.8	91.2	94.7						92.8	94.6	97.7	98.9	99.0	49.0	55.9	67.3	71.9	78.7
Indonesia	Palu	97.4	97.8	98.5	98.7	99.2						98.6	98.5	98.3	98.2	98.1	78.7	81.5	86.1	88.0	90.8
Indonesia	Bitung	89.7	91.3	94.0	95.1	96.8						73.6	79.2	88.4	92.1	97.7	83.2	85.3	88.9	90.3	92.4
Indonesia	Jaya Pura	86.3	90.6	97.8	97.8	97.8						47.1	67.2	94.1	94.1	94.1	82.7	85.1	89.2	90.8	93.2
Indonesia	Dumai	98.5	98.8	99.3	99.5	99.8						77.4	82.4	90.8	94.2	99.3	69.6	74.7	83.2	86.6	91.7
Myanmar	Yangon					93.0					44.8					95.3					81.4
Philippines	Metro Manila	78.4	78.8	79.4	79.6	79.9	65.8	67.8	71.0	72.3	74.3	83.8	86.8	91.9	93.9	96.9	85.6	88.9	94.2	96.4	99.6
Philippines	Cebu	51.1	54.1	59.0	61.0	64.0	63.1	66.5	72.2	74.5	77.9	66.5	73.8	86.0	90.9	98.3	79.0	81.0	84.5	85.9	87.9
Philippines	Cagayan de Oro	70.9	71.6	72.8	73.3	74.0	70.9	70.2	69.1	68.7	68.0	85.9	87.1	89.1	89.9	91.2	52.9	65.0	85.1	93.2	93.2
Philippines	Bacolod	41.8	46.3	53.9	56.9	61.4	69.6	70.1	70.8	71.1	71.6	84.2	84.2	84.2	91.3	91.3	56.0	62.3	72.8	77.0	83.2
Viet Nam	Ho Chi Minh City	99.2	99.4	99.7	99.9	99.9				70.5	77.4	99.9	99.9	99.2	99.0	98.6	89.6	91.5	94.5	95.7	97.6
Viet Nam	Ha Noi	95.1	96.3	98.4	99.2	99.2				75.7	82.2	96.2	97.1	98.7	99.4	99.4	58.4	67.8	83.3	89.6	98.9
Viet Nam	Hai Phong	98.1	98.0	97.9	97.8	97.8				84.3	95.9	98.0	98.5	99.0	99.2	99.0	1.0	1.0	69.3	81.7	100.0
Viet Nam	Da Nang	93.5	95.1	97.8	98.9	98.9				64.4	74.2	84.7	88.2	94.2	96.6	96.6	76.9	82.7	92.3	96.1	100.0
Western Asia																					
Armenia	Yerevan					98.9										99.4					93.6
Azerbaijan	Baku					99.4					88.2					91.6					85.3
Iraq	Baghdad										93.3					99.4					98.1
Iraq	Mosul										87.8					99.8					98.0
Iraq	Amara										88.8					93.1					88.8
Syrian Arab Republic	Damascus															99.7					99.1
Turkey	Istanbul	99.8	99.8	99.7	99.7	99.7	92.8	93.6	95.1	95.6	96.5	98.7	95.5	90.0	87.8	84.5	99.2	99.2	99.2	99.2	99.2
Turkey	Ankara	99.8	99.8	99.7	99.7	99.7	91.3	93.6	97.4	98.9	98.9	99.8	99.8	97.4	96.4	95.0	98.9	99.1	99.5	99.6	99.8
Turkey	Izmir	99.2	99.4	99.7	99.8	99.8	96.1	95.8	95.2	95.0	94.6	98.6	99.1	99.1	99.1	99.1	99.0	99.2	99.3	99.4	99.5
Turkey	Bursa	97.1	98.2	98.3	98.4	98.4	96.3	96.3	96.3	96.3	96.3	92.0	92.0	92.0	88.8	84.0	98.1	98.1	98.1	97.4	96.3
Turkey	Adana	99.8	99.8	99.8	99.8	99.8	80.5	83.6	88.7	90.8	93.9	97.7	98.6	98.6	98.6	98.6	99.9	98.8	97.1	96.3	95.3
Turkey	Gaziantep	99.8	99.8	99.8	99.8	99.8	70.0	68.0	64.7	63.4	61.4	96.8	96.8	96.8	95.5	93.6	99.4	99.4	90.4	86.8	81.4
Turkey	Kahramanmaras	78.8	78.6	78.3	78.1	78.0			82.6	87.1	93.8	88.6	92.9	92.9	92.9	92.9	95.4	85.7	69.6	69.6	69.6
Turkey	Antakya	99.5	99.5	99.5	99.3	99.0	89.5	91.5	94.8	96.1	98.1	92.7	92.7	92.7	89.8	85.4	99.4	99.4	83.9	77.6	68.3
Turkey	Aksaray										76.2					97.6					70.2
Yemen	Sana'a					91.0					65.9					93.9					77.9
Yemen	Aden					87.6					56.7					97.0					93.6
Yemen	Taiz					91.5					58.0					85.6					77.1

TABLE 5 PERCENTAGE OF MALNOURISHED CHILDREN UNDER FIVE AND UNDER-FIVE MORTALITY RATES

	PERCENTAGE OF MALNOURISHED CHILDREN UNDER FIVE YEARS (CHILDREN UNDERWEIGHT)							UNDER-FIVE MORTAITY RATES, DHS 1995-2003					
	Urban	Rural	Non-Slum	One Shelter Deprivation	Two Shelter Deprivations	Three+ Shelter Deprivations	All Slum	Urban	Rural	Non-Slum	One Shelter Deprivation	Two+ Shelter Deprivations	All Slum
Africa													
Benin	17.8	25.3	11.6	18.9	24.6	30.0	24.3	134.0	175.0	98.0	114.0	171.0	142.0
Burkina Faso	20.5	40.3	22.4	35.2	38.3	43.1	39.4	136.0	202.0	129.0	128.0	197.0	151.0
Cameroon	14.3	25.0	12.8	14.9	23.5	27.8	23.8	111.0	160.0	87.0	118.0	141.0	129.0
Chad	31.6	40.6	26.4	29.4	36.7	40.4	39.0						
Comoros	25.0	26.1	20.2	20.0	30.2	46.0	26.8						
Côte d'Ivoire	13.3	25.1	10.3	17.7	21.4	37.0	23.1	125.0	197.0	116.0	124.0	141.0	132.0
Egypt	6.8	9.6	7.9	11.2	9.3		10.7	42.0	63.0	41.0	76.0		61.0
Ethiopia	34.0	48.6	26.5	24.2	44.5	48.9	47.2	149.0	192.0	95.0	158.0	190.0	180.0
Gabon	10.0	16.8	8.2	10.7	14.6	20.8	14.1						
Ghana	14.9	25.2	13.8	21.2	26.7	26.4	24.0	93.0	118.0	95.0	99.0		91.0
Guinea	18.4	25.3	16.6	18.8	21.5	26.4	23.4	149.0	211.0	81.0	140.0	168.0	153.0
Kenya	12.6	21.4	8.1	11.9	19.3	24.1	21.0						
Madagascar	35.6	41.0	12.5	34.0	40.8	44.1	40.2						
Malawi	12.8	27.3	8.7	12.6	26.0	30.6	25.8						
Mali	20.6	37.2	15.5	20.5	35.9	36.9	34.2	185.0	253.0	131.0	155.0	229.0	197.0
Morocco	6.5	13.9	7.4	12.1	15.0	15.3	13.9	38.0	69.0	37.0	41.0	64.0	46.0
Mozambique	20.0	28.2	9.5	14.4	20.9	31.6	26.4	143.0	192.0	79.0	107.0	177.0	146.0
Namibia	16.5	25.9	13.5	21.0	27.8	26.7	26.5						
Niger	35.3	52.4	13.0	30.9	49.6	52.6	49.8						
Nigeria	22.4	31.7	12.1	23.1	30.0	38.4	31.0	153.0	243.0	81.0	141.0	205.0	174.0
Rwanda	15.6	26.1	8.8	15.4	25.9	26.5	25.3	141.0	216.0	105.0	147.0	162.0	156.0
Senegal								89.0	165.0	52.0	96.0	124.0	102.0
South Africa								43.0	71.0	38.0	62.0		58.0
Togo	16.1	27.9	16.9	21.8	26.7	29.4	25.5						
U. Rep of Tanzania	19.4	30.9	14.4	22.3	28.9	31.6	29.1	142.0	166.0	71.0	134.0	159.0	146.0
Uganda	12.4	23.6	7.4	16.0	23.2	28.1	22.8	101.0	163.0	51.0	87.0	136.0	107.0
Zambia	23.8	30.2	19.3	24.2	28.8	31.7	29.3	140.0	182.0	99.0	153.0	188.0	146.0
Zimbabwe	7.5	15.6	8.5	11.8	14.4	19.5	8.5	69.0	100.0	69.0	66.0	72.0	69.0
Latin America and the Caribbean													
Bolivia	4.8	10.9	3.1	5.0	10.7	12.9	9.0						
Brazil	4.6	9.2	3.2	5.6	7.3	19.1	7.5	49.0	79.0	34.0	55.0	90.0	64.0
Colombia	5.7	8.9	5.5	7.5	9.4	14.7	9.3	24.0	36.0	23.0	38.0		34.0
Dominican Republic	3.9	7.2	3.8	9.7	8.2	14.8	9.6						
Guatemala	15.6	29.1	11.4	22.5	30.9	32.7	29.0	58.0	69.0	60.0	46.0	62.0	56.0
Haiti	11.5	19.1	8.9	15.4	17.2	25.8	19.1						
Nicaragua	5.9	13.4	4.4	9.1	11.3	14.9	11.0						
Peru	3.2	11.8	1.5	4.8	9.1	12.4	9.4						
Asia													
Armenia	2.4	2.8	2.1	4.9		2.1	3.8	37.0	59.0	38.0	37.0	32.0	36.0
Bangladesh	39.8	49.2	24.7	25.2	48.9	53.6	42.5	97.0	113.0	80.0	76.0	130.0	105.0
India	38.2	49.3	33.0	49.8	52.8		50.2	65.0	111.0	48.0	88.0	80.0	87.0
Indonesia								42.0	65.0	38.0	47.0	54.0	48.0
Kazakhstan	4.8	3.9	5.5	7.9		.	4.0	50.0	73.0	36.0	95.0		75.0
Kyrgyzstan	5.9	12.5	1.9	9.0		1.9	8.2						
Nepal	33.1	49.5	27.6	26.8	42.2	43.7	37.0						
Pakistan	32.3	44.4	29.3	40.7	46.4		41.2	94.0	132.0	86.0	113.0	106.0	112.0
Philippines								30.0	52.0	26.0	36.0	47.0	39.0
Turkey	6.2	11.9	5.7	8.1	8.3		7.8	51.0	74.0	52.0	47.0	59.0	49.0
Uzbekistan	16.6	19.7	9.8	18.5	25.5	23.5	19.5	52.0	57.0	39.0	56.0	68.0	58.0
Vietnam								16.0	36.0	14.0	18.0	31.0	21.0
Yemen	24.5	30.7	21.3	29.8	31.4		29.5	117.0	141.0	93.0	124.0	161.0	141.0

TABLE 6 PERCENTAGE OF CHILDREN 12-23 MONTHS WHO RECEIVED MEASLES VACCINATIONS AND PERCENTAGE OF BIRTHS ATTENDED BY SKILLED HEALTH PERSONNEL

	PERCENTAGE OF CHILDREN 12-23 MONTHS WHO RECEIVED MEASLES							PERCENTAGE OF BIRTHS ATTENDED BY SKILLED HEALTH PERSONNEL						
	Urban	Rural	Non-Slum	One Shelter Deprivation	Two Shelter Deprivations	Three Shelter Deprivations	All Slum	Urban	Rural	Non-Slum	One Shelter Deprivation	Two Shelter Deprivations	Three Shelter Deprivations	All Slum
Africa														
Benin	75.3	64.1	87.4	72.7	68.0	55.1	65.1	79.9	59.0	94.9	78.0	64.6	44.7	62.1
Burkina Faso	73.1	53.3	69.7	66.9	52.6	49.6	54.2	87.7	30.5	86.3	57.0	30.6	22.4	32.4
Cameroon	67.6	49.3	74.0	74.7	53.3	39.5	50.5	82.1	44.8	85.2	84.0	50.8	34.6	49.8
Central African Rep	68.4	40.5	40.5	81.4	69.2	51.9	81.4	64.4	15.5	15.5	94.5	63.6	38.1	94.5
Chad	38.9	18.6	60.1	55.4	28.4	18.3	22.2	12.4	1.6	30.5	15.7	4.8	2.4	3.5
Comoros	63.0	63.5	85.7	67.9	57.0	37.5	60.0	64.9	35.9	72.8	53.3	27.9	22.1	38.0
Côte d'Ivoire	82.0	58.8	83.3	79.8	61.9	45.9	62.8							
Egypt	96.0	95.3	96.4	92.1	95.5	52.1	96.4	86.7	59.0	76.6	43.1	60.7	29.0	76.6
Ethiopia	63.1	22.3	88.6	71.0	37.5	21.6	26.4							
Gabon	61.1	37.1	65.8	58.1	47.0	31.8	48.7	89.6	63.1	93.0	84.9	74.7	59.5	76.2
Ghana	85.8	81.8	88.3	83.1	82.5	73.8	81.7	78.8	29.5	79.7	46.6	26.8	25.3	36.1
Guinea	66.9	46.7	80.1	65.8	60.1	41.0	51.5	16.8	6.7	32.9	16.0	11.5	5.4	8.8
Kenya	85.9	69.7	88.4	87.8	73.5	65.8	71.0	72.0	34.5	82.1	65.1	44.6	26.6	37.5
Madagascar	60.8	42.0	95.9	65.7	41.8	50.6	45.5							
Malawi	90.6	82.0	88.5	89.8	83.4	80.1	83.1	80.6	50.5	89.8	82.2	53.3	43.0	53.4
Mali	70.8	41.3	77.7	68.4	48.3	39.5	47.2	72.2	7.9	80.3	56.7	20.9	9.0	19.8
Morocco	94.2	85.9	93.8	87.5	84.8	81.9	85.4	85.3	39.5	80.5	49.5	38.1	21.3	39.1
Mozambique	93.0	47.1	95.1	92.3	76.7	43.0	57.0	80.6	30.1	96.7	85.6	59.5	25.5	40.2
Namibia	84.3	78.4	85.3	79.1	79.7	75.8	80.4	93.1	66.3	94.0	81.2	67.1	64.0	68.4
Niger	67.1	27.8	85.9	69.8	40.0	26.0	34.5							
Nigeria	52.1	28.5	76.5	46.6	35.1	14.2	31.3	57.0	25.0	83.9	51.5	30.6	12.1	28.8
Rwanda	89.9	86.3	90.7	91.4	86.1	86.2	86.7	65.7	20.0	82.7	57.0	24.9	18.4	23.9
Senegal								80.9	29.6	78.0	66.3	35.1	21.4	43.3
South Africa	85.1	79.3	85.4	84.1	81.1	68.0	80.3	93.4	75.5	95.0	84.7	76.6	62.0	77.5
Togo	58.0	38.2	71.7	47.5	41.1	32.8	41.2	72.3	17.9	84.5	43.7	21.9	13.6	28.0
U. Rep of Tanzania	90.3	75.3	96.3	88.7	78.5	73.0	77.9	76.8	26.4	85.0	77.5	31.6	25.2	35.0
Uganda	68.4	55.3	62.3	61.1	55.8	54.3	56.7	35.5	15.9	43.4	25.4	17.3	10.9	17.5
Zambia	85.5	83.9	85.9	88.3	88.2	80.1	84.2	78.7	26.1	90.9	74.9	37.5	20.8	36.2
Zimbabwe	86.2	75.7	82.5	73.1	74.3	84.1	82.5	89.4	64.2	85.7	69.8	66.3	58.7	85.7
Latin America and the Caribbean														
Bolivia	66.5	60.2	67.5	64.6	59.2	64.3	62.4	77.4	36.3	88.8	69.0	43.7	25.5	48.8
Brazil	90.2	76.5	91.7	89.2	78.5	73.8	83.9	92.0	72.7	95.3	88.9	78.4	55.1	81.6
Colombia								93.8	69.2	93.3	77.2	66.0	57.8	70.3
Dominican Republic	89.5	86.0	88.7	90.5	86.3	59.9	87.1	77.7	77.6	77.1	80.2	79.8	66.9	79.3
Guatemala	80.8	80.4	94.5	70.1	79.1	72.9	74.7	66.1	24.7	80.4	43.0	21.9	14.3	25.3
Haiti	60.9	50.2	70.7	54.8	51.4	37.7	48.9	49.3	9.2	50.6	25.3	10.2	5.5	13.7
Nicaragua	77.1	74.1	70.8	78.0	78.6	66.4	76.6	88.7	45.5	90.7	74.1	62.7	24.0	61.1
Peru	74.0	69.3	77.8	74.7	68.8	66.7	69.3	69.3	20.1	77.7	57.4	33.7	20.6	33.6
Asia														
Armenia	76.0	72.1	79.9	69.2		79.9	55.0	99.1	94.5	99.5	95.0			96.3
Bangladesh	80.7	68.9	91.2	85.2	76.9	73.8	79.3							
India	69.2	45.3	75.6	57.2	41.0	-	55.2	73.0	32.8	81.1	56.9	40.3		55.0
Indonesia	77.6	66.2	80.8	75.2	63.6	92.7	72.3	61.2	24.2	69.1	53.8	41.3	30.9	49.9
Kazakhstan	81.4	76.2	81.8	84.2	77.7	-	81.0	98.4	99.0	98.6	97.9	98.3		98.1
Kyrgyzstan	83.7	84.5	87.4	81.1		87.4	81.9	99.3	97.8	100.0	98.7			98.8
Nepal	80.6	69.9	86.1	75.8	75.6	86.0	76.3	50.3	8.1	69.0	53.9	32.2	9.8	37.3
Pakistan	64.6	43.6	67.8	55.0	59.0	N/A	55.4	40.3	6.9	47.6	21.8	14.7		21.0
Philippines	81.8	77.5	83.2	75.5	82.0	-	77.2	79.0	40.8	83.2	67.3	51.9	54.5	64.2
Turkey	82.2	72.7	84.7	72.9	80.4	41.8	72.7	87.7	68.8	89.1	83.5	82.4	80.5	83.2
Uzbekistan	84.2	94.7	89.1	84.0	62.1	-	82.4	100.0	96.4	100.0	100.0			100.0
Viet Nam	94.3	80.7	95.2	94.7	94.3	73.7	91.8	97.6	73.1	99.7	93.9	89.3	73.7	92.0
Yemen	68.6	37.1	72.6	59.6	63.0	-	63.5	45.8	10.0	55.4	38.5	31.0	20.7	35.1

TABLE 7 PERCENTAGE OF CHILDREN UNDER-FIVE WITH DIARRHEA AND ACUTE RESPIRATORY INFECTIONS (ARI)

	PERCENTAGE OF CHILDREN UNDER-FIVE WITH DIARRHEA, DHS 1995-2003							PERCENTAGE OF CHILDREN UNDER FIVE YEARS WITH ARI, DHS 1995-2003						
	Urban	Rural	Non-Slum	One Shelter Deprivation	Two Shelter Deprivations	Three Shelter Deprivations	All Slum	Urban	Rural	Non-Slum	One Shelter Deprivation	Two Shelter Deprivations	Three Shelter Deprivations	All Slum
Africa														
Benin	11.3	14.4	8.8	12.3	13.7	15.8	14.0	11.3	12.3	9.8	14.8	10.7	11.8	12.2
Burkina Faso	21.1	20.6	22.1	20.9	20.0	20.9	20.5	8.0	8.6	8.0	9.7	8.7	8.0	8.6
Cameroon	17.0	19.7	12.6	16.2	18.2	23.3	20.1	20.2	19.5	15.7	22.3	21.3		20.4
Central African Rep	19.5	24.8	24.8	13.6	21.8	24.2		28.1	28.2	28.2	24.8	30.8	28.9	24.8
Chad	22.2	21.3	16.7	21.0	21.8	21.5	21.6	12.6	12.7	10.9	12.1	11.4	13.2	12.7
Comoros	27.5	21.7	26.2	18.4	25.6	25.4	22.6	22.0	22.3	15.2	20.3	26.2		23.4
Côte d'Ivoire	16.9	23.6	17.9	23.4	22.0	20.0	22.0	14.6	17.2	14.7	14.3	15.6	23.3	16.6
Egypt	16.8	20.2	18.1	21.3	21.2		23.7	11.1	9.7	10.4	10.2	9.1	16.3	10.4
Ethiopia	16.7	24.5	6.2	13.4	23.8	24.1	23.7	16.3	25.4	2.8	16.8	21.7	25.7	24.5
Gabon	16.3	14.3	16.5	15.0	14.6	17.4	15.3	13.7	11.2	11.7	16.0	12.8	11.1	13.9
Ghana	13.6	16.1	12.3	16.7	14.3	19.0	16.1	8.9	10.6	8.1	11.1	10.3		10.6
Guinea	17.8	22.4	12.8	19.4	20.4	22.5	21.3	14.4	16.5	10.3	14.9	16.4	16.1	16.0
Kenya	17.0	15.8	10.7	15.4	16.5	16.8	16.5	16.4	18.9	10.8	15.6	17.9	20.8	19.2
Madagascar	30.0	26.3	29.9	27.6	27.1	24.9	27.0	21.1	24.6	8.6	17.4	24.9	25.7	24.0
Malawi	14.3	18.1	12.0	13.3	17.7	19.4	17.7	15.7	28.3	17.7	19.7	26.7	29.9	26.9
Mali	13.1	20.2	12.9	12.0	17.7	21.7	18.9	9.3	10.1	12.2	8.9	9.0	10.5	9.8
Morocco	11.5	12.4	10.9	13.6	14.6		13.4	11.3	12.0	12.1	10.0	11.8	11.7	11.1
Mozambique	30.6	18.0	18.0	25.0	17.8	21.1	20.7	15.7	10.8	15.3	16.7	12.4	10.7	11.8
Namibia	12.6	11.7	12.3	13.5	10.5	13.5	11.9	13.8	19.7	12.7	13.9	19.4	23.3	19.7
Niger	31.6	39.0	28.8	28.3	38.9	38.7	37.8	14.2	14.2	22.0	10.9	13.9	14.7	14.1
Nigeria	14.5	20.7	9.6	14.1	20.3	23.5	19.9	7.8	11.4	6.5	9.2	11.3	11.1	10.7
Rwanda	12.3	17.7	6.9	14.6	16.9	18.2	17.4	15.3	22.3	13.7	15.2	18.7	24.4	21.6
Senegal	13.9	15.7	14.9	14.8	15.0	15.9	15.1							
South Africa	10.8	15.7	9.8	15.6	14.6	16.5	15.5	18.9	19.6	19.5	19.2	18.7	19.7	19.1
Togo	27.0	32.3	17.7	30.0	32.1	33.6	31.7	18.8	20.7	19.3	17.1	20.7	24.6	20.3
U. Rep of Tanzania	9.8	12.9	6.1	9.1	13.7	11.9	12.5	12.2	14.3	10.8	14.3	14.8		14.0
Uganda	15.5	20.1	10.6	19.5	20.4	19.0	19.8	18.6	23.0	14.1	24.2	22.0	22.3	22.7
Zambia	21.1	21.2	14.7	23.6	21.6	21.6	22.0	13.8	14.9	8.2	14.7	16.6	14.9	15.4
Zimbabwe	11.9		11.9	16.2	14.0	15.2	11.9	11.3	18.1	11.5	18.9	18.0	18.2	11.5
Latin America and the Caribbean														
Bolivia	21.1	24.1	19.8	22.9	23.0	24.8	23.4	22.8	21.4	23.4	23.9	20.5	20.3	21.7
Brazil	12.6	14.6	11.0	13.5	15.8	16.4	14.6	24.0	22.9	23.0	24.3	22.9	28.5	24.2
Colombia	13.2	15.6	12.7	17.1		23.2	16.7							
Dominican Republic	13.5	14.9	13.2	15.0	19.1	20.7	16.6	19.3	20.1	19.3	20.4	19.8	26.5	20.6
Guatemala	12.8	13.6	12.6	11.1	13.5	16.4	13.6	17.6	20.1	16.4	19.0	20.3	21.2	20.2
Haiti	24.1	26.6	22.8	26.2	25.3	29.1	26.6	32.1	42.9	30.0	37.7	43.0	45.9	42.2
Nicaragua	11.7	14.4	9.8	13.3	14.2	14.2	13.9	28.5	33.2	27.0	27.1	34.5	35.1	31.8
Peru	13.6	17.6	10.8	16.0	18.6	16.8	17.4	19.8	20.6	18.0	20.2	21.8	20.9	21.1
Asia														
Armenia	7.8	7.8	7.8	10.6		7.8	8.4	11.5	11.4	11.8	7.3	15.8	11.8	9.5
Bangladesh	7.1	5.9	3.9	8.5	6.5	8.2	7.6	14.9	17.2	12.1	12.7	13.7	20.4	15.4
India	19.3	18.8	18.2	22.0	20.0		21.8	16.0	20.1	14.9	18.0	22.3		18.5
Indonesia	11.2	10.8	10.5	11.9	13.8		12.3	14.9	17.2	6.7	9.1	8.1		8.7
Kazakhstan	14.8	12.3	13.0	15.1	19.2		17.6	3.2	2.8	2.8	4.4	3.4		3.8
Kyrgyzstan	15.1	18.3	14.8	16.6		14.8	15.3	2.5	5.0	2.5	2.6		2.5	2.4
Nepal	16.6	20.7	15.4	15.7	17.3	25.4	17.4	23.8	22.7	24.4	24.5	22.5	25.5	23.5
Pakistan	15.0	14.2	14.6	15.5	21.6		16.1	13.8	16.8	12.9	16.2	16.8		16.2
Philippines	10.7	10.6	10.1	12.7	10.6	40.8	12.7	8.3	12.2	7.2	12.0	14.0		12.2
Turkey	26.1	35.7	24.5	30.6	32.5	48.4	31.7							
Uzbekistan	8.5	3.7	11.1	6.5	16.6		7.5	2.5	0.5	5.4	1.4	1.5		1.4
Viet Nam	3.5	13.0	3.7	1.4	9.2		2.9	14.0	20.7	14.4	11.4	22.5		13.1
Yemen	25.7	35.8	19.9	33.9	31.6	24.0	32.6	17.2	25.8	13.7	20.5	23.0	22.6	21.3

TABLE 8 EDUCATION; LITERACY RATES BY SHELTER DEPRIVATION

		WOMEN						
	Year	Total Urban	Total Rural	Non-Slum	All Slum	One Shelter Deprivation	Two Shelter Deprivations	Three Shelter Deprivations
Northern Africa								
Egypt	2003	73.7	42.9	75.0	60.2			
Egypt	2000	69.7	34.8	71.1	56.0			
Egypt	1995	65.9	30.0	69.7	31.6			
Morocco	2004	67.5	24.6	68.7	53.4			
Morocco	1992	57.6	12.0	60.7	35.9			
Sub-Saharan Africa								
Benin	2001	42.9	12.7	62.4	32.0	43.5	23.9	6.4
Benin	1996	40.5	9.5	58.5	30.8	44.2	22.2	5.1
Burkina Faso	2003	52.7	5.8	57.2	42.1	45.0	38.8	13.8
Burkina Faso	1999	47.0	3.6	82.7	45.2	49.6	32.5	33.3
Burkina Faso	1992	43.7	5.9	80.0	40.9	47.7	34.2	16.3
Cameroon	1998	80.4	53.9	88.6	72.9	80.6	69.7	44.3
Cameroon	1991	69.6	41.4	85.6	64.2	71.7	52.6	34.3
Côte d'Ivoire	1999	54.2	27.6	66.3	40.6	38.1	49.6	50.0
Côte d'Ivoire	1994	48.1	24.8	56.9	34.7	37.2	30.7	16.7
Ethiopia	2000	65.3	15.4	79.2	64.7	77.6	60.1	49.4
Ghana	1999	72.7	47.2	76.6	68.5	70.8	52.5	
Ghana	1993	62.3	30.8	68.8	58.1	61.8	46.4	
Guinea	1999	33.4	4.3	60.7	30.7	35.1	27.0	11.6
Mali	2001	35.8	5.7	47.7	30.4	35.2	21.4	
Mali	1996	31.2	4.1	69.4	29.6	41.5	27.4	
Mozambique	2003	64.9	21.6	94.2	62.1	78.6	55.8	31.4
Mozambique	1997	62.8	23.3	89.9	59.2	73.5	53.1	29.5
Nigeria	2003	67.5	37.9	89.3	58.9	67.2	55.1	22.7
Nigeria	1999	70.4	45.6	87.8	62.6	72.0	55.2	29.1
Nigeria	1990	66.2	27.2	88.9	58.4	64.5	49.5	38.5
Rwanda	2000	86.0	62.0	90.8	83.4	87.7	80.9	77.1
Rwanda	1992	81.4	59.6	93.8	78.7	88.2	75.7	67.6
Senegal	1997	50.8	11.4	69.2	43.6	45.5	37.7	23.5
Senegal	1993	46.4	6.3	62.0	39.3	42.4	26.3	17.1
South Africa	1998	96.2	87.9	97.1	93.0	93.3	91.5	
Uganda	2001	84.2	52.5	94.2	82.3	88.1	73.1	65.1
Uganda	1995	81.4	48.0	94.3	79.1	86.0	79.7	61.3
United Republic of Tanzania	1999	80.3	57.6	92.0	79.3	85.0	73.6	50.6
United Republic of Tanzania	1996	82.6	60.2	94.0	81.4	88.9	77.1	60.9
United Republic of Tanzania	1992	78.5	56.7	93.8	77.1	86.9	72.1	61.7
Zambia	2002	78.8	48.4	87.0	72.1	73.7	69.2	70.3
Zambia	1996	82.2	53.6	90.8	73.8	77.2	71.3	62.9
Zimbabwe	1999	90.6	63.3	91.1	85.2	85.9		
Zimbabwe	1994	83.2	53.4	85.2	72.2	78.3		
Latin America and the Caribbean								
Brazil	1996	94.2	80.8	96.2	91.6	92.8	88.2	82.4
Colombia	2000	88.0	53.2	89.4	73.2	75.6	63.5	43.8
Colombia	1995	97.4	88.6	98.0	89.2	93.4	82.7	61.9
Guatemala	1998	83.7	60.6	92.4	83.7	74.8	53.5	63.9
Asia								
Bangladesh	1999	62.2	41.7	89.8	56.8	74.0	63.1	41.2
Bangladesh	1996	58.3	33.5	82.3	49.4	73.9	48.8	30.9
India	1999	62.2	27.6	73.0	45.6	45.5	46.0	
Kazakhstan	1999	99.5	99.4	99.8	99.3	99.2		
Pakistan	1990	50.6	11.0	58.8	28.2	30.0	14.9	
Uzbekistan	1996	99.9	99.7	100.0	99.9	99.9		
Indonesia	2002	91.5	82.1	94.4	86.1	87.3	85.6	62.9
Indonesia	1997	92.8	77.9		92.8	93.6	88.9	
Indonesia	1994	89.8	72.5		89.8	91.9	82.2	72.2
Philippines	2003	97.8	93.8	98.5	94.6	94.8	93.5	
Philippines	1998	99.0	94.6	99.5	97.9	98.7	95.7	96.0
Viet Nam	2002	97.3	90.0	98.5	94.6	94.1	96.1	93.8
Armenia	2000	99.8	99.4	99.8	99.8	99.8	99.6	
Turkey	1998	88.0	77.6	88.2	88.2	87.9	86.9	63.4
Turkey	1993	82.1	64.5	83.5	79.5	82.4	73.4	57.4
Yemen	1991	37.4	6.2	48.2	24.0	29.6	17.4	12.9

TABLE 9 PERCENTAGE OF FEMALE AND MALE AGED 15-24 YEARS UNEMPLOYED BY SHELTER DEPRIVIATION

		FEMALE							MALE						
	Year	Urban	Rural	Non-Slum	One Shelter Deprivation	Two Shelter Deprivations	Three Shelter Deprivations	All Slum	Urban	Rural	Non-Slum	One Shelter Deprivation	Two Shelter Deprivations	Three Shelter Deprivations	All Slum
Africa															
Benin	2001	14.4	5.5	20.1	13.4	7.9	9.5	11.3	47.5	28.7	59.3	45.0	28.9	21.8	38.7
Burkina Faso	2003	13.8	1.4	14.2	13.0	12.6	8.3	12.7	2.7	4.7	2.8		24.5		2.7
Cameroon	2004	31.0	18.5	28.3	33.6	37.2	14.1	33.1	8.3	10.4					
Chad	1996	39.9	35.9	50.6	52.9	41.2	33.9	39.0	11.1	2.9	32.9	0.0	13.9	7.8	9.6
Comoros	1996	42.4	43.5	38.2	41.2	50.3	64.3	45.0	21.8	29.1	19.0	20.7	18.8	60.0	23.5
Côte d'Ivoire	1999	29.3	11.5	25.8	35.7	27.1		33.6	13.3	7.2	10.4	15.7	17.6	12.7	15.6
Gabon	2000	32.2	37.2	28.1	37.2	37.5	38.0	37.3	8.2	9.9	5.0	11.3	12.0	15.8	11.8
Ghana	2003	29.6	18.7	31.0	25.5	33.2	58.3	27.5	22.5	14.6	28.2	21.6	14.3		19.5
Guinea	1999	26.1	17.6	26.0	28.1	22.3	30.8	26.1	12.3	5.1	15.7	8.9	13.1	14.2	12.1
Kenya	2003	25.6	19.4	22.6	27.8	28.9	28.9	28.3	21.9	17.6	25.2	20.8	15.4	32.4	20.3
Madagascar	1997	25.4	17.3	22.3	27.3	27.7	21.5	25.6							
Mali	2001	34.8	38.9	21.6	31.1	32.4		34.3	2.4	0.8	4.9	2.5			1.3
Morocco	2004	49.6	78.3	47.7	61.5	73.1	87.5	62.46							
Mozambique	2003	46.1	19.1	20.4	49.2	49.7	40.9	49.0	22.0	24.5	13.3	21.1	22.7	27.4	23.3
Niger	1998	66.1	45.3	54.9	65.1	64.7	78.3	66.4	17.2	3.8					
Nigeria	2003	29.2	38.0	23.7	27.9	35.2	40.3	31.8	15.7	14.6	23.1	12.3	14.9	19.3	13.8
Rwanda	2000	35.6	8.7	15.1	50.4	32.4	26.6	39.3							
Senegal	1997	49.4	48.1	37.9	49.4	51.7	62.9	50.7							
South Africa	1998	26.9	31.0	25.0	33.3	31.6	33.3	33.0							
Togo	1998	14.3	16.1	13.5	15.5	11.7	14.5	14.8	11.0	9.8					
Uganda	2001	39.3	21.4	25.2	41.0	39.9	41.3	41.8	11.6	7.6	7.9	20.3	9.6	5.3	11.7
United Republic of Tanzania	1999	32.9	14.9	17.0	31.5	40.6	25.0	34.7	10.5	7.1	13.7	9.7	13.1	1.6	10.3
Zambia	2002	46.1	38.2	46.0	49.5	41.6	48.1	47.3	19.3	17.5	19.9	14.3	24.0		16.3
Zimbabwe	1999	40.1	36.5	38.8	36.0	52.8		43.4	30.7	21.4	34.1	4.0			6.1
Latin America and the Carribean															
Bolivia	2004	11.2	30.2	6.4	10.6	18.9	29.4	13.5	3.5	4.6	3.7	3.7	3.2		3.1
Brazil	1996	18.7	35.6	17.5	18.8	22.3	41.2	20.1	2.8	1.0	3.5	2.9	0.5		2.1
Colombia	2000	21.9	42.6	21.0	30.8	37.7	25.0	32.0							
Guatemala	1998	40.3	57.8	30.5	49.6	69.5	71.4	57.3							
Haiti	2000	29.1	33.4	23.3	32.8	40.6	39.2	34.7	14.9	12.5					
Nicaragua	2001	26.6	60.5	17.2	26.2	35.5	57.2	31.3							
Paraguay	1990	13.1	43.4	8.5	18.0	15.6	24.7	17.7							
Peru	2000	17.6	30.6	14.2	19.4	22.5	31.2	22.4							
Asia															
Kazakhstan	1999	27.7	38.9	19.6	31.6	30.3	39.2	32.5	19.7	37.4	15.5	12.1	31.0	43.7	22.7
Krygystan	1997	26.1	47.2	16.2	14.0	32.8	54.8	29.4							
Nepal	2002	45.2	22.7	61.0	45.3	34.7	0.0	35.1							
Philippines	1998	15.3	30.6	12.6	19.8	28.6	33.3	22.2							
Philippines	2003	17.6	31.5	23.5	14.3	23.3	34.0	25.9							
Turkey	1998	49.5	41.8	51.3	47.3	32.4	33.3	45.5	1.2	15.5	2.2				
Uzbekistan	1996	42.7	46.8	29.5	40.4	47.2	45.7	45.2							
Viet Nam	2002	16.7	10.4	18.8	12.5	16.7		12.5							

Index

Note: Page numbers in *italics* refer to maps, figures, tables and boxes. Those followed by 'n' refer to notes.

A

B

C

D

T

U

V

W

Y

Z